Denying Evolution

Creationism, Scientism, and the Nature of Science

Denying Evolution

Creationism, Scientism, and the Nature of Science

Massimo Pigliucci

University of Tennessee, Knoxville

SINAUER ASSOCIATES, PUBLISHERS
Sunderland, Massachusetts 01375

THE COVER

Original cartoon by Dave Granlund.

DENYING EVOLUTION:
Creationism, Scientism, and the Nature of Science

Sinauer Associates Inc.
23 Plumtree Road
Sunderland, MA 01375 USA

fax: 413-549-1118
email: orders@sinauer.com; publish@sinauer.com
www.sinauer.com

Library of Congress Cataloging-in-Publication Data

Pigliucci, Massimo, 1964-
Denying evolution : creationism, scientism, and the nature of science / Massimo Pigliucci.
p. cm.
Includes bibliographical references (p.).
ISBN 0-87893-659-9 (pbk.)
1. Evolution (Biology) 2. Creationism. 3. Scientism. 4. Science--Philosophy. I. Title.

QH366.2 .P54 2002
576.8--dc21 2002005190

To all my teachers,
for helping me develop my own way of thinking.

– CONTENTS –

- FOREWORD -

William B. Provine

Charles D. Alexander Professor of Biology,
Department of Ecology and Evolutionary Biology, Cornell University

In the past 20 years, evolutionists have published a host of books and papers directed at creationists of all stripes. Do evolutionists—or creationists—need one more, by Massimo Pigliucci? Yes, because both groups will benefit greatly from his criticisms and from the wide perspective he brings to the controversies.

The typical book written by an evolutionist is directed to the destruction of creationist arguments. Such books say nothing about the pedagogical benefits of inviting creationists to bring discussion of the evolution controversy into the controlled arena of the science classroom, and nothing about the limitations and failings of evolutionists to address the controversy within the confines of science education. Pigliucci's stimulating book is different.

Massimo Pigliucci is a prize-winning young evolutionary biologist who has studied deeply both history and philosophy; he is also a dedicated teacher. In this book, he bypasses the charged term "creationist" (in its many versions) and supplants it with the single phrase, "those who deny evolution." Then he evaluates the historical reasons for the existence of those in the United States who deny evolution, and uses his wide knowledge and uniquely broad perspective to dissect their arguments.

But Pigliucci's criticism of evolutionists is also unsparing. He faults them for placing near-total emphasis upon research at university levels, and for failing in many ways to promote the understanding of evolution. Evolutionists, he argues, hesitate to engage those who deny evolution in debate and forums, avoid participating in the construction of better textbooks for high school biology, and refuse to help revise and update the evolution sections of state examinations on biology. Most of all, he criticizes scientists for not thinking seriously enough about the teaching of evolution, and especially about the teaching of evolution to nonscience students.

No other book on the evolution–creation controversies has been so daring, and so challenging.

–ACKNOWLEDGMENTS–

This book would not have seen the light without the effort, counsel and support of many people. First and foremost, Barbara Forrest, who patiently went through the entire manuscript pointing out minor and major problems and peppering her editorial notes with witty comments about my (mis)understanding of philosophy of science. Norris Muth, one of my PhD students, also spent quite a bit of his time checking the manuscript, helping me improve it significantly. I am grateful to Andy Sinauer for asking me to do this and to Carl Schlichting and Doug Futuyma for suggesting my name to the publisher in the first place.

My appreciation of topics ranging from the philosophy to the history of science, from social movements to the implication of brain structure for education owes much to the following people: Shari Clough, Jonathan Kaplan, Ken Miller, Will Provine, Genie Scott, Michael Shermer, Elliott Sober, and the wonderful people of the Rationalists of East Tennessee (www.rationalists.org). I would like to thank several of my students, post-docs, and colleagues for helping out with the annual "Darwin Day" events that provided the springboard for writing this book: Josh Banta, Gordon Burghardt, Hillary Callahan, Mark Camara, Mitch Cruzan, Stan Guffey, Mike McKinney, Claudia Melear, Norris Muth, Carolyn Wells, and many others over the past several years. Thanks also to Amanda Chesworth, whose tenacity has made Darwin Day an international event with several dozen participating sites (http://darwin.ws/day/).

I am most especially grateful to my wife, Melissa Brenneman, not only for her patience in putting up with me (in general, as well as during the writing of this book), but for her continuous constructive criticism of everything I write.

Prologue: or, How I Got Into This Mess

Until a few years ago it had never occurred to me that one would have to defend science and evolutionary biology in the public arena. I am a professional evolutionary biologist, and I always minded my own business, concentrating my efforts on the arcane problems that—for whatever reason—strike me as interesting scientific puzzles (I work on interactions between genotype and environment, which is a fancy way to say nature versus nurture). Indeed, when I was still a college student living in Rome, Italy, I happened to see in a bookstore a copy of Douglas Futuyma's *Science on Trial* (translated into Italian), published by Sinauer Associates (the very same publisher of this book), and I remembered being puzzled. Futuyma was (and is) a prominent scientist; why was he wasting his time attacking myths about Noah's flood and a 6,000-year-old Earth? Was he that interested in medieval history?

Only much later did I fully understand why Doug and a few other scientists, philosophers, and educators devote a significant amount of energy to the creation–evolution "controversy." Soon after I came to the University of Tennessee as an enthusiastic young assistant professor in 1996, the state legislature considered an antievolution law. Things had apparently not changed much since the Scopes trial in 1925! Were these people serious? Didn't they realize that they were going to be the laughing stock of their constituents?

Alas, my naïveté had to give way to reality: In the United States, more than half the population believes in a more or less literal reading of the Bible, and the overwhelming majority of people (including a large proportion of high school science teachers!) reject the idea that humans evolved from "lower" forms of animals and that Earth is billions of years old.

Something needed to be done, so with a few students and colleagues I started the annual *Darwin Day*, a community education event intended to teach laypeople and schoolteachers about the scientific method and the real science behind evolutionary theory. It has been an incredible educational experience for me, too, as I've faced mistrustful undergraduate students, critical letters to the editor of the local newspaper, heated participation in radio talk shows, and even live debates against people who in all seriousness were saying that evolution is the theory that people come from bananas.

Then I slowly realized that all these people surely couldn't simply be stupid, as some of my colleagues occasionally hinted. It is not reasonable to blame only the public for something that it became more and more evident was an abysmal failure of our educational system, and hence of us as scientists and educators. It hit me that the roots of the problem go much deeper than most people on both sides of the debate realize. That is why I decided to write this book and to contribute to the debate from what I think is a different angle. Here, my interest is not so much in debunking creationist claims (although there is some of that, of course), but mostly in understanding the reasons for the problem itself.

I think creationism is more properly called *evolution denial*, which is why I titled the book after a suggestion of my friend Michael Shermer. Creationism is not a viable theory of anything, and it is certainly not a scientific theory. In the scientific community it ceased being a reasonable option for explaining life's diversity as soon as Darwin's *Origin of Species* became available to the public in 1859. Rather, creationism is really a form of denial, analogous to the denial of the Holocaust by some pseudohistorians, or the denial of environmental problems by so many pundits and special-interest groups.

In fact, while I was writing this book, I happened to be reading Paul and Ann Ehrlich's *Betrayal of Science and Reason*,[1] in which they analyze and debunk what they call the "brownlash" movement that is attempting to dismantle the few environmental protection measures hard-won over the course of decades because of the simplistic notion that there really isn't a problem after all. The parallels among evolution denial, the brownlash, and other antiscience positions are striking. Scientists (whether ecologists or evolutionary biologists) are alleged to somehow conspire to wreck our economy, worsen our style of life, and corrupt the souls of our children. Why scientists would want to do that is of course never explained, nor are the mountains of data supporting both environmental caution and evolutionary theory ever adequately accounted for.

In this book I suggest that there are many reasons for the evolution denial attitude of so many people in the United States and in other parts of the world (mostly countries where religious fundamentalism reigns supreme, such as many nations in the Middle East). I will discuss the many strands of anti-intellectualism that have plagued American society almost from its inception, as

[1] P. R. Ehrlich and A. H. Ehrlich, *Betrayal of Science and Reason: How Anti-environmental Rhetoric Threatens Our Future* (Washington, DC: Island Press, 1996).

well as the reasons for the failure of teachers to educate students about science as a method of discovery (instead of a list of facts as boring as a telephone directory). I will show the fallacies committed by scientists themselves when dealing with creationists, and examine the possibility that the human brain was simply not well designed (ironically, by natural selection) to think critically.

As in the case of the brownlash, part of the problem is an education gap in the general public, but part of it certainly comes from the shrewdness of a few ideologues who distort or ignore scientific findings, manipulate the media (who are often happy enough to be manipulated), and attempt to force their own ideological agenda on a nation. In the case of the brownlash, these people are conservative politicians and commentators, as well as the CEOs of major corporations whose bottom line would be hurt by environmental regulations. In the case of creationism, the sinister characters are associated principally with the so-called intelligent design movement and its Wedge strategy bent on literally destroying science as we know it and establishing a theocracy in the United States.

The creation–evolution debate, as I hope to make clear in this book, is not therefore a scientific debate—far from it. It is a particular instance of a broad cultural war between conservative and progressive forces, between a priori ideologies and the spirit of inquiry, between ignorance and education. As Carl Sagan once put it, science is a candle in the dark, and we need to make a Herculean effort just to keep it lit, if it is ever to become the bright light that many hope for (and others fear). That effort is most crucial because the ability to understand the world around us is perhaps the most precious thing we have.

– 1 –

Where Did the Controversy Come From?

History is a vast early warning system.

– Norman Cousins –

Life is infinitely stranger than anything which the mind of man could invent.

– Sherlock Holmes, in Arthur Conan Doyle's "A Case of Identity" –

When I was in high school I didn't appreciate history very much. I rather spent my hours reading about philosophy and, most often, science. It seemed to me that the past was just a repository of recorded events with no consequences for the present. Furthermore, it made no sense. It just seemed to be one damn thing after another.

I was profoundly mistaken, and today I spend a significant part of my time reading about history, in the hope to make up for what I missed by not paying attention in high school. The first time I realized my mistake was during my second year in college at the University of Rome. I was taking a mandatory course in genetics, which I loved. But the teacher was struggling between two alternative approaches of teaching the subject, reflected in radically different kinds of textbooks. On the one hand, the classical way to introduce students to genetics was to use a historical approach: One starts with Mendel and his experiments with peas and works one's way through the rediscovery of his work at the beginning of the twentieth century. One then learns of the first experiments aimed at uncovering the nature of the hereditary material, followed by Watson and Crick's discovery of the structure of DNA, the elucidation of the genetic code, and modern studies of biomolecules and intracellular machineries.

The alternative path was to ignore history altogether and start with the simplest genetic systems—bacteria—and then move to more complex phenomena, all presented from the standpoint of the most recent discoveries in the bur-

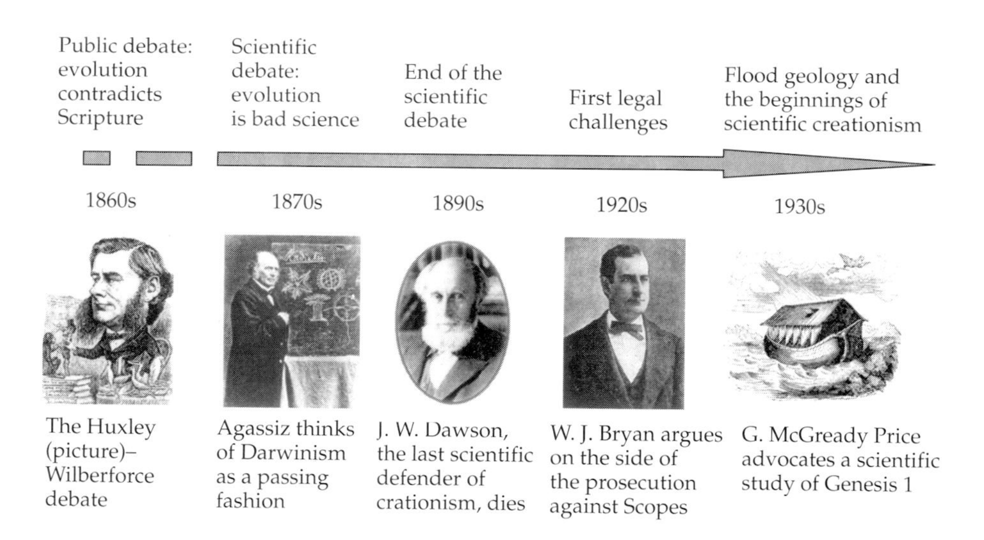

Figure 1.1 The progression of creationist tactics against evolution.

geoning field of molecular biology. Sadly, the trend in biology textbooks over the past few decades has been to move away from the historical perspective and instead to embrace the exciting science that is done here and now (I will come back to this topic in Chapter 8). It is the same mistake that I made in high school and one that is very costly to countless students for at least two reasons: First, they do not acquire a "memory" of their field of study. Second, and more importantly, they don't understand why people are asking certain questions now, and how science has arrived at answering the questions on which the current research builds. Science, then, becomes a matter of swallowing a lot of information (and more rarely a few concepts) with no appreciation of the *process* that has led us to rely on such information. And it is the process that is exciting, more so than the end result.

I do not wish to make the same mistake again while starting a book on the complex issue of creationism and evolution. This chapter is devoted to a brief (and rather idiosyncratic) history of what happened (Figure 1.1) because I believe that without such historical perspective, we cannot understand and appreciate the current state of the debate. I also believe that it is this lack of understanding among most of the people involved in the debate that hampers any progress toward a resolution. I feel reasonably confident that this debate will eventually go the way of the discussion between geocentrists and heliocentrists, but that it will take so much more time if we don't pay attention to its roots, both historical and conceptual.

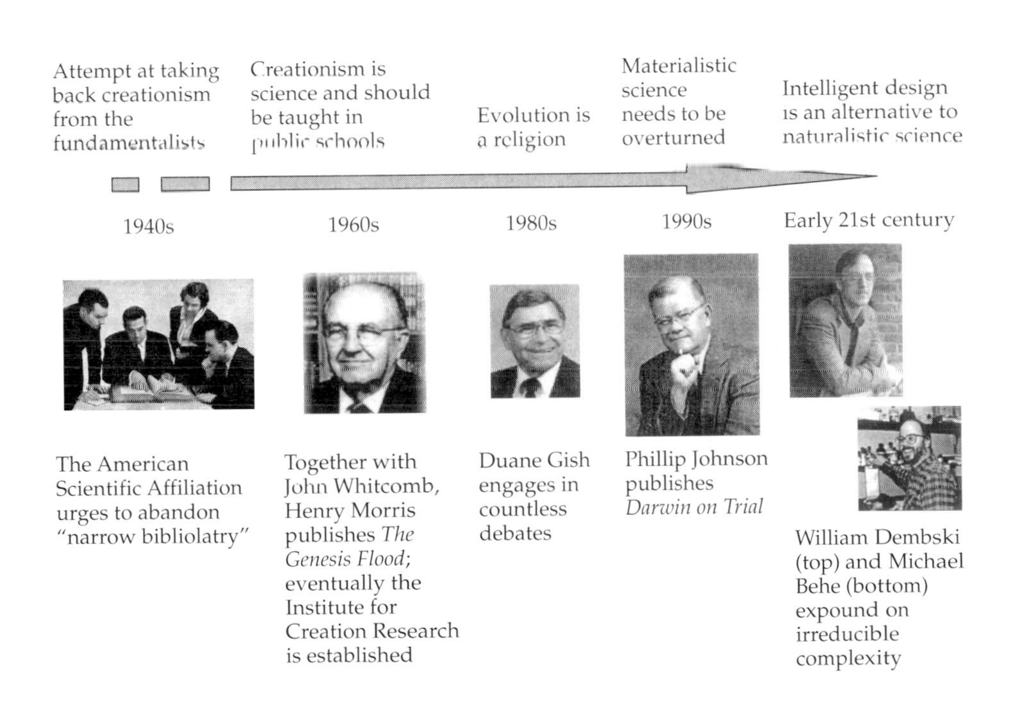

In the Beginning: The Great Debate That Never Was?

The standard account of the early stages of the Darwinian theory of evolution is well known. Darwin's voyage to the Galápagos and other areas of South America, the long hiatus between his return to England and the publication of *The Origin of Species* in 1859, and the fact that the latter was forced by the coincidence that Alfred Russel Wallace was getting ready to publish very similar ideas[1] are established items of the history of science. It is also generally understood that Darwin's book sold impressively fast for a "technical" volume of natural history, and that Darwinian ideas spread like wildfire, despite some resistance in both academic and theological circles. Before the end of the nineteenth century, the theory of evolution by natural selection was widely accepted not only in England but also in the United States, Russia, and the greater part of Europe.

What is less well known, but is important for understanding the complex history of evolutionary biology, is the active role that Darwin and two of his colleagues took in spreading their ideas among not only scientists, but the public at large. This less recognized part of the story is told by my colleague at the University of Tennessee, Edward Caudill, whose book is listed under "Additional Reading" at the end of this chapter. I will only briefly summarize the main points here.

Most people recognize Thomas Huxley as "Darwin's bulldog," a nickname he richly deserved for his vigorous attempts to defend the fledgling scientific theory. But Huxley was certainly not acting alone; his attempts were part of what was probably the first concerted effort at spreading scientific ideas while at the same time popularizing them. Botanist Joseph Dalton Hooker played an equally important part in furthering public acceptance of Darwinism, and Darwin himself was exceedingly active—albeit from his remote residence in Downe, outside of London.

Darwin, Huxley, and Hooker mounted a concerted publicity campaign in favor of *The Origin,* which included encouraging the publication of favorable reviews and letters to intellectual and more popular magazines, as well as keeping track of unfavorable commentaries. Darwin realized early on that the controversy was going to be on two fronts: On the one hand he could expect critiques by fellow scientists on the details of his theory (the major one came from one of the most respected naturalists of the time, Richard Owen, a comparative anatomist with connections to the royal family and the Anglican Church). On the other hand, Darwin knew that his doctrines would draw fire from people unwilling to accept the implications for humanity and religion, even though he was very careful in barely mentioning the import of his theory for our species.

---

[1] Wallace and Darwin presented a joint paper at the Linnean Society in 1858, a cooperative undertaking to which Darwin was forced by the urging of Joseph Hooker, who knew of the work of both scientists.

Huxley was also keenly aware of the theological implications of Darwin's work, and he embraced them while at the same time demonstrating an uncommon degree of political savoir faire. As we shall see, Huxley made no secret of his anticlericalism and had no reservations about pounding theologians who opposed Darwinism. But he also invented the term *agnostic* to deal with a delicate situation. He had been accused of atheism because of the publication of an article entitled "The Physical Basis of Life," but he labeled himself an agnostic to show that he had no inclination to accept or reject belief in the existence of a God given the lack of evidence one way or the other. Darwin also adopted the label, which to some extent shielded him, too, from the very same association that modern creationists seek to pin on all evolutionary biologists. And modern biologists—with some exceptions—by and large still adopt Huxley's move for the very same reasons. The connection between religion and science in the history of evolutionary biology was obvious to the protagonists of the story from the very first day, and this history is in part defined by the tactical moves on either side to either deflate or revamp that controversy.

Darwin kept track of "outsiders" and "insiders" in the debate beginning with the publication of *The Origin*, going to the extent of actually drawing up a list of opponents. Similarly, Huxley wrote to Hooker to enlist him in a plan for the two of them to write scientific articles for the *Saturday Review* favorable to Darwin's views. Indeed, in 1861 Huxley established a new magazine, the *Natural History Review*, with the express purpose of enlisting young scientists in support of the new theory.

Darwin is usually depicted as distant from the controversy and mostly absorbed by the continuation of his studies, as well as by his health problems. Yet in 1860 he asked Huxley what he thought about reprinting a favorable review of *The Origin* signed by the Harvard botanist Asa Gray. Eventually Darwin sent copies of the review to the *Natural History Review* and even paid for advertisements. Interestingly, Darwin wrote to Huxley, "I sometimes think that general and popular treatises are almost as important for the progress of science as critical work."

This concerted strategy (and the fact that Darwin's ideas were in fact scientifically valuable) paid off. By the mid-1860s Darwin had won over most British scientists and was tracking his progress in other European countries. By the end of the decade things looked even better. Hooker had become president of the British Association for the Advancement of Science in 1868, followed by Huxley in 1870. The book had already seen four British editions, two editions each in the United States, France, and Germany, and one each in Italy and Holland, plus a number in Russia.

Darwin, Huxley, and Hooker ushered in a new age for science, one in which concern for the public acceptance of scientific ideas was as important as their acceptance among peer scientists. These men understood—much like creationists understand today and scientists seem not to—that in a democracy, while it is important to come up with new ideas, it is equally crucial to defend and nurture them in the public's mind. The plan put together over the years and skillfully carried out by Darwin and associates does not explain why evolutionary theory became accepted in scientific circles. It did because it was good science. But it does explain why the acceptance came so rapidly and especially why the theory became so well known among the general public, perhaps for the first time in the history of science.

Naturally there were confrontations; the epic struggle between Huxley and Bishop Samuel Wilberforce immediately comes to mind. As a few modern commentators have pointed out,[2] however, the way things actually went is a bit unclear and may have been different from the myth that became common knowledge during the twentieth century. The problem is that both the classic account and the more recent attempts at revisionism are suspect because in both cases one is faced with a combination of few facts and a significant dose of ideological bias (on the revisionists' part, such bias may arise simply from the fact that if they didn't have a new story to tell, they would not have anything to say, and therefore to publish).

Let us first briefly examine the classical rendition of the Huxley–Wilberforce debate, moving then to some of the revisionist critiques, and finally to my personal assessment of the situation based on my own experience of participating in creation–evolution debates. The day that made "Darwin's bulldog," as Caudill puts it in his account, was Saturday, June 30, 1860, during the thirteenth meeting of the British Association for the Advancement of Science held in Oxford. The meeting had already seen a discussion of evolution, with a negative paper presented by Richard Owen and a rebuttal by Huxley. But Saturday was to be a special day because it included the equivalent of a long panel discussion, with about a dozen pro- and antievolutionists lined up to speak—among them Bishop Samuel Wilberforce (known by the nickname "Soapy Sam" because of the slickness of his rhetoric), Huxley, and Hooker.

Wilberforce spoke for a whole hour, by most accounts with great eloquence and wit, though equally clearly with little knowledge of the science behind evolution (he had apparently been hurriedly primed by Owen about the weaknesses of the Darwinian theory). One of the crucial moments that produced the legend allegedly came toward the end of his speech. Here is how Caudill puts it in his *Darwinian Myths:* "Apparently taken by the force of his own rhetoric, the bishop turned to Huxley and asked him if it was through his grandmother or grandfather that he claimed descent from a monkey. At that point, Huxley turned to the man seated next to him and whispered, 'The Lord hath delivered him into mine hands.'" The rest is—we are told—history. When the young Huxley took the podium, he told the audience that he had heard nothing scientifically valid against Darwin's theory, which he then proceeded to explain briefly. Finally, with a grave and quiet tone, he declared that he would

[2] Among them, Caudill in his *Darwinian Myths: The Legends and Misuses of a Theory* (Knoxville: University of Tennessee Press, 1997) and Stephen Gould in an essay entitled "Knight Takes Bishop?" published in *Natural History* in May 1986.

rather have an ape for a grandfather "than a man possessed of great means and influence who employed those faculties merely to introduce ridicule into serious scientific discussion." The house came down with cheers and applause, a woman fainted from the sheer power of the intellectual crisis, and Huxley became the white knight who defeats the evil bishop.

Or so many scientists like to think of the incident. But however much everybody needs heroes, it is unlikely that things went *exactly* that way. The story is too neat, and it smells of after-the-fact rearrangement. It has too much of a Hollywood flavor to it to represent real life accurately. Furthermore, we know from letters and newspaper articles that the debate was not considered very important at the time it happened, and that several commentators saw the arguments presented by Hooker, not Huxley, as dealing a more serious blow to the anti-Darwinian position. Darwin himself was ambiguous about this point (and he did not attend the debate anyway). He wrote to Huxley, jokingly asking whether he did not feel sorry for having attacked a "live Bishop . . . I am quite ashamed of you . . . By Jove, you seem to have done it well." Yet a few years later Darwin gave the credit to Hooker, writing that the botanist "did the Bishop so well at Oxford." Caudill suggests that Darwin was, as usual, piling praise on all his supporters, but of course it is also possible that he was reflecting the actual course of events because both Huxley and Hooker had attacked Wilberforce at the Oxford meeting.

Because of scant contemporary coverage by the newspapers, other details of the event are not clear. The room was full and noisy, so it is possible that part of the audience didn't even get to hear Huxley's rejoinder, and most likely the woman who fainted did so because of the heat, not the intellectual shock. What is definitely true is that *both* protagonists were convinced of having won the debate, and that they were reinforced in such beliefs by correspondence they exchanged with their respective supporters. Just one example will suffice to make the point. In 1887 the *Times* of London published a sharp exchange between Huxley and Wilberforce's son, Reginald. The latter complained about the tone of Huxley's response to a review of *The Origin of Species* and insinuated that Huxley was still hurt by the treatment he had received from the bishop in Oxford. Huxley replied with characteristic sharpness: "Those who were present at the famous meeting in Oxford, to which Mr. Wilberforce refers, will doubtless agree with him that an intellectual castigation was received by somebody. But I have too much respect for filial piety, however indiscreet its manifestations, to trouble you with evidence as to who was the agent and who the patient in that operation."

Of course, Huxley did in the end win in the broader sense of being on the side of evolutionary theory. As often is the case in history, the victors do have the privilege of—if not rewriting—certainly embellishing history. In particular, most of our modern perception of the debate was due to the efforts of Francis Darwin and Leonard Huxley, the sons of the two great evolutionists and hardly dispassionate observers of the unfolding drama in which their fathers had been protagonists. Furthermore, as both Gould and Caudill have pointed out, the whole story just fits so perfectly what liberals and scientists wished had happened: The young, self-educated, humble scientist and defender of the truth threw the old, privileged, conservative theologian into the dust. Humans do need myths, no matter how rational their general attitude toward life. Yet many myths are just the sanitized and enlarged versions of truths, and just as the Bible's flood was probably a local one, Huxley's victory was decisive only in the hindsight of history.

This whole story rings very familiar to someone like me who has direct experience of debating creationists. Indeed, I already have seen my little version of the Huxley–Wilberforce debate, mine having evolved over the span of just a year or so. In 2000 I debated creationist Kent Hovind in Knoxville and he used the very same tactics that "Soapy Sam" had used against Huxley: He knew nothing of science, but he was very witty and kept trying to throw ridicule on the other side. One of his rhetorical devices was to ask a child in the audience if he believed that humans came from bananas. The kid obviously replied no, and Hovind proceeded to accuse evolutionists of being denser than a 7-year-old in insisting on their absurd pronouncements.

Because I had a computer that had been used for my main presentation at the debate, and I had a few minutes to prepare the rebuttal, I quickly opened a new file and imported clip art of a banana, putting it next to a picture of Hovind. I then seriously explained to the audience that evolutionists do *not* believe that humans came from bananas and proceeded to jokingly insinuate (at the same time that I was showing the slide) that perhaps Hovind needed therapy to deal with his obsession about bananas (he had used the same exact tactic in a previous debate in Atlanta).

The interesting thing is that the audience laughed at both our antics and that we both received testimonials—verbally and in writing—from people who were at the debate affirming that we had "smashed" the opponent. Furthermore, the local newspapers made no mention of the banana incident in their coverage of the debate. More than a year later, at the time of this writing, the banana story has been retold by some of my supporters many times over, and I can already see the whole episode taking on larger proportions in my own mind, with some of the details being unwittingly distorted by people who report them (several of whom were not even present at the debate). I suspect that, on a much larger scale, this is pretty much what happened to the Huxley–Wilberforce exchange. Nothing like playing little Julius Caesar to understand what it felt like when the great general crossed the Rubicon. *Alea iacta est.*[3]

The Evolution of Creationism in America

Toward the end of the nineteenth century the action concerning the evolution–creation controversy moved from Europe to the United States, where it

[3] Latin for "the die is cast."

still is today (with the exception of fundamentalist Islamic countries). It is therefore interesting to follow this transition in some detail and to start by asking—as Ronald Numbers does in his *Darwinism Comes to America*—where the word *creationism* first originated.

Before the publication of *The Origin of Species,* the term *creationism* referred to the idea that each human fetus's soul was specially created, contrary to the alternative doctrine of traducianism, according to which souls were inherited from one's parents like genes (though at the time, of course, nobody knew anything about genetics). Despite this common usage, Darwin employed the label *creationists* to refer to people who opposed his ideas on descent with modification. Asa Gray, the chief American scientist championing evolutionary theory, wrote in 1873 of "special creationists" as those people who believed that species had been created supernaturally in their existing forms. Even so, prominent antievolutionists of the nineteenth century never referred to themselves as creationists. These included Louis Agassiz, who directed the Harvard Museum of Natural History until his death in 1873 and who in 1867 famously said, in reference to Darwinism, "I trust to outlive this mania."

According to Numbers, the chief reason why the label *creationism* was not consistently used by antievolutionists was pretty much the same one that is put forth today by proponents of the intelligent design movement who wish to distance themselves from young-Earth creationists (see Chapter 2): There is simply too great a diversity of opinion among evolution deniers, and the only commonality of these views is their objection (on moral grounds) of the perceived philosophical implications of Darwinism. For example, geologist Charles Lyell (who—ironically—had a great influence on Darwin) believed in various centers of creation where new species appeared as needed during the history of life on Earth. The Swiss scientist Arnold Guyot invoked only three miracles (for the separate creation of matter, life, and humans), while Agassiz himself advocated an infinite number of them.

After geologist John William Dawson died in 1899, there was not a single prominent scientist left in the United States to defend creationism, and the movement became entirely religious in nature—a position that, despite the protests of some of its most vocal proponents, it retains to this day.

The conservative, mostly evangelical, Christians who insisted on denying evolution at the end of the nineteenth century could still not rally around the banner of creationism for the simple reason that they had major disagreements about the interpretation of Genesis 1, a disagreement that has persisted through the twentieth century, as we shall see in Chapter 2. William Jennings Bryan, who we will encounter in a moment when examining the Scopes trial, endorsed the day-age theory of the interpretation of Genesis, according to which each day in the Scripture corresponded to a geological age. Bryan's position was very popular, rivaled only by the *gap theory,* which assumed a gap between different passages of Genesis to be able to reconcile a geological span of time with the classic six-day account. In fact, according to Ronald Numbers, pretty much the only people believing in a literal interpretation of the creation story were the followers of Ellen G. White, the members of the Seventh-Day Adventist Church. If this is a correct picture of the situation, it points to an interesting irony: late-nineteenth-century American Christians were in many ways much more accepting of science than their late-twentieth-century counterparts. The reason, ironically, can be found in the strong push toward evolutionism begun from within Christian circles at the turn of the nineteenth century.

The charge was initially led by the Adventists, and chiefly by their self-instructed geologist, George McCready Price. Price proposed the idea that scientific evidence was actually in agreement with the biblical story of the flood; with this idea, Price was anticipating modern "flood geology" and "scientific creationism" by more than half a century. Indeed, he even considered starting a magazine called *The Creationist,* a sign that the term was being adopted in the modern sense. One of Price's early converts, Dudley Joseph Whitney, created a society in the mid-1930s that would unite fundamentalists against evolution. His Religion and Science Association, however, died after a couple of years because—once again—its members vehemently disagreed on the interpretation of Genesis 1.

In 1941 a more progressive organization was established, the American Scientific Affiliation, whose members attacked flood geology as pseudoscience while at the same time defending their own version of creationism. Anthropologist James O. Buswell III even urged ASA members and other evangelicals to unite under the label of scientific creationists. Another primary exponent of ASA, Bernard Ramm, rejecting the ideas of a young Earth and a universal flood, told evangelicals to repudiate what he referred to as "narrow bibliolatry."

It is arguably this wave of attempts at shifting the creationist movement toward a more progressive position, close to the idea of theistic evolution that we will discuss in Chapter 2, that provoked the inevitable backlash. The reaction crystallized in the publication of John C. Whitcomb and Henry M. Morris's *The Genesis Flood* in 1961, which eventually led to the establishment of the current Institute for Creation Research in the 1970s.

At the beginning of the twentieth century, the creationist battle, initially not clearly identifiable with any particular area of the United States, increasingly saw the South as the uncontested leader, both in terms of the legal actions (attempted or effected) and of the number of prominent exponents among the evolution deniers. Although Numbers wisely warns that the reality was more patchy and complex than just "the South against the rest of the world," it is certainly true that some of the most relevant episodes of our story took place in that region of the country. Ironically, the argument has been proposed (by Numbers) that it was the very early success of Darwinian ideas in the South that generated the backlash against evolution from which we still suffer today. Of course, it remains to be explained why such a backlash did not initially take place in other regions of the United States—though recent statistics from the

National Center for Science Education do attest to the fact that the problem has now spread almost everywhere.[4]

First we need to understand that—contrary to a belief popular in certain circles—the South was neither more nor less populated by scientists and academicians than any other part of the country. Before the Civil War, the South and surrounding regions had the same number of leading scientists per 100,000 white residents as the states in the North and West (although it is not clear if similar statistics apply to other measurements of scientific involvement and literacy). Although many southerners before the publication of *The Origin* had already pushed for a reinterpretation of the Genesis story in light of modern geological findings, the Presbyterian theologian Robert Lewis Dabney complained about what he considered the continuous encroaching of science on the Scripture, referring in particular to the ideas of an old Earth, a local flood, and the nebular hypothesis of the origin of the solar system.

Perhaps the most famous and symptomatic episode was what has come to be known as the *Woodrow affair,* after what happened to James Woodrow, a former student of the antievolutionist Agassiz and the first occupant of the Perkins Professorship of Natural Science in Connection with Revealed Religion established at the Columbia Theological Seminary in South Carolina in 1861. The goal of the professorship was "to forearm and equip the young theologian to meet promptly the attacks of infidelity made through the medium of the natural sciences," and Woodrow was hired because of his profession of absolute belief in the inerrancy of the Bible.

The problem was that Woodrow also had an open mind, and by 1884 he had seen and read enough about evolution to come to the conclusion that it was probably true. He had the temerity of actually saying so in front of an assembly of alumni of the seminary, which immediately catalyzed a vehement attack against him. Woodrow was eventually fired and even tried (and convicted) for heresy by the presbytery of Augusta. Only newspapers in the North came to Woodrow's assistance, with the *Presbyterian Journal* of Philadelphia labeling the whole affair "the ecclesiastical blunder of this generation" and the *Interior* of Chicago calling it "a thundering fiasco."

The Woodrow affair was by no means a common outcome of the clash between moderate and conservative creationists, but it wasn't an isolated case either. Another instance saw the expulsion of geologist and theistic evolutionist Alexander Winchell from the Methodist Vanderbilt University in Nashville, Tennessee—though in that case Winchell's abolitionist positions probably played an even larger role than his views on evolution.

[4] As we shall see in Chapter 3, I think the more recent upsurge of creationism nationwide goes back to the swing toward conservative politics beginning with Richard Nixon, followed by Ronald Reagan's courting of the Religious Right at the same time that Reverend Jerry Falwell formed the Moral Majority. This change in the political climate, combined with public illiteracy in science and the increasing political militancy of the Religious Right, made an upswing in creationism inevitable.

The beginning of the twentieth century looked a bit better for evolution supporters, mostly because of the backlash against the Woodrow case. Henry Clay White was able to teach evolution at the University of Georgia and even planned a special celebration in 1909 for Darwin's hundredth birthday—though he had to host the event at his home because of concerns of propriety raised by the chancellor of the university. In the same year, the dean of the University of South Carolina acknowledged progress since the time of Woodrow, but also warned that the point had hardly been reached where one could make the subject of evolution prominent in the school's curriculum.

Things started picking up for the evolution deniers after World War I, when fundamentalists increasingly blamed the teaching of evolutionary theory for society's deep problems—a tactic that has not abated to this day. This increase in momentum led to what Numbers refers to as a witch hunt against evolutionary biologists in Southern Baptist colleges during the 1920s. Simultaneously, many southern legislatures considered antievolution laws, with Arkansas, Florida, Mississippi, Oklahoma, and Tennessee actually passing them in one form or another (Florida, for example, declared the teaching of Darwinism "improper and subversive"). While several major southern newspapers fought against the fundamentalist onslaught, fundamentalist preachers such as J. Frank Norris gained enough influence to fire college professors, as in the case of a sociologist at Baylor University in Waco, Texas (the current institutional home of neocreationist William Dembski[5]).

During the last decades of the nineteenth century and the first ones of the twentieth, pockets relatively favorable to Darwinism were to be found in scientific circles, although the Charleston Circle in South Carolina included many more critics than defenders of evolution. But outside such circles, the anti-Darwinian sentiment was brewing and was soon to explode into a full-fledged crusade that led to the historic confrontation in Dayton, Tennessee—the Scopes "monkey" trial.

Yet Another "Trial of the Century"

So much has been written on the Scopes trial that I will limit myself to a brief summary here. On the other hand, this episode should not be overlooked, as it is not only historically important but provides a good example of the ideological foundations of the creation–evolution controversy. (A good in-depth history of the trial can be found in Ed Larson's book listed under "Additional Reading" at the end of this chapter.)

The monkey trial took place in the summer of 1925 in Dayton, Tennessee, and saw the participation of some of the major legal minds of the time, par-

[5] Dembski himself ran into problems with his home institution, but he did not lose his job. The trouble was generated by the fact that he set up a center for intelligent design "research" within Baylor University without knowledge of many of the faculty, an unorthodox procedure that would get anybody into trouble, creationist or not.

Figure 1.2 The courtroom in Dayton, Tennessee, during the Scopes trial. Clarence Darrow (defense) is standing with his fist raised. (©Bettman/Corbis.)

ticularly of controversial secularist Clarence Darrow on Scopes's defense team (Figure 1.2) and of populist religionist William Jennings Bryan prosecuting the state's case. Both Darrow and Bryan volunteered their services because of the large impact they thought the dispute was going to have on issues of religion and academic freedom.

The story inspired the famous (or infamous, depending on whom one asks) play and Hollywood movie *Inherit the Wind* (the original production featured Spencer Tracy, Fredric March, and Gene Kelly; there are now three cinematographic versions available). As many recent commentators have pointed out, the movie is a bit far from the reality of events—not surprising, given the fact that it is a Hollywood production. Of course Scopes did not fall in love with the fiery preacher's wife, and Bryan did not die in the courtroom at the end of the trial (but a few days later, while still in town). However, I also think that recent revisionist critics of the play/movie go too far when they assert that the production gives a completely distorted view of the events. On the contrary, most of the essential facts are there, much of the court dialogue is original or close to it, and the movie certainly succeeds in reconstructing the circus atmosphere of the original events.[6]

[6] I have actually been in Dayton during the annual Scopes Festival, visited the courthouse (still in use), and even attended a play based on a reconstruction of part of the actual court proceedings. (I kept feeling that Spencer Tracy was going to show up at any moment.)

What is true about the dramatic rendition of the trial is that it was not intended as a comment on the events of the time, but was rather the anti-McCarthyism message of playwrights Jerome Lawrence and Robert E. Lee (the play was published in 1955) and of movie director Stanley Kramer (the movie was released in 1960), a message promoted by other writers and filmmakers in that period (e.g., Arthur Miller in his 1952 play *The Crucible,* about the witch hunt in Salem, Massachusetts, that occurred in 1692).

An important point to understand about the Scopes trial is that, to a certain degree, it was actually planned. The then-fledgling American Civil Liberties Union actively advertised for a local teacher to challenge the Tennessee law against the teaching that humans descended from a lower rank of animals, and some of the prominent people in Dayton talked Scopes into volunteering for the trial, hoping to bring publicity to the town and thereby "put it on the map." This backdrop to the events, however, does not alter—contrary to what several modern commentators and creationists maintain—the genuine fact that there was an antievolution law on the books and that it was legally challenged. The events that led to the trial were staged to some extent, but the legal concerns and proceedings were independent of any such staging and were themselves serious and significant.

The details of the exchanges during the trial are interesting as much for their historical significance as for the light they throw on the positions and beliefs of the various participants. But the most important thing to consider here is what the outcome was and why it came about. Scopes was in fact convicted, and justly so given that he had acknowledged teaching evolution when it was illegal to do so (there are doubts that Scopes really did teach anything like evolution in his class, but that issue also is only marginally relevant to the general principle of teaching evolution discussed here). The judge refused to hear the testimony of scientists brought in to explain evolution because he considered the matter irrelevant to the issue at hand—and from the narrow viewpoint of a local court he was certainly right. Even the famous (or, again, infamous) cross-interrogation of Bryan by Darrow on the Bible was immaterial to the problem at hand (i.e., Scopes's alleged violation of a state law), even though it provided a revealing look at the complex psychologies and ideologies of both protagonists.

Indeed, the outcome of the proceedings was exactly as the ACLU had hoped. The ACLU was planning to bring the matter eventually to the U.S. Supreme Court to challenge the *constitutionality* of the Tennessee law, regardless of the fact that Scopes had broken it. In fact, the ACLU did not see Darrow's participation in the Scopes trial as a good thing because they wanted to move to the level of higher courts as soon as possible, and the chance that Darrow's abilities would actually succeed in getting Scopes acquitted was a constant nightmare during the trial. Unfortunately for the ACLU, the judge made a technical error in deciding the fine that Scopes had to pay (instead of leaving that to the jury), so the conviction was overturned by the Tennessee Supreme

Court on a technicality. Because the state decided not to pursue another potentially damaging trial, the ACLU had to wait a lot longer for its aim to be realized. One of the consequences was that the Butler Act (as the relevant law was known) remained on the books in Tennessee until it was repealed in 1967.

Although legally the Scopes trial was a victory for creationism, the developments over the next few decades revealed themselves as a mixed blessing for both sides. On the one hand, evolutionists were able to declare moral victory because of the shame that most of the press piled upon Dayton and the South during and after the trial itself (H. L. Mencken's writings on the episode for the *Baltimore Sun* still make for highly entertaining or irritating reading, depending on which side one is on). This moral victory translated to some extent into a quiet acceptance of the teaching of evolutionary ideas throughout the South. On the other hand, the underlying tension never disappeared, eventually erupting in the modern creation science movement of the 1970s, which I will consider in Chapter 2. Tennessee and much of the rest of the country are still in a situation of stalemate when it comes to the creation–evolution issue, and the Scopes trial hasn't helped to move the situation in one direction or the other.

Bryan's Last Speech

Before we leave the Scopes trial to consider more recent events in the long history of evolution denial, it is worth our while to take a close look at William Jennings Bryan's last speech, which was delivered immediately after the end of the trial itself. It is a rare, in-depth example of the mind of a creationist, and it speaks to one of the main goals of this book, which is to understand why we have a problem in the first place.[7] It is difficult to find the entire transcript of the speech, so it is reproduced in Appendix B of this book for the reader's convenience. Here I will refer to some of the main sections of the speech and comment on what they tell us about the creationist mind.

The speech starts with a bit of rhetoric that was probably to be expected, given the occasion. Bryan's nickname, "The Great Commoner," was well deserved, given his history of public service, and he does start out by praising "the stern virtues, the vigilance and the patriotism" of the class from which the jury is drawn.

Interestingly, he then immediately goes on the defensive, assuring his audience that he is not about to make a speech against academic freedom. As we will see in this section, academic freedom is often invoked by both creationists and evolutionists, albeit in a rather asymmetrical fashion. Bryan, however, also reminds us that teachers are paid by the state, and they should be teaching what the state tells them to teach. This is a very delicate issue indeed, because

[7] For another commentary on Bryan's piece, see M. Shermer, "Why Creationists Fear Evolution," *Skeptic* 4(2) (1996): 88–89.

it is certainly true up to a point, while it is also true that the state cannot pretend to have hired robots that blindly follow what they have been told to present in class. The crucial point as Bryan saw it was that the Tennessee statute was not meant to force any particular religion on anybody (though this is in strident contrast to the repeated references to the sanctity of Christianity throughout the speech), and that the majority has to be able to defend itself from the attacks of an "irresponsible oligarchy of self-styled 'intellectuals.'" This is a clear case of anti-intellectualism, one of the deep roots of the whole controversy (see Chapter 3), and Bryan's position is based on the mistaken idea that a democracy means a rule of the majority not only in political matters, but in all others as well (including science and science education).

Bryan defines evolution by evoking the words of Howard Morgan, a 14-year-old boy who was taught by Scopes, and proceeds to "show" that the notion itself corrupts the young mind, claiming that there is no need for expert testimony. This is a tactic still common among creationists today because the whole point, as far as they are concerned, is not really to establish the scientific credibility of evolutionary theory, but to prevent its alleged moral implications from being accepted by the population. Bryan was conscious that he was fighting a battle in front of the whole world, not just for the citizens of Dayton, Tennessee, and he wanted people to realize that the state of Tennessee has "a high appreciation of the value of education."[8]

Bryan says, apparently with the force of full conviction, that Christianity is well known for welcoming the truth, no matter what its source. It appears that he was not aware of the trials of Giordano Bruno or Galileo Galilei, or of the clearly documented fact that the official positions of the Christian Church (Catholic or otherwise) have consistently been based on the denial of any advances in the natural sciences until the latter are so evident that ridicule piles up too high at the church's door. Pope John Paul II, for example, waited roughly 350 years before "pardoning" Galilei.

As we shall soon see (Chapter 5), one of the mantras of modern creationists is that evolution is not a truth—which no scientist claims it to be—but a "mere" hypothesis, by which they seem to mean a guess or hunch. "Merely an hypothesis" is how Bryan was already describing the theory of evolution in 1925. He went on to suggest that this theory did not explain the origin of species—contrary to the title of Darwin's masterpiece—and that it would soon follow the fate of the British naturalist's other great theory, sexual selection, which in Bryan's time was temporarily out of vogue among biologists.

Ironically, not even many biologists realize that *The Origin of Species* does explain what its title purports to explain, by the radical suggestion that species, in a deep biological sense, do not really exist; a species is just one step along an

[8] I actually live and teach in the state of Tennessee, which, at least since I moved here, has shown a remarkably callous attitude about higher education, slashing funds for it repeatedly, even during times of economic prosperity. I am told this is a trend that has not significantly changed from a time long before my arrival.

almost seamless continuum of biological forms. This idea is still under intense debate among evolutionary biologists. What is clear by now, however, is that Darwin was largely correct in proposing a major role for sexual selection as an explanation for a variety of strange things, such as the absurdly large and colorful tail of the peacock because of female sexual choice.[9]

Bryan shows himself to be well versed in scientific controversies when he mentions that the theory of evolution is demonstrated to be impossible by modern chemistry, which he considered the queen of sciences. Indeed, one of the obstacles that Darwin had to deal with was Lord Kelvin's contention—based on the understanding of physical phenomena of the time—that the sun hadn't been around long enough to permit the long time spans that evolution needed. Kelvin, as we now know, was wrong because he could not have known that nuclear, not chemical, energy fuels the sun. But modern creationists such as Duane Gish still invoke the "queen of sciences" (today, clearly physics) to "demonstrate" that evolution is impossible—though no physicist today is on record as attempting to repeat Kelvin's mistake. (For a discussion of physics and evolution, see Chapter 6.)

Another attack against evolution employed by Bryan is—again—a staple of modern scientific creationism. He complains of the inability of evolutionists to produce all the intermediate links between extant species, a task that is clearly impossible in principle and not necessary to test the theory in a scientific context (for a discussion of the alleged incompleteness of the fossil record, see Chapter 6). With a sudden shift of emphasis, however, Bryan claims that the "facts" of religion—such as that souls are drawn to heaven—are as well established as the fact that all matter is attracted to the center of Earth. This is a great example of the persistent double standard in matters of burden of proof that is typical of creationists.

The Great Commoner seems occasionally to be confused and to defend different versions of his doctrine. For example, at one point he says that the Tennessee statute did not prohibit the teaching of evolution up to but excluding the origin of humans, a position similar to the one expressed by Pope John Paul II, who, however, limited his restriction to the origin of consciousness (if that is what one could mean by God's instilling the soul directly into primitive humans). Bryan wants to make the point that the law under discussion at the Scopes trial—far from being coercive—was actually quite conservative if one considers that the good citizens of Tennessee probably held (and still largely hold) much more radical biblicist views of biology. Bryan even pleads for fairness to the human race, claiming that it is simply reasonable to make a funda-

[9] It is interesting to note that although some feminist theorists have accused Darwin of being a male chauvinist, his idea of sexual selection was remarkably revolutionary by the standards of Victorian England, which also explains why it took biologists so long to accept it. Victorian white males had no trouble relating to the idea of males fighting against each other for the possession of a female—as Darwin also showed happens in many species—but they were extremely recalcitrant to admit that *females* could choose their male partners, instead of the other way around. See Geoffrey Miller's *The Mating Mind* (New York: Doubleday, 2000).

mental distinction between humans and every other animal on Earth. What justifies this distinction, other than pure self-conceit, is something that psychologists, philosophers, scientists, and even science fiction writers have discussed ever since. But the conclusion that Bryan draws is that even if one could make the case that evolution should be taught, most certainly children should be spared such ignominy.[10]

After the first third of his speech, Bryan gets to the heart of the matter—the only reason why fundamentalist religionists have a problem with evolution: "Evolution not only offers no suggestion as to a creator but tends to put the creative act so far away to cast doubt upon creation itself." In other words, Darwin was sweeping humanity from the pinnacle of creation, much in the same way that Copernicus had swept Earth from the center of the universe, and that modern cognitive sciences are threatening to reduce human consciousness to the intricate, marvelous, but mechanical workings of a machine. This is really the crux of the matter, which is why the evolution–creation controversy is not about science, but about philosophy and religion. Indeed, Bryan immediately warns his audience to pause and consider carefully before "accepting a new philosophy" built on materialistic foundations.[11]

At times Bryan verges on the hysterical, as when he says, "It is all animal, animal, animal, with never a thought of God or religion." Why *should* we hear about God and religion in a science class? This is yet another reflection of the different worldviews that clash in the controversy: For a religious fundamentalist, religion not only *is* reality, but is the *only* reality there is, which means that it has to permeate every aspect of private and public life. This is the root of the perpetual conflict with the authority of the state and with the discoveries of science.

The second half of Bryan's speech is built around five "indictments" against evolution, which I will examine in turn. The first indictment is that evolutionary theory disputes the account of creation as reported in the Bible. This obsession with using the Bible as a book of factual truths about the universe as opposed to spiritual insights marred the history of Western religion throughout the Middle Ages and Renaissance. With the advent of the Enlightenment, the Catholic Church and the mainstream Protestant denominations have progressively retreated from this increasingly untenable position. But it remains at the heart of fundamentalist doctrine.

The second indictment is that evolution allegedly puts the faithful on a slippery slope. Once one questions the literal truth of the Bible, Bryan asserts, one slides progressively into agnosticism and finally into the opprobrium of atheism. It is interesting here to pause for reflection and observe that Bryan is partially

---

[10] Notice that Bryan here was adopting the same attitude that a possibly apocryphal story attributes to a respectable Victorian lady upon hearing Darwin's theory for the first time: "Oh, let us pray that it is not true, but if it is, let us hope that the news will not spread!"

[11] On the philosophical implications of the theory of evolution, see the excellent article by Barbara Forrest, "Methodological Naturalism and Philosophical Naturalism: Clarifying the Connection," *Philo* 3 (2000): 7–29.

right and partially off the mark. It is certainly true—as Bryan himself remarks a bit later when commenting on the work of psychologist James Leuba—that there is an inverse relationship between literal belief in a personal god and degree of education (not just of acceptance of evolution). In this sense the most educated among us do tend to be more skeptical of a young Earth or a six-day creation. On the other hand, it is also true that there are many people who accept the advances of science, including evolution, and still call themselves Christians (or members of whatever other religion they belong to). This seems to indicate that while the probability of questioning Christianity increases with education, it surely does not follow logically that educated people reject religion.

In fact, Bryan himself quotes Darwin to say precisely as much: "Science has nothing to do with Christ, except insofar as the habit of scientific research makes a man cautious in admitting evidence." Darwin then goes on to say that he himself does not believe in any revelation. But as historians have pointed out, this conclusion may have had less to do with his formulation of the theory of evolution than with the early loss of his daughter to disease. In any case, one cannot draw on a single example, no matter how quintessential it is thought to be, to arrive at the sort of sweeping conclusion that Bryan wants to reach. Indeed, Darwin did not consider himself an atheist, as is clear by another passage quoted in the Great Commoner's speech: "The mystery of the beginning of all things is insolvable by us; and I, for one, must be content to remain an agnostic." But Bryan's interpretation is quite different: "Here we have the effect of evolution upon its most distinguished exponent; it led him from an orthodox Christian, believing every word of the Bible and in a personal God, down and down to helpless and hopeless agnosticism."

Bryan is of course correct in saying that "belief in evolution can not bring those who hold such belief any compensation for the loss of faith in God," but this is yet another example of his refusal to see that evolutionary theory is science, not philosophy. Why *should* an understanding of nature bring any relief (or discomfort, for that matter)? Do we expect moral guidance from understanding how the sun burns its fuel, or from the discovery that continents move on tectonic plates? But Bryan goes even further, suggesting that scientists *consciously* undermine morality and the belief in God: "What pleasure can they find in robbing a human being of the hallowed glory of that creed?" he asks, clearly baffled.

The work of psychologist James Leuba had, and still has, an important impact on the creation–evolution dispute. His results are cited in Bryan's speech as proof that evolution leads to atheism. Leuba showed that more than 50 percent of all scientists (not only biologists) do not believe in a personal god or life after death, a conclusion that has not changed much according to recent research by Larson and Witham.[12] Worse—from Bryan's standpoint—Leuba's

[12] See two articles by E. J. Larson and L. Witham: "Scientists Are Still Keeping the Faith," *Nature* 386 (1997): 435–436; and "Leading Scientists Still Reject God," *Nature* 394 (1998): 313.

work also indicated that young adults' beliefs are affected by their level of education: Unbelief increased from 15 percent in a typical freshman class to 40 to 45 percent by the time students graduated from college. Of course, this part of Bryan's speech clearly indicates that his battle is not against evolution per se, but against higher education; that is, despite his protestations of valuing education, he is the quintessential anti-intellectual.

Bryan then turns his guns against his direct opponent in the Scopes trial: Clarence Darrow. Bryan vehemently attacks Darrow's previous defense of two youngsters, Leopold and Loeb, accused of murdering another child. Bryan criticizes Darrow's defense strategy in that case because it was based on the contention that the two boys were essentially mentally disturbed adolescents who had come under the spell of the philosophical ideas of German philosopher Friedrich Nietzsche, twisting these ideas to justify their own insane acts.

It is crucial here to understand not only what Bryan is saying, but especially what Darrow and Nietzsche meant to say. For Bryan this is yet another example of how modern philosophy corrupts the mind and, if left unchecked, will eventually destroy our sense of humanity. But it is also quite clear that Bryan did not understand Nietzsche, since he accuses him (as many did afterward and are still doing) of nationalism and incitation to war, two concepts that the German philosopher clearly abhorred. Most revealingly, Bryan entirely misses the point of Darrow's defense and speech.

Darrow was a much subtler thinker than Bryan, and he needs to be understood in context. Darrow goes through a chain of reasoning according to which he initially blames Nietzsche for the influence on the two boys, then shifts the blame to the university for having taught the philosopher to such impressionable minds, and finally expands the fault to the parents themselves for the carelessness with which they had sent two mentally disturbed individuals to learn about complex philosophies. While Bryan takes all of this at face value and calls for prohibiting the teaching of any "dangerous" philosophy to the young, Darrow realizes that the problem is more complex. In the first place, almost any philosophy can become lethal in the minds of people who are already prone to abnormal behaviors for other reasons. The proof of this is that most people read Nietzsche and go on with their lives without turning violent. Second, Darrow blames the university not for teaching per se, but for teaching a large number of students with few faculty, which means that education becomes impersonal and that faculty are in no position to realize the potential dangers brewing in a particular youth's mind. This, alas, is still very much a problem in contemporary society (see Chapter 8), where education at large universities—and even high schools—is not really education at all, but rather a regurgitation of material read from a textbook, accompanied by standardized tests that hardly test for comprehension of such material. Darrow concludes that what happened in the specific case must be considered an acceptable risk because of the benefits of education for the overwhelming majority of people. As he puts it, "You can not destroy thought, because forsooth, some brain may be deranged by thought."

After dealing with Darrow, Bryan moves to his fourth indictment, which is that evolution allegedly paralyzes the hope for the reform of humankind, thereby discouraging those who labor for the improvement of the human condition. Bryan here (and in other places) is committing the naturalistic fallacy described by philosopher David Hume: He is assuming that there is a direct connection between what is and what ought to be without doing the work of justifying such an assumption. Since evolution depicts a world of struggle and slow change, if this is the only possible world our hopes and efforts are destined to be frustrated. But this is clearly a non sequitur, and many philosophers and evolutionary biologists have pointed out that knowing how things are in nature provides a stark contrast with what humans have been able to achieve through passionate commitment and intellect, so that we should be somewhat uplifted, rather than depressed, by the comparison.

Yet Bryan defines evolution as a cold and heartless process, and he sees as its logical consequence the establishment of a program of genetic breeding of humans directed by "self-appointed, supposedly superior, intellectuals." Of course, the rise of eugenics during that time—and the fact that the founder of the movement, Francis Galton, was Darwin's cousin—provided the Great Commoner with plenty of good reasons to worry, but it also suggests that Bryan (and Galton, for that matter) could not see what should have been obvious: Scientific findings—for example, in the field of genetics—ought to be judged on their own merits, regardless of the ethical connotations some people might see in them. Ethical choices, on the other hand—while they should certainly be informed by the best science available—are too important to be left only in the hands of scientists. The theory of evolution is no danger to the welfare of humanity as long as we understand that it tells us how living things have changed and come to be as they are now; it is not a guide to how we should proceed with our lives, nor was it ever intended as such by Darwin.

This confusion between the purposes of science and religion is of course based on the fundamentalists' misunderstanding of their sacred scriptures as not only books on how to live, but also descriptions of how the universe works. By the same token, then, scientific discoveries must describe not only how the world is, but how it *should* be. This is perhaps the single most tragic mistake repeatedly made by both sides of the debate, though much more often by the religious side than the scientific side.

The fifth and final indictment that Bryan raises against evolution is actually not distinct from the previous one: If used as a philosophy, evolution "would eliminate love and carry man back to a struggle of tooth and claw." But of course evolution is *not* a philosophy, though knowledge of science may have implications for people's philosophical viewpoints. Yet these implications can develop in very different directions. For example, during Bryan's time most philosophically inclined commentators accepted some version of Social Darwinism, the idea that natural selection should be extended to the way we run our society, leading to the social equivalent of the "struggle of tooth and claw" and to a naturalistic justification of unbridled capitalism. Indeed, Bryan quotes extensively—toward the end of his speech—from precisely these sorts of thinkers. But other secular commentators have taken the evolutionary message to be something completely different: Most recently, Peter Singer[13] and Elliott Sober and David Sloan Wilson[14] have emphasized the fact that humans are *social* animals, and that within societies natural selection actually favors cooperation and a reduction of interpersonal conflict (at least, within each group). These later authors are much more aware of the naturalistic fallacy and do not claim that evolutionary theory therefore tells us that cooperation is better than selfishness. But they do point out that evolution can go a long way toward explaining precisely those features of human empathy that struck Bryan, with his distorted and narrow view of evolutionary theory, as so incompatible.

Of course, it is another example of Bryan's doublethink that he can actually quote a contemporary author as suggesting that "the plea for peace in past years has been inspired by faith in the divine nature and the divine origin of man; men were then looked upon as children of one father and war therefore was fratricide. But now that men are looked upon as children of apes, what matters whether they were slaughtered or not?" It is well known that religion, economics, and the egomaniacal attitude of a few leaders have always been (and still are) the common causes for war, but this same argument is used today by creationists to blame all the maladies of humanity on evolution—never mind that most of these problems have been around much longer than the evolutionary theory has.[15]

Bryan then cites another contemporary source with which most scientists today would perfectly well agree: "Unless the development of morality catches up with the development of technique, humanity is bound to destroy itself." This statement leads Bryan to claim that science needs religion, but in fact religious hatred and intolerance can be armed by technology as well, as has been repeatedly demonstrated throughout the twentieth century and into the beginning of the twenty-first. What we really need is exactly what the quoted author said: a better-developed sense of morality. Whether this will come from religion is an open question, and certainly secular philosophy can add positively to the discussion as well. Bryan's declaration that science *fails* to address moral concerns is a profound misunderstanding of what science is about. One can accuse science of failing to solve problems it has set out to solve—such as explaining the origin of life—but one cannot accuse science of not solving prob-

---

[13] In *A Darwinian Left: Politics, Evolution and Cooperation* (New Haven, CT: Yale University Press, 1999).

[14] In *Unto Others: The Evolution and Psychology of Unselfish Behavior* (Cambridge, MA: Harvard University Press, 1998).

[15] A similar attitude was recently expressed by a U.S. senator, who declared evolution immoral because a popular rock song includes verses to the effect of "you and me, baby, we're just mammals, let's do it like mammals do it"—as if promiscuous sex became a social "problem" only after the publication of *The Origin*.

lems that are by definition outside its purview. That would be like blaming sports for not telling us the meaning of life.

As important as it was historically and ideologically, the Scopes trial was only the first legal battle waged by creationists. It was also the only one they won. Next we will briefly examine the series of subsequent legal challenges mounted by creationists throughout the twentieth century and observe the repeated appearance of the themes raised by Bryan's speech in 1925.

A String of Legal Defeats

There was a long hiatus between Scopes and the next important case, which did not occur until 1968. In that year the U.S. Supreme Court struck down an Arkansas law because it violated the First Amendment's stipulation that no state can require that teaching be constrained by adherence to the principles of any religious doctrine.

More than another decade passed relatively quietly, but the 1980s saw an eruption of important cases. In 1981 the Sacramento Superior Court decided in favor of the state of California, dismissing the plaintiff's contention that teaching evolution in public schools curtailed his children's freedom of religion. The court in fact directed the state to expand the application of its antidogmatism policy—which included discussions of how science reaches certain conclusions and the teaching of hypotheses about origins in a nondogmatic fashion—to all areas of science, not just evolutionary biology.

The following year a federal court issued a very significant ruling against a statute proposed by the Arkansas Board of Education. The board had sought to require equal treatment of evolution and creation science in public schools, but Judge William R. Overton delivered his decision after careful examination of what constitutes a science, an examination that led him to reject creationism as unscientific. This conclusion implied that the statute under examination did not have a secular purpose and was therefore in violation of the U.S. Constitution. Furthermore, Judge Overton noted the creationists' emphasis on the problem of the origin of life and its characterization as an intrinsic part of the theory of evolution. The judge realized that evolutionary theory does not actually address the problem of origin at all because it deals with whatever happened *after* the appearance of the first forms of life on Earth (see Chapters 2 and 5).

The arguments put forth by the Arkansas judge are worth examining in a bit more detail before we proceed because they sparked an interesting controversy among philosophers of science, whose professional business is precisely to understand what science is and how it works.

The controversy was published in the pages of *Science, Technology and Human Values* in 1982, and it featured Michael Ruse—who testified at the proceedings—and Larry Laudan. To understand and discuss Ruse's rebuttal, we must start with Laudan's attack on Judge Overton's ruling. Notice that neither of these men was trying to make an argument for creationism to be taught in public schools. However, they realized that the judge in this case explicitly ventured onto open philosophical ground, and they wished to examine how that ground was covered in what still is a landmark legal decision.

Laudan points out that Overton offered five criteria by which to separate science from other activities (what is known in philosophy of science as the *demarcation problem,* so defined by Karl Popper): "1) It is guided by natural law; 2) it has to be explanatory by reference to natural law; 3) it is testable against the empirical world; 4) its conclusions are tentative, i.e., are not necessarily the final word; 5) it is falsifiable."

Laudan points out further that it is simply not true that creationism does not make empirically falsifiable claims (points 3 and 5). I agree with him on this point. It is true that *some* creationist claims (such as that humans were created *ex nihilo,* from nothing) are indeed unfalsifiable as a matter of principle. But other parts of the creationist doctrine, such as the occurrence of a worldwide flood about 4,000 years ago, can in fact be subjected to empirical test: They are falsifiable, and they just happen to be false. Laudan is therefore correct that Overton has taken the teeth out of science by arguing that all of creationism is untestable and unfalsifiable. The best reason not to teach creationism as science is that it is no better than the flat-Earth theory: Regardless of what ideology inspired the theory, the theory itself is wrong.

Laudan also remarks that science includes some nonfalsifiable statements as well in its framework, such as, for example, the fact that there is a natural world made of matter and energy. As we shall see, this is exactly the basis on which some neocreationists like Phillip Johnson attack science and claim that it is not distinct from a religious position (see Chapter 2). However, Laudan may be conceding too much here. Although we all necessarily have to make assumptions about the world in order to live our lives, some assumptions represent small and others large leaps of faith, and science is distinguished by the attempt to make those leaps as small as possible—in fact, no larger than any person of common sense would make while engaged in the process of buying a used car, for example. It is one thing to assume the existence of a physical world explained by natural laws, yet quite another to posit the reality of an entirely intangible supernatural realm that we cannot inquire about and that continuously affects the physical realm. Perhaps one of the distinctions between science and religion is in fact to be found in the breadth of the respective leaps of faith that each takes in framing an understanding of the world.

Another observation made by Overton concerned the fact that the conclusions of science are tentative, unlike the dogmatism of religion (point 4). Laudan wants to argue that this also is not true because modern-day creationists espouse different positions from those of their nineteenth-century counterparts. I doubt that these shifts are as significant as Laudan implies, since creationist positions seem to *oscillate* rather than shift, with more or less literal interpretations of Genesis (or several alternative "literal" interpretations) temporarily finding themselves on the upswing only to be replaced by former positions. In other words, it is difficult to argue for progress within creationism in the same sense in which historians of science see progress in the scientific enterprise.[16]

I also think that Laudan is mistaken in his attack on Judge Overton's insistence that science deals only with natural explanations (points 1 and 2). Laudan counters that there are plenty of cases in which good science has been carried out when no known explanation was available for the phenomena under study. The case of Mendel comes to mind, in which the Benedictine monk essentially invented the science of genetics in the 1860s without any understanding of what a gene was and how it worked (understanding that came much later and culminated in the 1953 discovery of the structure of DNA by Watson and Crick). But Overton referred to a *naturalistic* framework in general, not to specific explanations in particular. Mendel never invoked God to explain his results; he just assumed (reasonably, as it turned out) that a physical phenomenon was responsible for the patterns of inheritance he was documenting. The problem with creationism is that it rejects a priori the possibility of entirely naturalistic explanations of the world. This is different from the case of science, which rejects the possibility of supernatural explanations not as a matter of principle, but of *methodology:* What kind of research would one do, what kind of methodology would one use, if the premise were that God can do whatever He pleases whenever He wishes to do it?

Laudan is correct in pointing out that the requirements of testability, revisability, and falsifiability (which are actually three aspects of the same thing) are much too weak to define what science is, and therefore to solve the demarcation problem. Indeed, philosophers of science have long recognized this point and have moved beyond the simple answer originally given by Popper[17] in terms analogous to those chosen by Judge Overton.

Ruse's response to Laudan makes most of the points I just raised but also includes some additional arguments that we need to consider briefly. For example, Ruse says, "Whatever the merits of the plaintiff's case, the kinds of conclusions and strategies apparently favored by Laudan are simply not strong enough for legal purposes." This statement essentially argues that we *know* creationism is wrong (which we do) and so we have to concoct the strongest possible legal argument, regardless of how unsound it is philosophically. Such a move comes dangerously close to being unethical and should not be pursued under any circumstances, no matter how grave the danger is perceived to be.

[16] This is most certainly not to say that I believe in a simple vision of science as marching steadily toward higher and more encompassing truths. On the contrary, science has seen and will always see—by its very nature—plenty of dead ends and blind alleys. However, I have little patience also for extreme philosophical positions that deny the very meaning of progress in our understanding of the world. The atomic bomb, countless cures for diseases, and the very computer I am using to write this book are ample proof that twentieth-century science made significant progress (for good or for evil) over its nineteenth- and eighteenth-century counterparts.

[17] See K. R. Popper, *Conjectures and Refutations: The Growth of Scientific Knowledge* (New York: Harper & Row, 1968).

Ruse goes on to say that Laudan's strategy would lead to the conclusion that creationism is very weak science, but since the U.S. Constitution doesn't prohibit the teaching of weak science, only the teaching of religion, creationists would win the day. And yet the argument can be easily made that creationism is *both* very weak science *and* religion, and therefore should not be taught in public schools.

Ruse uses an interesting example to make a point against Laudan, a point that I think is well taken. Ruse quotes from Charles Lyell's *Principles of Geology*[18] in which the British scientist, after defending his principle of uniformitarianism—the idea that we can understand past geological events in terms of currently acting natural causes—edges toward an exception for humans: "We are not, however, contending that a real departure from the antecedent course of physical events cannot be traced in the introduction of man." Ruse claims that there are two possible interpretations of what Lyell is saying: Either science (in the form of Lyell's *Principles*) has sometimes played with going beyond the laws of nature, or Lyell mingled science and nonscience. Ruse prefers the second explanation, and I agree. This explanation does not excuse creationism, because *by definition* it has to mingle the natural and the supernatural, contrary to what Lyell was doing with his geology (i.e., he didn't *have* to invoke the supernatural but could have limited himself to saying that we know little of the appearance of humans in the geological record). In fact, argues Ruse—again correctly—even going for the more charitable (to the creationists) first interpretation and allowing that science has confused the two realms in the past would simply confirm that science has evolved gradually from natural philosophy with heavy theological influences (as it was in the eighteenth and part of the nineteenth century) to the full-fledged, independent entity it is today. Since Judge Overton's talk about modern science and creationism, the distinction between the two based on the unacceptability of supernatural explanations in science has held.

Ruse does, I think, make a mistake when he attacks the falsifiability of creationism and says, "Their publications (and stated intentions) show that . . . there is no way they will relinquish belief in the Flood, whatever the evidence. In this sense, their doctrines are truly unfalsifiable." Here Ruse is confusing creationists as individuals with creation*ism* as a doctrine or hypothesis about the world, a mistake he did not make in distinguishing Lyell's writings as a scientist from the process of science in general. The fact that dowsers refuse to let go their belief that they can find water with a wooden stick, even when every double-blind test performed so far has shown that they can't, demonstrates only that dowsers are particularly stubborn and attached to their beliefs, not that the practice of dowsing itself is not falsifiable.

The take-home message from the Overton decision and the ensuing debate is that the matters relevant to this discussion are indeed complex and involve

[18] C. Lyell, *Principles of Geology* (London: J. Murray, 1830–1833).

not only science and religion, but deep and contentious philosophical and legal arguments. No wonder most people feel confused by the whole parade of arguments and counterarguments!

Let us now close the digression on Overton's decision and consider the third major ruling of the 1980s, handed down by the U.S. Supreme Court while striking down the Louisiana Creationism Act. The motivation for the decision was, again, a violation of the constitutional separation between state and church: The act provided for the teaching of evolution only when creation science was also taught. But the Court found this stipulation to damage science education and to endorse religion directly in the public schools by the very implication of the term *creation "science."*

Four more court cases resulted in legal defeats for creationists throughout the 1990s. In 1990 the Seventh Circuit Court of Appeals pushed the debate back into the area of academic freedom (where it was during the Scopes trial) when it decreed that a school district can, on the basis of the establishment clause of the First Amendment, prohibit a teacher from presenting creation science in the classroom. Similarly, the Ninth Circuit Court of Appeals ruled that the Capistrano School District had a legal right to require one of its teachers to teach evolution because it is a scientific theory, and rejected the teacher's contention that there was such a thing as a religion of evolutionism.

A particularly important decision again concerned Louisiana. In 1997 the U.S. District Court for the Eastern District of that state rejected a provision requiring teachers to read aloud a disclaimer about evolution. The idea of the disclaimer was allegedly to further critical thinking. The court didn't buy it, noting that not only were students already encouraged to think critically, but the disclaimer singled out evolution as a religious position and presented it as a view inherently opposed to that of other religions. Furthermore, for the first time the court found that *intelligent design* (see Chapter 2) is the same as creation science and is therefore subject to the same limitations. Importantly, the ruling was confirmed by the Fifth Circuit Court of Appeals, and the U.S. Supreme Court refused to hear the case, so the lower court's ruling stood.

Finally, in 2000, another decision against the idea that requiring teachers to present evolution and not creationism violates their right to free speech was delivered by District Court Judge Bernard E. Borene, who dismissed a case brought against Independent School District 656 of Minnesota. This judge, like others before, simply noted that a vast case history upholds the right of a school to have its teachers follow the curriculum (which in that case included evolution), and that the teacher's right to free speech does not extend to overriding the curriculum.

It is important to pause and reflect on the meaning of the legal battles that creationists and evolutionists waged during most of the twentieth century, especially considering that recent news of new bills introduced in several state legislatures indicates that if anything, the clash will continue with renewed energy during the first part of the twenty-first century.

The first obvious thing to note is that creationists have repeatedly shifted their tactics to adapt to previous legal decisions. Whereas the attitude during the Scopes trial was to drive evolutionists back to where they came from and to affirm the principle that the majority has the right to decide what is taught in publicly funded schools, goals have become more modest since. Creationists first shifted to the idea of demanding equal time on the pretense that creationism is a science. Failing that, they have increasingly insisted that evolution is not a science either and—if it has to be taught because it is so entrenched in academic culture—at least it has to be preceded by warning labels in the manner of potentially toxic products like genetically engineered foods and cigarettes. Proponents of intelligent design are now taking the tack of demanding that the views of scientists who disagree with evolutionary theory be taught, adopting the slogan "Teach the controversy!" in the name of critical thinking.

Particularly sad is the fact that often pro-creationist legal actions have been started by science teachers, who supposedly should know better. I think it says a lot about the sorry state of education in this country (see Chapter 8) if educators themselves don't understand the distinction between science and religion, or science and pseudoscience.

The contrast between some science teachers' views and the court's repeatedly held position on academic freedom is also very interesting, and it is one of the subtleties of the debate that most often enrages creationists. They claim that Scopes was defended by evolutionists on the ground of academic freedom, but that such freedom somehow does not apply to other science teachers who wish to present a more "balanced" view. They are right in noticing the asymmetry, but such asymmetry stems from two very reasonable differences in the two cases.

First there is the matter of the First Amendment's clause on the separation of state and church: Because creationism is a religion and evolution is not, the teaching of the former is prohibited in public schools, while that of the latter is protected. Second, academic freedom protects the right of a teacher to teach *a given subject matter* the way she thinks best, but does not extend to that teacher's ability to ignore the school's curriculum and teach whatever she pleases. Parents would certainly, and rightly, become upset if their children where taught Pythagorean philosophy instead of geometry in a math class, and the situation is no different for creationism and biology.

For all our discussion of the history of creationism, we still have not addressed a most urgent question: What exactly *is* creationism? As we shall see in Chapter 2, it is many things to many people, and as our discussion continues, the landscape of the debate will become even more complex.

Additional Reading

Alters, B. J., and S. M. Alters. 2001. *Defending Evolution: A Guide to the Creation/Evolution Controversy.* Sudbury, MA: Jones and Bartlett. A must-read for teachers and other educators, complete with a guide explaining how to answer specific questions from students in the classroom setting.

Caudill, E. 1997. *Darwinian Myths.* Knoxville: University of Tennessee Press. A delightful book by a journalist and historian. A scholarly and yet entertaining debunking of so many silly notions normally associated with "Darwinism."

Larson, E. J. 1997. *Summer for the Gods: The Scopes Trial and America's Continuing Debate over Science and Religion.* New York: Basic Books. A deserving winner of the Pulitzer Prize, the best account of the events and cultural milieu that led to the Scopes trial in Dayton, Tennessee, in 1925.

Numbers, R. L. 1998. *Darwinism Comes to America.* Cambridge, MA: Harvard University Press. A lively history of the first steps of Darwinism and creationism in the United States.

– 2 –

Evolution–Creationism 101

The arguments of these fundamentalist missionaries often involve tortured logic, a stubborn denial of the evidence, a shallow understanding, or a reckless disregard for the truth.

– Zoologist Tim Berra –

What harm would it do, if a man told a good strong lie for the sake of the good and for the Christian Church . . . a lie out of necessity, a useful lie, a helpful lie, such lies would not be against God, he would accept them.

– Martin Luther –

There is one theory of evolution, just as there is one theory of general relativity in physics. True, there are different schools of evolutionary thought that emphasize distinct mechanisms to explain organic evolution, and creationists have tried to capitalize on these differences to show that the whole field is in disarray. Yet differences among scientists are the bread and butter of scientific progress. It is through the empirically driven resolution of theoretical disagreements that science at its best progresses and yields a better understanding of the natural world. To an outsider, the distinction between neo-Darwinian gradualism and Gould and Eldredge's theory of punctuated equilibria (see Chapter 5), or between individual and multilevel selection, may create the impression that biologists talk past each other in the midst of complete chaos. But the realities of scientific practice and discourse are far from what creationist propaganda claims. Furthermore, the idea of a monolithic and unchangeable science is a dangerous myth—one that scientists and science educators should work toward eradicating.

On the other hand, even a superficial look at creationism itself clearly shows that creationists have gone to a great deal of trouble to construct what looks deceptively like a unified front to naïve outsiders. As we shall see, the difference between proponents of intelligent design theory such as William Dembski and young-Earth creationists like Duane Gish spans a theological and scientific abyss. One of the few things that these people have in common is their hatred for what they perceive as a materialistic, scientific worldview that leaves no space for God and spirituality. Intelligent design defender Phillip Johnson has proposed the idea that creationists can win by driving a wedge into what he thinks is a small but crucial crack in the edifice of science (I will discuss the Wedge strategy later in this chapter). I would like to suggest to scientists and educators that the crack in the *creationist* camp is much wider and easier to exploit, if only we stop being on the defensive and initiate a counterattack. In this chapter I will first briefly discuss the astounding variety of creationist positions, focusing in particular on the two most popular ones as embodied by the Institute for Creation Research and by the Discovery Institute and its Center for the Renewal of Science and Culture. As a counterpoint, I will then explain what evolutionary theory is really about in order to provide the reader with the minimum scientific foundations necessary for understanding the arguments in the central part of the book.

The Many Forms of Creationism

Perhaps the best classification of positions on the question of *origins,* as the broader conception of evolution is often referred to, has been proposed by Eugenie Scott of the National Center for Science Education[1] and is summarized in

[1] See E. C. Scott, "Antievolution and Creationism in the United States," *Annual Review of Anthropology* 26 (1997): 263–289.

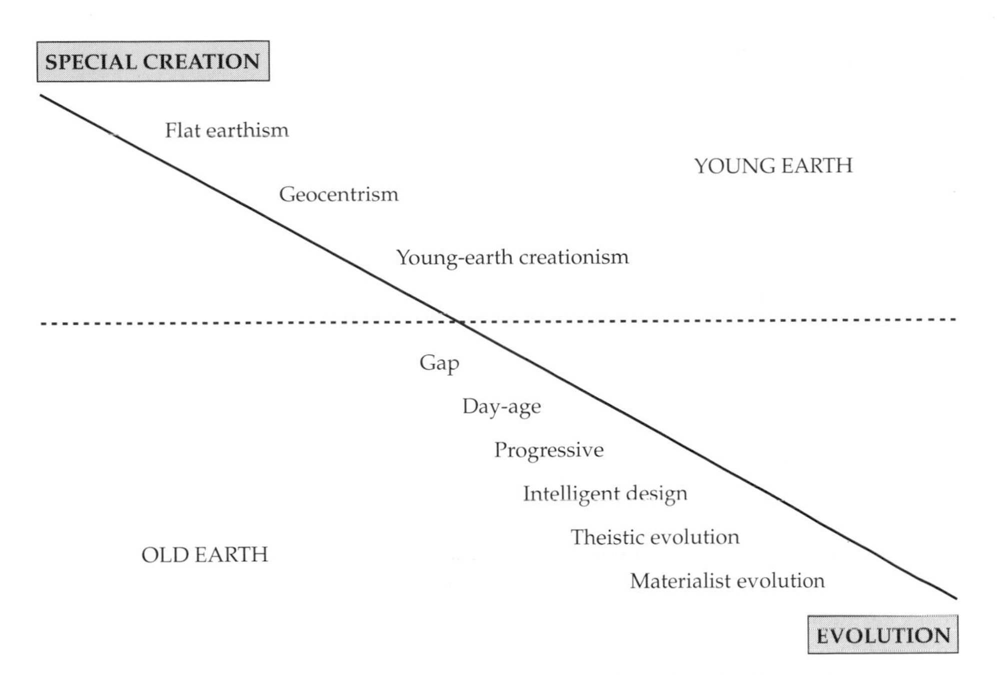

Figure 2.1 The various positions on the question of origins. (Redrawn from a summary prepared by Eugenie Scott of the National Center for Science Education)

Figure 2.1. As Scott points out, there is a panoply of ideas to choose from, and they differ in degree as to what they accept from science on the one hand and from the Bible on the other.

At one extreme we find people believing that Earth is actually flat. While these are certainly a minority even among creationists, their thoroughly literalist position is arguably the most biblical of them all: In addition to rejecting evolution, they believe not only that Earth is 6,000 years old, but that it is flat and is the center of the universe, precisely as the Bible says.[2]

A slightly different position is the geocentric one, which accepts the idea of a spherical Earth but squarely rejects astronomy after Copernicus and Galilei. This is also a minority view, but it is instructive—together with the flat-Earth position—because it is indicative of an interesting aspect of mainstream creationist thinking. On many occasions I have had conversations or have exchanged letters with young-Earth creationists (the next group in Scott's classification) who vehemently deny being gullible, antiscience individuals. They claim that the *scientific* evidence is definitely against evolution and in favor of a young Earth. I usually politely point out that the only reason they think so is that they believe in the inerrancy of the Bible in scientific as well as spiritu-

What?!

[2] See Daniel 4:11, where the possibility of seeing all the kingdoms of Earth at once is presented, and Matthew 4:8, where a mountain is high enough to see all the kingdoms of Earth. Both situations are possible only if Earth is considered flat, as it was in the Jewish tradition of the time.

al matters. I then occasionally ask why they don't believe that Earth stands still while the sun moves around it and that our planet is flat, since both notions are also present in the Bible.[3] In fact, one could argue that these two positions are much more clearly defined in the Christian scriptures than the age of Earth, which has to be calculated on the basis of assumptions concerning the life span of the lines of descent mentioned in the book.[4] The astonishing creationist response is to deny that the Bible makes claims either about a flat Earth or in defense of geocentrism. But this goes clearly against not only the existence of creationists who see and defend both claims, but also against the historical evidence: For most of Western history, Christians have espoused both views precisely on biblical grounds! It is not for nothing that both Copernicus and Galilei got into trouble with the Church of Rome.

Young-Earth creationists, however, seem to be able to live with this internal contradiction; they actually represent the majority of creationists in the United States (according to a 1999 Gallup poll, 45 percent of Americans believe that God created human beings "pretty much in (their) present form at one time or another within the last 10,000 years"). For them the story in Genesis[5] is to be taken literally: The world was created 6,000 years ago, and most humans and animals died in a worldwide flood that occurred about 4,000 years ago. As we shall see in more detail in Chapter 5, several interesting fallacies underlie this position, and young-Earth creationists are the epitome of what happens when science education fails completely. It simply makes no sense—given the evidence that we have today from a variety of fields, such as geology, paleontology, ecology, physics, and astronomy—to deny that Earth is billions of years old and that while mass extinctions certainly occurred, they were not due to floods and no such event happened on a worldwide basis so recently in Earth's history.

The next category in Scott's classification of theories of origins marks a fundamental theological, if not yet scientific, divide from the positions discussed so far: We are now entering the realm of *old-Earth creationism*, which therefore at least acknowledges modern geology as it has been conceptualized since Sir Charles Lyell and his *Principles of Geology* (1830–1833). It is also in agreement with the scientific research in this field conducted throughout the twentieth century. Within old-Earth creationism, the idea often referred to as *gap theory* is by far the most peculiar. Supporters maintain that there is a large temporal gap be-

[3] On the revolving of the sun, see Joshua 10:12–14.

[4] Bishop Ussher famously calculated that the world began on the morning of October 22, 4004 B.C.E., although since the genealogies in the Bible are given in whole years, not days or minutes, this is a good example of doubtful accuracy.

[5] Or rather, the two—quite different—stories in Genesis 1 and 2. Genesis 1 presents the following sequence of events: Day 1: sky, Earth, light. Day 2: water, both in ocean basins and above the sky. Day 3: plants. Day 4: sun, moon, stars. Day 5: sea monsters (whales), fish, birds, land animals, creepy crawlers (reptiles, insects, and so on). Day 6: humans (both sexes at the same time). Genesis 2, however, has a quite different sequence: Earth and heavens (misty); Adam, the first man (on a desolate Earth); plants: animals; finally Eve, the first woman (from Adam's rib).

tween the first and second chapters of Genesis in the Hebrew Bible, suggesting the existence of a pre-Adamic Earth that was destroyed and replaced by a second creation, when God started over and (re?)made Adam and Eve. This interpretation obviously solves the well-known problem posed by the discrepancy between the two accounts of creation in Genesis, but literalists are clearly less than happy with such a solution because it is obtained at the theologically costly price of introducing a scenario (two successive creations) of which there is no hint in the Bible itself. This is the much-dreaded "slippery slope" of interpretation of the sacred Scripture that, even though it is adopted in one fashion or another by most practicing Christians, is seen as very dangerous by fundamentalists, who believe that the word of God should be a clear and universal message, not subject to the whims and fashions of human explanations.

An even more liberal interpretation of the Bible is adopted by people espousing the next category of old-Earth creationism in Scott's taxonomy: the day-age system. According to this idea, each "day" referred to in the traditional six-day account of creation is comparable to a geological age, so it literally took tens of millions of years to create stars, planets, and life on Earth—in convenient agreement with the evidence from astronomy and geology. This solution still suffers some obvious shortcomings from a scientific standpoint, most egregiously the incompatibility between the chronology of events in Genesis (e.g., whales before land animals) and the data from the fossil record. But it also makes religious fundamentalists very unhappy because it proceeds further down the slippery slope of arbitrary interpretation of the Bible: What is there to stop the believer from even accepting a substantial amount of evolution?

Not much, as is clear from a cursory examination of the next position: so-called progressive creationism. A typical proponent of this version of old-Earth-ism is Hugh Ross of Reasons To Believe ministries, which, as the name implies, is based on the idea that someone can accept Christianity on the basis of *reason*, not just on faith.[6] Progressive creationism is a peculiar and idiosyncratic blend of creationism and science that accepts, for example, the Big Bang and many other scientific conclusions, even within the biological sciences, but limits the power of evolution. For example, evolution is said to occur, but only within the basic "kinds" of organisms originally created by God. Although many young-Earth creationists, such as Duane Gish, also allow what they refer to as *microevolution* within kinds, progressive creationists—because they accept long spans of geological time—at least don't find themselves in the awkward posi-

[6] The conflict within Christianity between apologists, who believe that reason can aid the true religion, and people who reject any appeal to reason as contrary to the very idea of faith has been going on for a long time, and it is reflected by the variety of modern creationist positions. Thomas Aquinas (1225–1274) was among the first proponents of logical reasons to believe in his Five Ways, or proofs, of the existence of God. Augustine of Hippo (354–430 C.E.), however, had already made a separation between faith and philosophy, claiming that the latter cannot possibly suffice to reach the truth because of the limits intrinsic in human reason. According to Augustine, the best reason could do was to elucidate things already accepted by faith as divine revelation.

tion of having to concede more evolution than even the most ardent evolutionist would feel comfortable with (the number of "kinds" was limited by the size of Noah's Ark for a young-Earth creationist, so tens of millions of species had to evolve from a few thousand in as little as 4,000 years: an *astronomic* evolutionary rate by any standard!).

Working our way through Scott's useful classification, we finally arrive at intelligent design (ID) theory—to which I will return in more detail later in this chapter. This is the idea—originally formulated in some detail by the ancient Greeks—that the universe is the result of some kind of supernatural plan evidently constrained by forces that even the gods cannot entirely control. Plato, in the *Timaeus,* presents us with the idea of a god (later called the *demiurge,* literally "the craftsman") which makes the universe "as best as it can be"[7] made within the constraints imposed by contingency.

In 1831, William Paley published *Natural Theology: or, Evidences of the Existence and Attributes of the Deity, Collected from the Appearances of Nature*, perhaps the most famous premodern defense of ID. Paley's memorable image of a watchmaker-creator was evoked in this fashion: "In crossing a heath, suppose I pitched my foot against a stone and were asked how the stone came to be there, I might possibly answer that for anything I knew to the contrary it had lain there forever. . . . But suppose I had found a watch upon the ground, and it should be inquired how the watch happened to be in that place, I should hardly think of the answer which I had before given, that for anything I knew the watch might have always been there." Essentially, Paley said that nobody would necessarily invoke a supernatural designer to account for the existence of simple rocks, but complex and marvelously functional objects such as eyes beg for an explanation that transcends natural laws. If there is a watch, there was a watchmaker; ergo, if there is an eye, there must have been an intelligent designer of that eye.

Unfortunately for Paley, the famous skeptic philosopher David Hume had already sharply criticized this sort of argument more than 50 years before Paley's formulation. In his *Dialogues Concerning Natural Religion,* Hume left it to his legendary character, Philo, to concisely explain what is wrong with the argument from design: "The world plainly resembles more an animal or a vegetable than it does a watch or knitting-loom. Its cause, therefore, it is more probable, resembles the cause of the former. The cause of the former is generation or vegetation." That is, the universe is a natural, not a supernatural, creation. It is interesting that the argument from design is still the most popularly cited reason for why people believe in God.[8]

---

[7] "God desired that all things should be good and nothing bad, so far as this was attainable" (Plato, *Timaeus*, 30a). This and other passages clearly seem to indicate that the demiurge had some limits imposed upon him, which explains why this is not the best of all conceivable worlds, but only of all *possible* worlds.

[8] See M. Shermer, *How We Believe: The Search for God in an Age of Science* (New York: W. H. Freeman, 2000).

Before we get to the contemporary version of ID later in this chapter, it is important to examine more closely the structure of Hume's critique and understand where exactly the intelligent design argument falls flat on philosophical grounds. In the exposition that follows, I will add my commentary and examples to clarify each point, given that Hume's language is at times obscure and obviously not informed by our current knowledge of the physical universe (see Appendix A for a more detailed commentary of the original text).

One can discern six objections to the argument from intelligent design in a complete reading of Hume's *Dialogues:*

(1) The analogy between the universe and human artifacts is weak. In the preceding quote, Hume does not think that the universe resembles a complex machine at all. Although the regularity of the laws of nature may superficially inspire the analogy, human artifacts are always clearly designed for a preconceived function. It often takes quite a bit of imagination to see any purpose in some aspects of the universe. Along similar lines, biologist J. B. S. Haldane once famously answered a reporter who asked what his study of genetics told him about God, "He must have an inordinate fondness for beetles," referring to the tens of thousands of species of these insects existing for no apparent purpose other than their own reproduction.

(2) Intelligence is only one of the active causes of pattern in the world. Many natural phenomena obviously do not require intelligence to occur. Tides, for example, would hardly make a good choice for Paley because even after a cursory examination, one realizes that their explanation in terms of simple gravitational interactions does not require any intelligent design.

(3) Even if intelligence is everywhere operative now, it does not follow logically that this intelligence is responsible for the origin of the universe. This concept can be illustrated in modern terms if we imagine that somebody one day demonstrates that life on Earth was seeded by a race of extremely intelligent extraterrestrials. Such discovery, of course, would not make them gods, nor would it provide an explanation for the origin of the extraterrestrials or for the universe as a whole. In fact, humans may someday do something of this sort, without being elevated to divine status because of it (other than perhaps in the minds of the possibly superstitious products of our own experiments).

(4) The origin of the universe is a unique case, so analogies are pointless. This is a subtle but very good point: Whereas we have plenty of natural objects, organisms, and human artifacts from which to construct analogies, we have only one universe. Science can derive meaningful analogies by comparing two or more populations of objects or entities. But whereas we may compare and contrast the attributes of rocks, eyes, tides, and watches, to what shall we compare the universe? Anything we might think of would be comparing only a part of the whole to the whole itself, and we are unable to find another self-contained whole for comparison.

(5) The analogy between the human and the divine mind is clearly anthropomorphic. Nature resembles a mindless organism rather than a purposeful and intelligent one. This is another way to put the first objection, this time by highlighting the parochialism of a theology that would pretend to understand the mind of God simply as a version of the human mind writ large.

(6) The fruit of anthropomorphic thinking is a finite God. Here Hume goes on the counterattack by showing that if the argument from design is taken seriously, one has to conclude that the god acting in the universe is very different from the Christian one. Because there is no independent argument for the perfection of the designer in the analogical argument from design, we have to judge its ability and character from what we see of the universe. And our observations of the universe we live in are not friendly to the concept of an omnipotent designer god. To paraphrase Bertrand Russell, if *I* had millions of years of time and infinite power and had come up with the universe as we know it, I should be ashamed of myself.

The last step in this long series of creationist positions is usually referred to as *theistic evolution.* This is the position more or less implicitly accepted by the majority of Christians, especially in western European countries: Simply put, God works through the natural laws and processes that He created, and there is no reason to think that natural selection is an exception. This is also essentially the official position of the Catholic Church after two writings of Pope John Paul II. Previous popes, such as Pius XII[9] had taken a stern position against evolutionary theory: "Some imprudently and indiscreetly hold that evolution . . . explains the origin of all things. . . . Communists gladly subscribe to this opinion so that when the souls of man have been deprived of every idea of a personal God, they may the more efficaciously defend and propagate their dialectical materialism."[10] John Paul II (the pope who, without being a paragon of liberalism, still managed to pardon Galilei, albeit after a few centuries of delay), on the contrary, wrote a much more sober letter in 1997 to the Pontifical Academy of Sciences[11] stating that "new knowledge has led to the recognition of the theory of evolution as more than a hypothesis," although he couldn't help adding, "However . . . the moment of transition to the spiritual cannot be the object of this kind of observation," thereby attempting to reserve an untouchable realm of knowledge for the church quite outside naturalistic explanations.

Several scientists were more than happy to go along with the pope's separation of "magisteria," as Stephen Gould called it,[12] and John Paul II himself further elaborated it in a much weightier document, an encyclical this time, en-

[9] In the encyclical *Humani Generis,* published in 1950.

[10] Notice that—ironically—this is exactly the position of the father of contemporary intelligent design theory, the Protestant Phillip Johnson, who seems convinced that he is the discoverer of the connection between evolution and materialistic philosophy.

[11] For the full text of the pope's letter and varied commentaries from evolutionists presenting different philosophical positions, see "The Pope's Message on Evolution and Four Commentaries," *Quarterly Review of Biology* 72 (December 1997): 381–406.

[12] See S. J. Gould, "Nonoverlapping Magisteria," *Natural History* March 1997: 16–22. See also my critique of Gould's position: M. Pigliucci, "Gould's Separate 'Magisteria': Two Views," *Skeptical Inquirer* 23(6) (1999): 53–56.

titled *Fides et Ratio*,[13] in which he essentially maintained that faith and reason are complementary avenues to the truth and cannot contradict each other. While one may quibble with both John Paul II and Stephen Gould for philosophical sloppiness, most scientists surely drew a sigh of relief, not only at realizing that they were not, after all, being condemned to hell because of their work on evolution, but because the pope's position meant that hundreds of millions of Christians worldwide would not have any doctrinal reason to give them a hard time in the classroom. (Of course, this does not apply to the many Christian denominations in the United States that do not recognize the pope's authority.)

The last entry in Scott's classification is the only one that does not involve any creationist component at all, even of the mild type accepted by the Catholic Church and most mainstream Protestant denominations: the much-dreaded (by all creationists) materialistic evolution. This is a *philosophical* (as opposed to scientific) position maintaining that there is no reason whatsoever to invoke supramaterial causes for any natural law or process, including evolution. Not even the most ardent Christian would submit that God works directly through the law of gravity by supervising the motion of every single object; analogously, biologists tend to think that natural selection is just that—a *natural* process with no need of supervision.

Two things are important to realize in connection with materialistic evolution. First, and perhaps most importantly for our discussion and for the creation–evolution controversy, even materialistic evolution does not automatically imply atheism. There are several more possibilities that permit acceptance of both materialistic evolution and belief in a god.[14] This is true both in theory and in practice. In theory, for example, one can be a deist—that is, somebody who believes that God created the universe and its laws but then refrained from any further direct intervention in His creation. In practice, Theodosius Dobzhansky, the evolutionist responsible for one of the sentences most hated by creationists ("nothing in biology makes sense if not in the light of evolution"[15]) was himself a devout Christian. Second, it has always been commonly accepted that most scientists are materialists and do not believe in a personal god.[16] This turns out to be quantitatively accurate. In two interesting articles, Larson and Witham[17] have published the data of two surveys showing that

[13] The full text of the encyclical is available online at http://www.cin.org/jp2/fides.html.

[14] For a classification and discussion of a variety of positions on science and religion, see M. Pigliucci, "Personal Gods, Deism, & the Limits of Skepticism," *Skeptic* 8(2) (2000): 38–45.

[15] T. Dobzhansky, "Nothing in Biology Makes Sense Except in the Light of Evolution," *American Biology Teacher* (1973): 125–129.

[16] For example, see the last essay in Richard Feynman, *The Meaning of It All: Thoughts of a Citizen Scientist* (Reading, MA: Perseus, 1998).

[17] See the amusingly contradictorily titled articles by E. J. Larson and L. Witham: "Scientists Are Still Keeping the Faith," *Nature* 386 (1997): 435–436; and "Leading Scientists Still Reject God," *Nature* 394 (1998): 313.

such is the belief (or more properly lack thereof) of a majority of "average" scientists and of almost every "top" scientist (as measured by their membership in the National Academy of Sciences).

So we are left with three major conclusions: (1) There is a panoply of creationist positions to choose from, characterized by varying degrees of contradiction with the scientific evidence; (2) reassuringly (for the creationist), materialistic evolution does not have to and does not necessarily lead to atheism; and (3) troublingly (again for the creationist), most scientists are atheists or agnostics, or at the very least do not believe in a personal God who listens to prayers and intervenes in the material world. It is to the first conclusion, and the trouble it brings for creationists, that I wish to turn my attention now.

Trouble in the Creationist House

Given the much greater heterogeneity of opinion regarding important conceptual issues within the creationist camp compared to the evolutionist camp, it is amazing that creationists can present such a unified front that includes young-Earth proponents such as Duane Gish and mainstream Catholics who believe in an old Earth and quite a bit of evolution, as is the case of biochemist Michael Behe. Or is this unified front just a matter of appearances, thinly disguising the vat of trouble that is brewing among creationists?

This question was brought to my attention in particularly vivid form during my debate against Jonathan Wells of the Discovery Institute (more on him and his ideas in Chapter 8). The setting was the University of Tennessee's Clarence Brown Theatre, and the occasion was a special public event before a performance of *Inherit the Wind* (the play—really about McCarthyism—that was loosely based on the 1925 antievolution Scopes trial; see Chapter 1). As a UT faculty member, I was playing on home turf in front of an unusually friendly crowd. This notwithstanding, a majority of the attendees were clearly sympathetic to Wells's positions. We went on for a considerable amount of time discussing the fine details of developmental biology and the pertinence to evolutionary theory of the question of the origin of life, until it came time for the final statements, before the question-and-answer session open to the public.[18] At one point, Wells—apparently in an attempt to demonstrate to the audience his ideological neutrality, or at least open-mindedness—claimed that he went into his Ph.D. in molecular biology accepting Darwinian "doctrines" and that it was only a careful study of biology that made him see the light and turn to intelligent design theory.

That claim turned out to be a capital tactical mistake. I had done my homework and found an article by Wells, posted on the Reverend Sun Myung Moon's Unification Church Web site, in which my opponent clearly states that

---

[18] Complete audiovisual coverage of the debate is available at http://burns.tns.utk.edu/research/cb/evdebate.htm, and a play-by-play summary at http://fp.bio.utk.edu/skeptic/Debates/Wells_play_by_play.html.

he went into molecular biology under precise instructions from Moon, whom he calls "Father," to acquire the tools of biology so that he could more effectively "destroy Darwinism" and the evil it carried.[19] I had not planned to use the article, but I couldn't believe that Wells, in making such a false statement, was handing me his head on a silver platter. I promptly moved to chop it off. I told the audience about the article, essentially pointing out that Wells was lying about his motives. The audience was visibly taken aback. Wells tried to portray my revelation as an *ad hominem* attack (a no-no during public confrontations, except in politics), but of course, it was not. If one's debating opponent has just made a false claim of open-mindedness to an audience, hoping to score points by doing so, it is only fair to point out that the opponent is not open-minded at all.

Regardless, what was surprising to me was *why* the audience was taken aback and why even ardent creationists reacted with an audible gasp to the news I had just brought into the debate. It wasn't the fact that Wells was lying that apparently bothered most people (after all, Luther did say that a good lie in the name of the Lord is not a bad thing). And certainly they were not moved by my rational arguments pertaining to the actual subject matter of the debate. No, it was the revelation that Wells belonged to the Unification Church that did the trick. UC members, although they consider themselves Christians, are actively despised by most fundamentalist Christian denominations. I was reaping the benefits not of superior logic and evidence, but of the intertribal warfare among Christians themselves!

I am certainly not advocating here that an evolution proponent skip the intellectual debate and go straight for the creationist's jugular whenever possible. After all, scientists and educators are involved in this debate because they care about science education, not about winning an ideological war. In this, I can safely say, they are different from creationists. Even though most creationists are sincerely convinced of their positions, they are interested only in winning the ideological war. If evolutionary theory had no theological implications (say, like atomic theory), there would be no debate. This point is strangely missed by scientists, who continue to behave as if creationists were either lunatics (which by and large they are not) or as if they needed to be rebutted on solid scientific grounds, after which they would go away.

The reason it is important to explain science to the general public is that it is important for our society that people have a scientific understanding of the world. Without it, as a recent report of the National Science Foundation remarked,[20] people are likely to make bad decisions in the voting booth, as members of a jury, or in their private lives when considering an insurance policy or a moneymaking scheme. But to refuse to accept that this particular debate is

---

[19] Wells's article on his true motivations for getting a Ph.D. in biology, as narrated by him, is available at http://www.tparents.org/library/unification/talks/wells/DARWIN.htm.

[20] See "Science Indicators 2000: Belief in the Paranormal or Pseudoscience," *Skeptical Inquirer* January/February 2001: 12–15.

about ideology rather than science is foolish and largely accounts for the lack of progress we have made since the Scopes trial. It is therefore urgent that scientists and educators be in a position to counterattack and to point out the internal inconsistencies in the creationist camp. As uncomfortable as this may be, this is bound to make much more of an immediate impact than any esoteric explanation of the second principle of thermodynamics (see Chapter 6).

Let me offer two examples of the degree of division within the creationist camp that can and should be used to point out to creationists that their position is not the self-consistent and problem-free monolith they believe it to be. The first case is the negative reactions that even the most biblically oriented of creationist institutions, the Institute for Creation Research, can elicit among fundamentalist Christians. The second case is at the other end of the spectrum in the ongoing discussion between moderately liberal and ultraliberal Christians about the scope and implications of Darwinism.

The Institute for Creation Research (ICR) was established in 1977 as a branch of Christian Heritage College by young-Earthist par excellence Henry Morris (currently honorary president after the succession of his son John to the presidency). Morris's right arm has always been Duane Gish, a prolific writer and debater for the ICR.[21] The ICR's statement of educational philosophy is one of the most amusing and at the same time irritating "academic" documents that I have come across.[22] It starts by stating that the institute presents to its students standard, factual, scientific content analogous to that taught in secular institutions, but it immediately adds that such content is enriched by complementary material "presented in accordance with the distinctive ICR mission and beliefs." The mission is to promote the word of Jesus Christ as taught in the Bible, and the belief is that such word is the word of God. There is nothing wrong with either the mission or the belief, but both clearly are in strident contrast with the teaching of science, which, by definition, is independent of any religious tenet.

The situation gets worse. The ICR requires *from its faculty* and students what it refers to as a "unique" statement of faith, which incorporates the "basic Christian doctrines in a creationist framework." In other words, if the conclusions of a scientist's "research" put him at odds with the Bible, he had better change the interpretation of his results accordingly.[23] The "tenets" of the ICR are expressed very clearly and in perfect accord with the fundamentalist religious ideology espoused by the institute's founders. For example, the insti-

[21] I have had the dubious pleasure of debating Duane Gish five times so far. A more or less complete account of one of our debates is found in Chapter 11 of my book *Tales of the Rational* (Atlanta, GA: Freethought Press, 2000).

[22] The Institute for Creation Science statements referred to in this chapter can be found online at the ICR Web site, at http://www.icr.org/.

[23] This reminds me of the *Hitchhiker's Guide to the Galaxy* by the immortal and much-missed Douglas Adams. The guide (a fictional editorial product, of course) comes with a disclaimer aimed at avoiding lawsuits of galactic proportions: "In any case in which the Guide and reality are at odds, reality must be wrong."

tute's Web site says that the creation record is considered factual and historical, which means that any theory invoking evolution must *by definition* be false. This is a remarkable admission for an institute devoted to "research," since it means that all conclusions of the alleged research are already known and that all their "scientists" need to do is to gather evidence to back them up. Such a statement is, of course, made possible by an obvious advantage that ICR scientists have over their secular colleagues: God has already told them the end of the story, so they don't have to engage in any "whodunit" speculation without knowing the entire script. If only normal science would work that way, it would be a much more straightforward (albeit rather boring) activity.

But things are not that easy for the ICR even within fundamentalist circles. Plenty of creationists consider Morris's institute not quite fundamentalist enough and have voiced repeated criticism of it. In fact, the ICR actually originated out of a schism within its original incarnation, the Creation Science Research Center (CSRC). The CSRC was organized in 1970 by Henry Morris, together with two housewives upset by the 1962 Supreme Court decision prohibiting officially sanctioned school prayers in public schools—Jean Sumrall and Nell Segraves—as well as Segraves's son, Kelly.[24] Morris wanted to build a creationist movement from the bottom up, through grassroots education, but his cofounders preferred a top-down approach based on passing legislation favoring the teaching of creationism. Morris was seen as too "soft" for a true defender of the word of God. In fact, the Christian Identity movement (whose members believe that blacks and other "inferior" races were originally created as "beasts") considers the ICR unscientific and accuses Morris and colleagues of being crypto-evolutionists by conceding limited change within kinds of organisms. Fundamentalist Gary North also opposes the ICR on the grounds that the latter's acknowledgment of the second principle of thermodynamics (one of the most fundamental and undisputed laws ever arrived at by scientists) is an implicit statement that the ultimate triumph of *sin* and *unbelief* (seen as synonyms with *disorder* and *entropy*) is inevitable (I am sure this will shock most physicists). Furthermore, hyper-Calvinist John Robbins (of the Trinity Foundation) considers the whole idea of scientific creationism—the basis of the ICR's activities—a fraud, a deception, and a cowardly surrender to the anti-Christian forces. The reader should now be in a position to appreciate what I meant when I said that this is an ideological, not a scientific, war.

More trouble has been brewing even within ICR's own walls. The institute has a graduate school (recognized but unaccredited by the state of California) that dispenses master's degrees in astro-geophysics, biology, geology, and science education. But the dispensation of such degrees has created internal schisms that have often resulted in the sudden dismissal of some ICR scientists

---

[24] The story of the ICR, as well as an interesting virtual tour of its museum, can be found in T. McIver, "A Walk through Earth History: All Eight Thousand Years," *Skeptic* 4(1) (1996): 32–41. This is also a main source for my description of the trouble internal to the ICR (for further reading, see the references in McIver's article).

and administrators. For example, a 1986 thesis titled "A Classical Field Theory for the Propagation of Light" was signed by Duane Gish and Thomas Barnes, but not by the third committee member, ICR's coordinator of research Gerald Aarsdma. Apparently Aarsdma had objected to the fact that the thesis was based on pre-Einsteinian physics and invoked the long-dismissed existence of ether. Then again, perhaps one should not be surprised at the content of work coming out of an institution at which the dean, Harold Slusher, got his Ph.D. from an unaccredited correspondence school and used to rail against post-Newtonian physics and particularly against relativity and quantum mechanics (before leaving the ICR under strained circumstances).

In this ideologically charged climate it must be difficult to hire (and retain) people, as was clear when Morris considered (and rejected) the possibility of welcoming Arleton Murray to the ICR. Murray, otherwise known as "Mr. Fossil," believes in the gap theory briefly described earlier—that is, that millions of years passed between the times referred to in Genesis 1 and Genesis 2. According to Murray, the creation described in the Bible was actually a *re*-creation; before that Satan had ruled the world of dinosaurs. This was not in line with the tenets to which the ICR demands adherence from its faculty, and Morris did not further consider Murray's candidacy for a post at the institute.

The point I wish to make is that these sorts of sharp internal fights are not what one would expect within a scientific community devoted to the pursuit of truth. I am certainly not naïve enough to suggest that academic squabbles and ideological wars are not found within the scientific establishment. They are, and they make excellent material for philosophers and sociologists of science, who relentlessly (and justly) document them. Yet the degree of "ideological" separation within the scientific community is much less broad, and it is usually settled—in the long run at least—by appeal to an external source of evidence: the natural world. It may take decades, but the school of thought that is closer to the truth will eventually emerge. No such appeal is possible for creationists, given the complete lack of external criteria outside the Bible, which can notoriously be interpreted (like any literary text) in countless ways. In fact, it is this lack of objective criteria that makes the divisions within creationism so large. The appearance of a unified front is made possible in part by a lack of publicity of these internal debates and, specifically within the "scientific creationists," by a complete lack of hypothesis testing, fully one-half of the scientific process. We will consider shortly Phillip Johnson's idea of breaking the scientific establishment's hold on academia and public funding by driving a wedge in a small crack he thinks he has uncovered. If scientists and educators were bent on winning this debate, they would have no difficulty driving a bulldozer into the much larger abyss separating different factions of creationists.

This feeling of internal uneasiness and sharp disagreement is reinforced when one considers the writings of less rabid antievolutionists, such as many mainstream Christian commentators. An example is provided in an article by John Wilson aptly entitled "Your Darwin Is Too Large," which appeared in

Christianity Today in May 2000. Wilson rails against a then recently published book by John Haught entitled *God after Darwin: A Theology of Evolution*,[25] an example of the theistic evolution position (i.e., at the opposite extreme of the creationist spectrum compared to the Institute for Creation Research, if we don't count the few Californians who belong to the Flat Earth Society).

Wilson starts out by complaining that "every five minutes" a book is published arguing that God is an outmoded concept among rationally thinking people. (He could have fooled me, since I have exactly the opposite impression every time I stop by my local bookstore—but that just goes to show how one's ideological commitments color one's perception of the world.) Wilson goes on to state—quite inaccurately—that Darwinism is being invoked to explain everything from the formation of galaxies to the eating habits of teenagers. This is a typical mistake that is important to recognize. Evolution is *not* the theory that everything changes all the time, an idea found in ancient times in Heraclitus's philosophy (he was the fellow who said that one can't step into the same river twice, because the river constantly changes). Although the public, and even some scientists, sloppily use the word *evolution* just to indicate change (as in the evolution of galaxies, or the evolution of cars), this is *not* Darwinian evolution, which is a specific theory about the patterns and mechanisms of change throughout the history of life. Although the mechanisms of Darwinian evolution—a source of variation (such as mutation) and natural selection—can be meaningfully extended to other realms of application (e.g., the creation of sophisticated computer algorithms), they certainly do not apply to galaxies or cars, and very likely not even to the eating habits of teenagers. To think of this issue in other terms, the theory of gravity also certainly applies to everything that exists in our universe, but it does not *explain* everything in that universe—and explaining is what a theory is supposed to do.

Wilson attacks Haught for giving too much ground to Darwin because Haught writes statements such as, "Any thoughts we may have about God after Darwin can hardly remain the same." But Haught is simply making a reasonable philosophical—and historically accurate—statement here. He is not making a scientific proposition at all, or defending an imaginary claim of Darwinists to explain everything. Haught is saying that knowledge of the Darwinian theory of biological evolution inevitably alters anyone's conception of what God is and does. This is a truism, given the very existence of the evolution–creation controversy, the intervention of the pope, and the panoply of books and articles written on the subject.

Haught again: "By expanding the horizons of life's travail, Darwin gives unprecedented breadth to our sense of the tragic." Well, this is hardly controversial. If one denies this, one is simply not paying attention. Evolutionary biology documents endless examples of the intricacies of the battle of every living creature for existence and reproduction. Although the Darwinian theory in it-

[25] J. Haught, *God after Darwin: A Theology of Evolution* (Boulder, CO: Westview, 2000).

self does not have any moral or emotive implications, and therefore could not as a science have anything to do with a sense of the tragic, certainly any human being considering what we now know of the way life evolved and works every day cannot avoid feeling a sense of tragedy heightened by the discovery. I do not see how this in any sense gives too much space to Darwinism.

Haught can—and perhaps should—be criticized on theological grounds, on what he derives about God from the philosophical implications of evolutionary theory. But this would not be a criticism of the science of evolution; it belongs in a debate in which a scientist—in her capacity of scientist—is utterly uninterested. Wilson does not like Haught's position that God put restraints on Himself because supernatural beauty depends on God's not knowing how the unfolding of His creation will all play out. The universe is viewed by Haught as an ongoing work of art, and by analogy with the work of an artist, although the general scheme is known to the artist in advance, the details will be figured out as one goes on with the work. According to Haught, "All the moments of an evolving world are harvested into the divine experience in an ever intensifying aesthetic pattern." Whatever that means, it is certainly no scientific statement and has nothing to do with evolution.

The point is that Wilson manages to attack both Haught and evolution—Haught for his heretical and ultraliberal version of Christianity, and evolution for failing to yield answers to questions about evil and divine action. As if that were the proper scope of evolutionary biology to begin with! Wilson makes the same mistake that Henry Morris and Duane Gish make: He conflates the (real or alleged) philosophical-theological implications of evolution with the science. The former are part of an ongoing discussion that does not always take account of empirical evidence (though any respectable philosophical viewpoint should at least be consistent with empirical evidence); the discussion on the latter has been settled by empirical evidence for more than a century.

Let me now turn to a more in-depth analysis of the content of two major creationist positions: scientific creationism and intelligent design theory. This analysis will be the basis for our discussion of specific creationist fallacies in Chapter 5 and for the contrast with what evolution really is (and is not) at the end of this chapter.

Creationism 101: Scientific Creationism

"Neither evolution nor creation qualifies as a scientific theory," asserts Duane Gish in a pamphlet published by the Institute for Creation Research aimed at discussing the controversy about the teaching of creationism and evolution in the public schools.[26] One wonders, then, what all the fuss is about. One would also be more than justified in asking why Gish's institute has the words *Creation Research* in its title and claims to be the reference point for a movement

[26] See D. Gish, *Creation, Evolution, and Public Education* (Institute for Creation Research, no date).

called *"scientific creationism."* I will discuss in Chapter 5 why Gish is correct about creationism's not being a science and wrong when he makes the same pronouncement about evolution. For now, let us consider an overview of what scientific creationism actually is, according to its own proponents.

Gish defines creation as the supernatural origin of the basic kinds of plants and animals (though I'm sure he did not mean to exclude other life forms, such as bacteria, protozoa, and fungi), plus *at least* one worldwide catastrophe (to be studied by a special branch of scientific creationism appropriately known as *flood geology*). So scientific creationism, as the pledge of the ICR claims, is about scientific support for the truth of Genesis. In fact, the first hall of the ICR Museum of Creation & Earth History near San Diego, California, is called Science and Faith, and it claims that *true* science confirms a literal biblical account of creation. As for the flood, an exhibit on the ark at the museum goes into quite a few details, meant no doubt to convey the idea that ICR scientists have given the matter some considerable thought. The ark, as we know from the Bible, was 450 by 75 by 45 feet, and the challenge is to explain how 30,000 "kinds" of animals could be stored on it and fed for 371 days.

Scientific creationism includes other very peculiar statements that are worth pondering from a scientific standpoint. For example, the history of the universe has to follow the biblical version. Besides the fact that the whole process allegedly took only six days—contrary to everything we know about the age of the universe—the sequence in Genesis 1 includes light being created on day 1, plants on day 3, and sun, moon, and stars on day 4. This implies that plants (and light) existed before there was a sun. Genesis also claims that birds and marine mammals appeared before land animals. The fact that this doesn't square with anything we can see in the actual geological record seems to bother the creation scientist not at all.

To be sure, some reconciliation of Scripture with reality is attempted, albeit in a tortuous way. For example, ICR scientist Harold Slusher explains that light actually travels through non-Euclidean and non-Einsteinian shortcuts to arrive on Earth in moments as opposed to years, as established by standard cosmology. There is, of course, no evidence to support this claim, and the only reason that creationists accept it is their need to reconcile a priori biblical notions with reality. Once again, with creationists the conclusion comes first; then they selectively look for the evidence to back it up. Failing that, they must invent a far-fetched scenario to save the day.

Perhaps one of the most peculiar conclusions of scientific creationism is that the Fall of Adam is essentially the same thing as the second principle of thermodynamics. In Chapter 6 I will devote considerable space to the endless problem creationists seem to have with the second principle, but it should strike anybody as at least odd to connect a fundamental law of physics with an ancient tale about morality and obedience. As I mentioned already, some creationists are harshly critical of the ICR for making this connection to begin with, so Gish must feel under attack on two fronts in this regard.

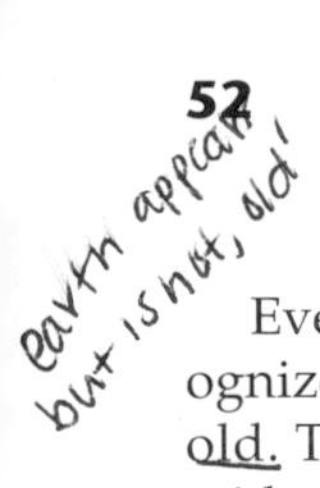

Even Duane Gish's ability to deny reality, however, has a limit. He does recognize, for example, that Earth's life forms come with the *appearance* of being old. This, in his worldview, is no different from Adam having been created with a belly button, even though he had no mother or birth. (How Gish or anybody else knows that Adam had a belly button is beyond my abilities to comprehend. Since Adam was created in God's image, does this imply that God, too, has a belly button? Why would God have a belly button?) Of course, strictly speaking, Gish is correct. Historical scientists (and evolution is largely about the history of life on Earth) can only point to the appearance of things, what are technically referred to as patterns, and then deduce the most likely processes that led to the formation of such patterns.

Quite aside from considerations of philosophy of science and epistemology (which we will discuss in Chapter 4), one would feel justified to ask what on earth was on God's mind when He created the universe with the appearance of being old. Was He trying to test the faith of His creatures? In that case, why endow them with a sense of critical thinking and with an ability for rational thought? In the movie *Dark City,* the main character discovers that the place where he lives and the people who inhabit it indeed show only the appearance of having a past. In fact, the entire city is an experiment run by aliens to understand human behavior, but the scenario is rather sinister and surely would not fit with the popular image of the Christian God. Yet Gish and his ICR associates don't seem to even think about the questions raised by their cavalier interpretation of the *appearance* of life on Earth. To paraphrase the defense counsel in the fictional version of the Scopes trial, *Inherit the Wind,* it frightens me to think of the low level of driving curiosity that characterizes creation "scientists."

Some of the historical claims to be found in the ICR museum are also stunning and show how easily ideology gets the better of accuracy. Apparently Darwin was not the first to think of evolution; such dubious honor is attributed to Nimrod, who was instigated by Satan to build the Tower of Babel, with the infamous results we all know about. Jumping forward a few thousand years, the German philosopher Friedrich Nietzsche—always a favorite target of antimaterialists—is said to have been an evolutionist who helped bring about World War II. Never mind that Nietzsche despised nationalism and militarism and never correctly understood the idea of Darwinian evolution. Wernher von Braun, on the other hand, is considered a hero by the ICR merely because he was a scientist who spoke against evolution—a position consistent in its dishonesty with the other unsavory aspects of Von Braun's life and work. Von Braun was the man that single-handedly helped the Americans recover from the initial stunning advantage that the Soviets had gained in the space race. Yet he also used slave labor as a scientist in Germany during the Nazi regime and designed the lethal V-1 and V-2 bombs that blasted London for most of what writer Kurt Vonnegut calls "civilization's second failed attempt at suicide." But what are a few good lies about Von Braun's heroism in the general scheme of serving the Truth? Again, Martin Luther would be proud of the ICR staff.

The positions endorsed by the Institute for Creation Research, and common to other brands of creationism, go well beyond the issue of evolution and its perceived direct applications. For example, the explanatory panels in the ICR's museum suggest that "evolution-based" environmentalism is really a ploy to promote New Age religions and other forms of paganism, even though the actual connection between environmental activism and paganism is very tenuous, and few environmentalists know anything about evolution. Regardless, ICR exhibits make a point of the fact that Hitler was a vegetarian and opposed to vivisection (he was also nominally a Christian, but apparently that detail is of minor importance).

Along the same lines, of course, global warming is considered improbable, and the depletion of the ozone layer a "ploy." According to the ICR, our planet can easily sustain a population of 100 billion people using conventional agriculture, though what data or models have been used to produce such an astonishing result are not revealed. Perhaps most ironically, creationists at the institute are not worried about extinction because species can re-evolve (!) from the stock represented by their given kind. (So much for bumper stickers claiming that extinction is forever.) It is precisely this sort of reasoning that is most frightening and shows why scientific literacy in general is much more important than the particular creation–evolution controversy. It was one of Ronald Reagan's top aides who claimed—no doubt cheered by creationists of various creeds—that he wasn't worried about the state of the environment because the Second Coming was near anyway.[27]

All these incongruities notwithstanding, even a cursory examination of creationism clearly leads to the conclusion that the real issue for creationists is not the flood, carbon dating, evolution, or the Big Bang. The issue is the perceived *moral* conflict between two worldviews: scientific materialism and Christian supernaturalism—which brings us to the other major branch of creationism, the so-called intelligent design theory and its detailed program for the destruction of science as we know it.

Creationism 102: The Three Weapons of Intelligent Design Theory

Intelligent design (ID) theory represents a relatively new brand of creationism that reached prominence on the scene in the mid-1990s. The neocreationists who are proponents of intelligent design, as we have seen, largely do not believe in a young Earth or in a too-literal interpretation of the Bible. Although neocreationism is still propelled mostly by a religious agenda and financed by mainly Christian sources, such as the Templeton Foundation and the Dis-

[27] James Watt, Ronald Reagan's nominee for Secretary of Interior and supporter of "Natural Law" said he would run the department "according to scripture." When asked how he would manage things for the long term, he replied, alluding to the Second Coming, "I am not sure how many more generations there will be."

covery Institute, the intellectual challenge that it poses is sophisticated enough to require detailed consideration.[28] As we shall see immediately, the level of the debate shifts from the almost ludicrous to the philosophically sophisticated. At least ID theorists give philosophers (if not scientists) a good run for their money.

Among the chief exponents of ID is William Dembski, a mathematical philosopher and author of *The Design Inference*[29] (among other books). In that book Dembski attempts to show that there must be an intelligent designer behind natural phenomena such as evolution and the very origin of the universe.[30] One of Dembski's favorite arguments is that modern science ever since Francis Bacon has illicitly dropped from consideration altogether two of Aristotle's famous four types of causes, and in particular the idea of a final cause, thereby unnecessarily restricting its own explanatory power. Science is thus incomplete, and intelligent design theory will rectify this sorry state of affairs, if only closed-minded evolutionists will allow Dembski and company to do the job. Let us briefly see what the philosophical fuss is all about.

Aristotle said that one can ask four fundamental questions about any given thing: (1) What is its material cause—that is, what something is made of? (2) What is its formal cause—that is, the form, pattern, or structure embodied in the object or phenomenon? (3) What is its efficient cause, the maker or activity producing a phenomenon or object? (4) What is its final cause, the function or purpose of the object we are investigating? For example, let's say we want to investigate the "causes" of the Brooklyn Bridge. Its material cause consists of the physical materials that went into its construction. The formal cause consists of the blueprints drawn by engineers, the blueprints themselves preceded by the idea of what the engineers wanted to build—a bridge rather than a skyscraper. The efficient causes are the labor of men and machines that assembled the physical materials and put them into place. The final cause of the Brooklyn Bridge is the necessity for people to connect two landmasses separated by a large body of water, thereby enabling them to travel from one side to the other.

Dembski maintains that Bacon and his followers did away with both formal and final causes (the so-called teleonomic causes, because they answer the question of why something is, based on Aristotle's concept of *telos*—that is, a thing's aim, goal, or purpose, indicated by its function) in order to free science from philosophical speculation and ground it firmly in empirical observation. Dembski interprets Aristotle's concept of *telos,* or purpose, in a way that

[28] On intelligent design, see T. Edis, "Darwin in Mind: 'Intelligent Design' Meets Artificial Intelligence," *Skeptical Inquirer* 25(2) (2001): 35–39; and D. Roche, "A Bit Confused: Creationism and Information Theory," *Skeptical Inquirer* 25(2) (2001): 40–42.

[29] W. A. Dembski, *The Design Inference: Eliminating Chance through Small Probabilities* (Cambridge, England: Cambridge University Press, 1998).

[30] For a critique of *The Design Inference,* see M. Pigliucci, "Chance, Necessity, and the New Holy War against Science," *BioScience* 50(1) (2000): 79–81.

favors his own belief in divine purpose. He overlooks historical developments in scientific methodology and epistemology in order to preserve this concept because it is an integral part of his Christian worldview. However, the way science was done, and correspondingly the understanding of scientific methodology, changed in important ways with the work of Charles Darwin.

Darwin was addressing a complex scientific question in an unprecedented fashion: He recognized that living organisms are clearly "designed," in a manner of speaking, to survive and reproduce in the world they inhabit; yet as a scientist he worked within the framework of naturalistic explanations of such apparent design. The important point here is *why* he did so. He did so because the then current supernatural explanations—explanations that were distinctly teleological in the Aristotelian sense, the concept having been absorbed into the Christian metaphysics and through the latter into pre-Darwinian, nineteenth-century scientific thought—had proven unworkable. They simply could not explain the data that Darwin and other contemporary scientists had observed. Darwin *had* to construct his explanation of the data from within a naturalistic framework in the absence of any other *workable* explanation.

Darwin found the answer to the question of apparent design in his well-known theory of natural selection. Selection, combined with the basic process of genetic mutation, makes design possible in nature without the need for supernatural intervention; the reason is that selection is definitely nonrandom and therefore has *creative* (albeit nonconscious) power. It is initiated by the particular mutations that occur and subsequently guided by the nature of the environment within which those mutations appear. And the environment may well be one in which mutations are preserved because they are advantageous, in which case organisms with those mutations proliferate through successful reproduction. What Dembski calls "design" is actually the *result* of an organism's adaptation to the particular ecological niche that it occupies. And such adaptation is a continual process. Creationists usually do not understand this point and think that selection can only eliminate the less fit, but Darwin's powerful insight was that selection is also a cumulative process—analogous to a ratchet—that can build things over time despite the lack of a plan.

Darwin has made it possible to put all four Aristotelian causes back into science, but only in an importantly different sense with respect to both formal and final causes. For example, if we were to ask what the causes of a tiger's teeth are within a Darwinian framework, we would answer in the following manner: The material cause is provided by the biological materials that make up the teeth; the formal cause is the genetic and developmental machinery that makes a tiger's teeth different from any other kind of biological structure; the efficient cause is brought about by natural selection, which promotes some genetic variants of the tiger's ancestors over their competitors; and the final cause is provided by the fact that having teeth structured in a certain way makes it easier for a tiger to procure its prey and therefore to survive and reproduce—the only "goals" of every living being.

Therefore, (the appearance of) design and Dembski's "missing" causes are very much a part of modern science, at least whenever there is a need to explain a structure that has an obvious function (such as the parts of living organisms). However, the "design" of an organism that serves as an adaptive advantage is an advantage conferred *after* the fact of natural selection; it is *not* the fulfillment in the original Aristotelian sense of *telos,* or purpose, guiding the development of the organism. For Aristotle, *telos* is the immutable, even though immanent, essence of the organism. In a Darwinian sense, contrary to Dembski's preferred Aristotelian interpretation, *telos* is wholly natural and certainly not immutable. In fact, the very essence of evolution is change rather than immutability. Hence, all four Aristotelian causes may be reinstated within the realm of scientific investigation, though with important differences in how formal and final causes are understood, and science is not therefore "incomplete" in a philosophical sense. What, then, is left of the argument of Dembski and of other proponents of ID? They, like William Paley well before them, make the mistake of ignoring the distinction between natural design and intelligent design by rejecting the possibility of the former and concluding that any design must be by definition intelligent.

One is left with the lingering feeling that Dembski is being disingenuous about ancient philosophy. It is quite clear, for example, that Aristotle himself never meant his teleonomic causes to imply intelligent design in nature in any sense analogous to the deliberate, consciously creative activity of Dembski's intelligent designer. Aristotle's mentor, Plato (in the dialogue *Timaeus*), had already concluded that the designer of the universe could not be an omnipotent god, but at most what later commentators called a *demiurge* (literally, "a craftsman"), a lesser god who merely fashions the universe, imposing form upon chaos, and with mixed results. Aristotle believed that the scope of God was even more limited, essentially to the role of prime mover of the universe, with no direct, creative interaction with the cosmos itself.

In *Physics,* where he discusses the four causes, Aristotle treats nature itself as a craftsman, but clearly devoid of forethought and intelligence. A tiger develops into a tiger because it is in its nature to do so, and this nature is due to some essence given to it by its father (although Aristotle was not thinking of physical essences, we would call it DNA, and we would also acknowledge that the mother had at least an equal contribution to the matter), which starts the process. Aristotle makes clear this rejection of a god as a final cause when he says that causes are not external to the organism (such as a designer would be) but internal to it (as modern developmental biology clearly shows).[31] In other words, the final cause of a living being is not a consciously enacted plan, intention, or purpose, but simply intrinsic in the developmental changes of that organism. This means that Aristotle identified final causes with formal causes

[31] Modern evolutionary theory does include external causes, of course—primarily the physical environment in which organisms find themselves.

as far as living organisms are concerned. He rejected chance and randomness (as do modern biologists) but did not invoke an intelligent designer in its place, contra Dembski. We had to wait until Darwin in order to reinstate in any genuinely scientific sense Aristotle's conception of the final cause of living organisms, and until modern molecular biology to achieve an understanding of their formal cause.[32]

There are two additional arguments proposed by ID theorists to demonstrate intelligent design in the universe: the concept of *irreducible complexity* and the *complexity specification* criterion. *Irreducible complexity* is a term introduced in this context by molecular biologist Michael Behe in his book *Darwin's Black Box*.[33] The idea is that the difference between a natural phenomenon and an intelligently designed one is that a designed object is planned in advance, with forethought. Although an intelligent agent is not constrained by a step-by-step evolutionary process, the latter is the only way nature itself can proceed, given that it has no planning capacity (this may be referred to as incremental complexity). Irreducible complexity then exists whenever all the parts of a structure have to be present and functional simultaneously for it to work, indicating—according to Behe—that the structure was designed initially at that level of complexity and could not possibly have been gradually built by natural selection.

Behe's example of an irreducibly complex object is a mousetrap. If one takes away any of the minimal elements that make the trap work, it will lose its function; in addition, a natural process cannot produce a mousetrap because it won't work until the last piece is assembled. Forethought, and therefore intelligent design, is necessary. Of course it is. After all, mousetraps as purchased in hardware stores are indeed human products; we know that they are intelligently designed.[34] But what of biological structures? Behe claims that, although evolution can explain much of the visible diversity among living organisms, it is not sufficient to produce complexity at the molecular level. The cell and several of its fundamental components and biochemical pathways are, according to him, irreducibly complex.

The problem with this statement is that it is contradicted by the available literature on comparative studies in microbiology and molecular biology, which

[32] It is worth noting here that Aristotle's god is a final cause in an ultimate sense. The prime mover as ultimate final cause, which gives purpose to the cosmos as a whole and by extension to every individual thing, is the idea Dembski understands and is trying to exploit. But it is also precisely what had to be discarded from Aristotle because it is beyond empirical verification. The idea of god as ultimate *telos* is, for Aristotle, a logical necessity because he needs (or so he thought) an ultimate principle of explanation. And ultimate explanations are beyond the scope of science. Dembski well knows this, of course, but this is what he cannot admit because it would not serve the religious use to which he wants to put Aristotle.

[33] M. J. Behe, *Darwin's Black Box: The Biochemical Challenge to Evolution* (New York: Free Press, 1996).

[34] It is amusing to note that somebody has actually gone through the pains of showing that mousetraps are actually *not* irreducibly complex, and that the problem resides only in Behe's limited imagination. For examples of mousetraps with fewer and fewer pieces (down to one!), see http://udel.edu/~mcdonald/mousetrap.html.

Behe conveniently ignores. For example, geneticists are continuously showing that biochemical pathways are partly redundant.[35] Redundancy is a common feature of living organisms in which different genes are involved in the same or in partially overlapping functions. Although this may seem a waste, mathematical models show that evolution by natural selection has to produce molecular redundancy because when a new function is necessary, it cannot be carried out by a gene that is already doing something else, without compromising the original function.[36] On the other hand, if the gene is duplicated (by mutation), one copy is freed from immediate constraints and can slowly diverge in structure from the original, eventually taking over new functions. This process leads to the formation of gene *families,* groups of genes that clearly originated from a single ancestral DNA sequence and that now are diversified and perform a variety of functions (e.g., the globins, which vary from proteins allowing muscle contraction to those involved in the exchange of oxygen and carbon dioxide in the blood). As a result of redundancy, mutations can knock out individual components of biochemical pathways without compromising the overall function—contrary to the expectations of irreducible complexity.[37]

To be sure, in several cases biologists do not know enough about the fundamental constituents of the cell to be able to hypothesize about or demonstrate their gradual evolution. But to consider this current absence of knowledge to be evidence in ID's favor is an argument from ignorance, not positive evidence of irreducible complexity. William Paley advanced exactly the same argument to claim that it is impossible to explain the appearance of the eye by natural means. Yet today biologists know of several examples of intermediate forms of the eye, and there is evidence that this structure evolved independently several times during the history of life on Earth.[38] The answer to the classic creationist question, "What good is half an eye?" is, "Much better than no eye at all!" However, Behe does have a legitimate point concerning irreducible complexity, although not in the sense he intended. It is true that some structures simply cannot be explained by the slow and cumulative processes of natural selection.

---

[35] For examples of and discussion about molecular redundancy, see F. B. Pickett and D. R. Meeks-Wagner, "Seeing Double: Appreciating Genetic Redundancy," *Plant Cell* 7 (1995): 1347–1356; and A. Wagner, "Redundant Gene Functions and Natural Selection," *Journal of Evolutionary Biology* 12 (1999): 1–16.

[36] For a discussion of the role of gene duplication in evolution, see M. D. Ganfornina and D. Sanchez, "Generation of Evolutionary Novelty by Functional Shift," *BioEssays* 21 (1999): 432–439.

[37] Notice that creationists, never ones to yield a point, have also tried to claim that redundancy is yet more evidence of intelligent design, because an engineer would produce backup systems to minimize catastrophic failures should the primary components stop functioning. Although very clever, this argument once again ignores the biology: The majority of duplicated genes end up as pseudogenes, literally pieces of molecular junk that are eventually lost forever to any biological utility. See E. E. Max, "Plagiarized Errors and Molecular Genetics: Another Argument in the Evolution–Creation Controversy," *Creation/Evolution* 9 (1986): 34–46.

[38] On the evolution of the eye, see D.-E. Nilsson and S. Pelger, "A Pessimistic Estimate of the Time Required for an Eye to Evolve," *Proceedings of the Royal Society of London B* 256 (1994): 53–58.

From Behe's mousetrap, to Paley's watch, to the Brooklyn Bridge, the irreducible complexity of manufactured objects is indeed associated with intelligent design—*by humans*. The problem for ID theory is that there is no evidence so far *in nature* of irreducible complexity in living organisms.[39]

William Dembski uses an approach similar to Behe's to back up creationist claims in that he also wants to demonstrate that intelligent design is necessary to explain the complexity of nature. His proposal, however, is both more general and more deeply flawed. In *The Design Inference* he claims that there are three essential types of phenomena in nature: "regular," random, and designed (the last of which he assumes without further discussion to be products of intelligence). A regular phenomenon would be a simple repetition explainable by the fundamental laws of physics—for example, the rotation of Earth around the sun. Random phenomena are exemplified by the tossing of a coin. Design enters as an explanation whenever two criteria are satisfied: complexity and specification (which I will consider in a moment).

There are several problems with this neat scenario. First of all, temporarily leaving aside design, the remaining choices are not limited to regularity and randomness. Chaos and complexity theory have established the existence of self-organizing phenomena, situations in which order spontaneously appears as an emergent property of complex interactions among the parts of a system. And these kinds of phenomena, far from being only a figment of mathematical imagination as Behe (and, interestingly, Gish) maintains, are real. For example, certain meteorological phenomena, such as tornadoes, are neither regular nor random but are the result of self-organizing processes. Furthermore, there is now evidence of self-organization in precisely one of Behe's favorite examples of irreducible complexity: the assembly of bacterial flagella.[40]

But let us go back to complexity and specification and take a closer look at these two fundamental criteria, allegedly capable of detecting the action of intelligent (supernatural) agency in nature. Following one of Dembski's examples, if SETI (Search for Extraterrestrial Intelligence) researchers received a very short signal that could be interpreted as encoding the first three prime numbers, they would probably not rush to publish their findings just yet. The reason is that even though such a signal could be construed as due to some kind of intelligence, it would be so short that its occurrence could just as easi-

[39] Of course, there are examples of irreducible complexity in living organisms, such as genetically engineered organisms. The problem for Behe is that they are all due to human agents and are readily identifiable as such.

[40] On the self-assembly of the bacterial flagellum, see T. Surrey, F. Nedelec, S. Leibler, and E. Karsenti, "Physical Properties Determining Self-organization of Motors and Microtubules," *Science* 292 (2001): 1167–1171. My students and I once had an online discussion with Michael Behe (transcripts available at http://fp.bio.utk.edu/skeptic/Chats/behe.html), and we got him to agree that if the evolution of the bacterial flagellum were to be explained satisfactorily, he would give up his support of the ID movement. We are getting excitingly close to that moment, and I am curious to see if Behe will follow through with his promise and have the intellectual honesty (and courage) to recant his *Darwin's Black Box.*

ly be explained by chance. Given the choice, a sensible scientist would follow Occam's razor (see Chapter 4) and conclude that the signal does not constitute enough evidence for extraterrestrial intelligence. However, also according to Dembski, if the signal were long enough to encode all the prime numbers between 2 and 101, the SETI people would have to open the champagne and celebrate all night. Why? Because such a signal would be too complex to be explained by chance and would also be specifiable, meaning that it would not just be a random sequence of numbers, but rather an intelligible message.

The specification criterion needs to be added because complexity by itself is a necessary but not sufficient condition for design. To see this, imagine that the SETI staff receives a long but random sequence of signals. That sequence would be very complex, meaning that it would take a lot of information to archive or repeat the sequence (one would have to know where all the zeros and ones were), but it would not be specifiable because the sequence would be meaningless.

Dembski is absolutely correct that many human activities, such as SETI itself, investigations into plagiarism, or the detection of encryption, depend on the ability to detect intelligent agency. Where he is wrong is in assuming only one kind of design: For him design equals intelligence, and even though he admitted on an Internet forum that such an intelligence might be an advanced extraterrestrial civilization, his preference is for a god—namely, of the Christian variety.

The problem is that natural selection, a natural process, also fulfills the complexity specification criterion, thereby demonstrating that it is possible to have unintelligent design in nature. Living organisms are indeed complex. They are also specifiable, meaning that they are not random, but rather systematic assemblages of organic compounds, clearly formed in a way that enhances their chances of surviving and reproducing in a changing and complex environment. What, then, distinguishes organisms from the Brooklyn Bridge? Both meet Dembski's complexity specification criterion, but only the bridge is irreducibly complex. This fact has important implications for design theory.

In response to some of his critics, Dembski claims that intelligent design does not mean optimal design. The criticism of suboptimal design has often been advanced by evolutionists who ask why God would do such a sloppy job with creation that even a mere human engineer could easily determine where the flaws were. For example, why is it that human beings have hemorrhoids, varicose veins, backaches, and foot pain? If one assumes that we were "intelligently" designed, the answer must be that the designer was either incompetent or had an evil sense of humor—something that would hardly please a creationist (though it wouldn't bother Plato, as we have seen). Instead, evolutionary theory has a single and simple answer to all these questions: Humans evolved bipedalism (walking with an erect posture) only very recently, and natural selection has not yet fully adapted our body to the new condition (Figure 2.2).[41]

[41] For an amusing look at what better-designed humans would look like, see S. J. Olshansky, B. A. Carnes, and R. N. Butler, "If Humans Were Built to Last," *Scientific American* 284(3) (2001): 50–55.

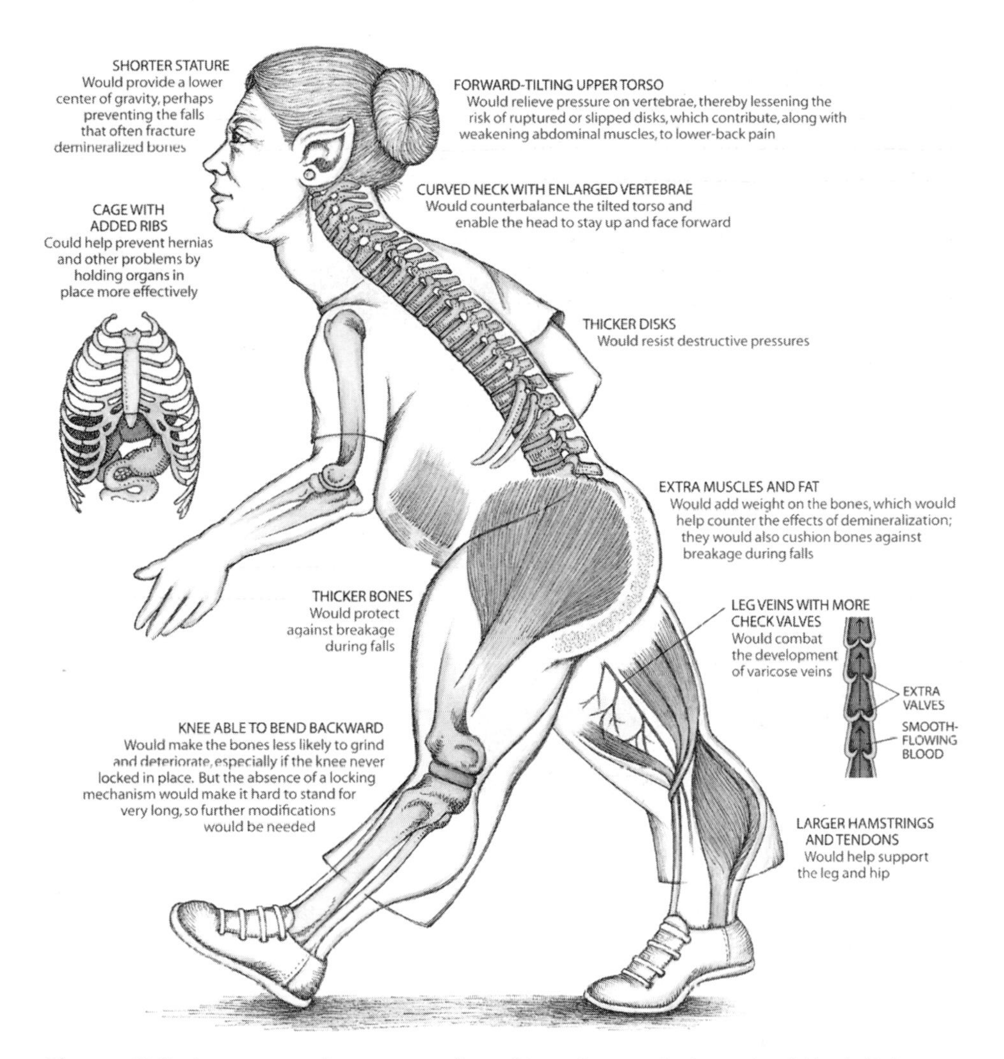

Figure 2.2 A conceptual representation of how human beings *should* look if they were better adapted to their environment, and in particular if they were to cope better with advanced age. (From S. J. Olshansky, B. A. Carnes, and R. N. Butler, "If Humans Were Built to Last," *Scientific American* 284(3) (2001): 50–55)

Dembski is of course correct in saying that intelligent design does not mean optimal design. As much as the Brooklyn Bridge is a marvel of engineering, it is not perfect, meaning that it had to be constructed within the constraints and limitations of the available materials and technology, and it still is subject to natural laws and decay. The bridge's vulnerability to high winds and earthquakes, and its inadequacy to bear a volume of traffic for which it was not built can be seen as similar to the back pain caused by our recent evolutionary history. However, the imperfection of living organisms, already pointed out by Darwin, does away with the idea that they were created by an omnipotent and omnibenevolent creator, who surely would not be limited by laws of physics that he himself made up from scratch.

Given these considerations, I propose a framework here that includes both Behe's and Dembski's suggestions, and shows why they are both wrong in concluding that we have evidence for intelligent, supernatural design in the universe. Figure 2.3 summarizes my proposal. Essentially I think there are four possible kinds of design, which, together with Dembski's categories of "regular" and random phenomena, and the addition of chaotic and self-organizing phenomena, truly exhaust all known possibilities. Science recognizes regular, random, and self-organizing phenomena, as well as the first two types of design described in Figure 2.3. The other two types of design are possible in principle, but I contend that there is neither empirical evidence nor logical reason to believe that they actually occur.

The first kind of design is *nonintelligent-natural*, exemplified by natural selection within Earth's biosphere (and possibly elsewhere in the universe). The results of this design, such as all living organisms on Earth, are not irreducibly complex, meaning that they can be produced by incremental, continuous

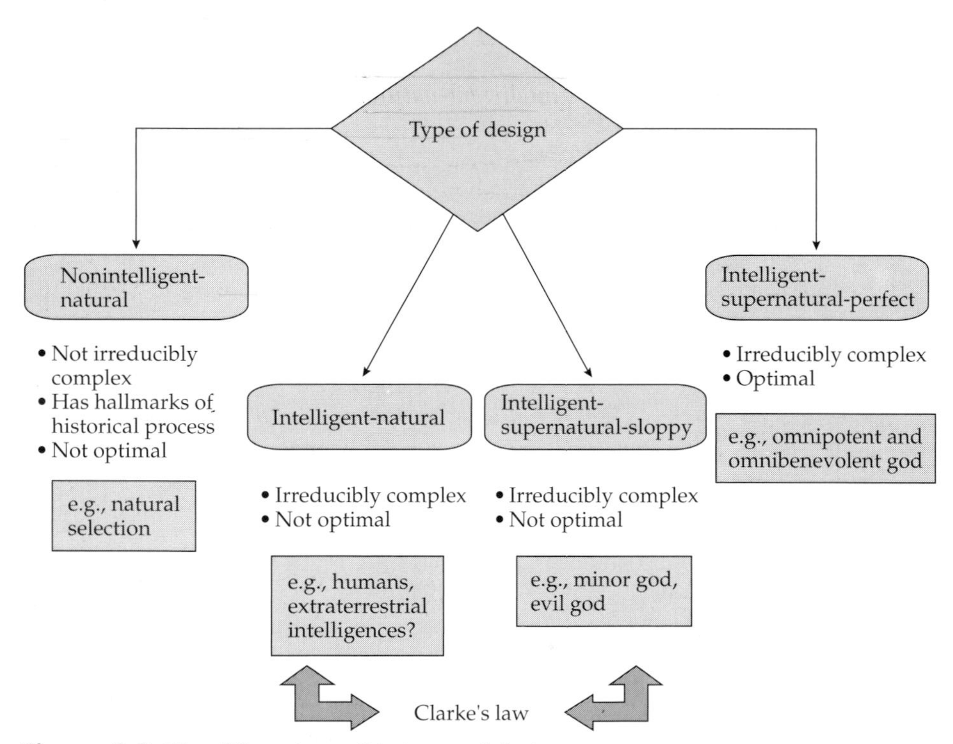

Figure 2.3 The different possible types of design.

(though not necessarily gradual) changes over time. These objects can be clearly attributed to natural processes also for two other reasons: They are never optimal (in an engineering sense), and they are clearly the result of historical processes. For example, they are full of junk—nonutilized or underutilized parts—and they resemble similar objects occurring simultaneously or previously in time (see, for example, the fossil record). Some scientists and philosophers of science feel queasy at considering this "design" because they equate the term with intelligence. But I do not see any reason to embrace such limitation. Design simply means an underlying pattern or system that determines how a thing functions. If something is shaped over time—by whatever means—such that it fulfills a certain function, then in that sense it manifests design, and the question is simply how such design happened to materialize.

The teeth of a tiger are clearly designed to efficiently cut into the flesh of its prey and therefore to promote survival and reproduction of tigers bearing such teeth. This is what misleads so many people to believe in intelligent design. Because of our anthropomorphic tendencies, it is difficult to talk about design at all without the logical implication of a designer. The teeth of the tiger function in a way that is very good for the tiger, but they did not arise in the genetic makeup of the tiger *in order* for tigers to more efficiently bite into their prey. The tiger's ancestors simply happened to have mutations that modified their teeth in a way that was advantageous for their offspring, and such mutations were retained by natural selection.

The second type of design is *intelligent-natural*. Artifacts of this type of design are often irreducibly complex and are designed by natural beings with intelligence, such as a watch designed by a human. They are also not optimal, meaning that they clearly compromise between solutions to different problems (trade-offs), and they are subject to the constraints of physical laws, available materials, expertise of the designer, and so on. Humans are not the only ones to generate these objects. Just consider bird's nests, or the twig tools designed by chimpanzees, or the possibility of artifacts from any extraterrestrial civilization.

The third kind of design, which is difficult, if not impossible, to distinguish from the second, is what I term *intelligent-supernatural-sloppy*. Objects created in this way are essentially indistinguishable from human or extraterrestrial intelligence artifacts, except that they would be the result of the ancient Greeks' demiurge, a minor god with limited powers. Alternatively, they could be due to an evil, omnipotent god that just amuses himself with suboptimal products. The reason that intelligent-supernatural-sloppy design is not distinguishable from some instances of intelligent-natural design (but it is by no means indistinguishable from all) is Arthur C. Clarke's famous third law: From the point of view of a technologically less advanced civilization, the technology of a very advanced civilization is essentially indistinguishable from magic (such as the monolith in his *2001: A Space Odyssey*). At the moment, I do not see a way around Clarke's law.

Finally, we have *intelligent-supernatural-perfect* design, which is the result of the activity of an omnipotent and omnibenevolent god. Artifacts of this type of design could be irreducibly complex and optimal. They would not be constrained by either trade-offs or physical laws (after all, God created the laws themselves). Although this is the kind of god many Christian fundamentalists believe in (though some do away with the omnibenevolent part), it is at least questionable that such a god exists, given the realities of human evil, natural catastrophes, diseases, and back pain. Dembski recognizes this difficulty and, as I pointed out earlier, admits that his intelligent design could be due to a very advanced extraterrestrial civilization, and not to a supernatural entity at all.

In summary, it seems to me that the major arguments of intelligent design theorists are neither new nor compelling: (1) It is simply not true that science does not address all Aristotelian causes whenever design needs to be explained. Science can address Aristotelian causes because it has subsumed and improved upon their function, a fact showing that Aristotle in important ways anticipated modern scientific explanation. (2) Although irreducible complexity may indeed be a valid criterion to distinguish between intelligent and nonintelligent design, living organisms do not appear to be irreducibly complex.[42] (3) The complexity specification criterion is actually met by the products of natural selection and therefore cannot provide a way to distinguish intelligent from nonintelligent design. (4) If supernatural design exists at all (but where is the evidence or compelling logic?), it is certainly not of the kind to which most religionists would likely subscribe, and it is indistinguishable from the technology of a very advanced civilization.

Therefore, Behe's, Dembski's, and other creationists' claims that science should be opened to supernatural explanations and that these should be allowed in academic as well as public school curricula is unfounded and based on a misunderstanding both of design in nature and of what the neo-Darwinian theory of evolution is all about.

A Brief, Indirect Q&A Session with a Neocreationist

At this point it might be a good idea to summarize some crucial arguments pro and con intelligent design as a serious possibility to be considered within the framework of science. An article by Todd Moody that appeared in *Philosophy Now* provides a good framework for such a summary. Moody's piece (interestingly entitled "Intelligent Design: A *Catechism*")[43] is constructed around a series of short questions and answers concerning ID. I shall summarize Moody's relevant answers and then attempt to show why they are wrong.

[42] See, for example, N. Shanks and K. H. Joplin, "Redundant Complexity: A Critical Analysis of Intelligent Design in Biochemistry," *Philosophy of Science* 66 (1999): 268–282; and B. Fitelson, C. Stephens, and E. Sober, "How Not to Detect Design—Critical Notice: William A. Dembski, The Design Inference," *Philosophy of Science* 66 (1999): 472–488.

[43] My emphasis. The article was published in *Philosophy Now* 31 (March/April 2001): 31–33.

(1) What is intelligent design? According to Moody, ID is the theory that there is empirical evidence for the claim that some things in nature are the result of intelligent agency. There are two problems with this statement. First, ID is not a scientific *theory,* it is simply the statement of a possibility. To achieve theory status it would have to be articulated in a way so as to be potentially falsifiable, which so far ID "theorists" have utterly failed to do. Second, no scientist would deny that certain objects are the product of intelligent design. Surely the laptop I am using to write this book was the result of intelligent agency. Furthermore, and contrary to what ID supporters claim, scientists have seriously entertained the possibility that some interesting scientific puzzles might be solved by judicious invoking of intelligent design. For example, Francis Crick—the Nobel Prize–winning codiscoverer of the structure of the DNA molecule—has suggested that life on Earth might have originated from extraterrestrial intelligences that were "seeding" different planets, a theory known as *directed panspermia.*[44] Although biologists have considered this proposal with some skepticism (as any scientist should consider any radically new proposal), they have not ridiculed it or excluded it a priori. What science necessarily has to exclude is the possibility of *supernatural* agency because such a hypothesis is utterly untestable by definition (given that a god can do whatever it pleases without having to follow natural and understandable laws).

(2) Isn't this just creationism? According to Moody, it is not creationism because ID makes no claims as to the identity of the designer, it does not claim that the design is supernatural, and it accords no priority to revelation or to the reconciliation of Scripture and science. As we have seen, *all* of these claims are inaccurate. Although Dembski has been forced to admit that the designer *might* not be supernatural, that is his conclusion in his major book, *The Design Inference* (and Dembski speaks freely of his belief in the Christian God as designer when he is before friendly audiences who share his religious beliefs). We will see from our analysis of the Wedge strategy (see the next section) that the ID think tank, the Center for the Renewal of Science and Culture, very much accords priority to Christian religious doctrine, a position substantiated by the curious observation that the overwhelming majority of ID theorists belong to conservative Christian denominations.[45]

---

[44] On the idea of directed panspermia, see F. Crick, *Life Itself* (New York: Simon and Schuster, 1981). The original suggestion of (undirected) panspermia was formulated by the great Swedish chemist Svante Arrhenius in his book *Worlds in the Making,* first published in 1907.

[45] Sometimes skeptics who point out the uncanny relationship between creationism and fundamentalist Christianity are accused of committing the *genetic fallacy*—that is, arguing that something is wrong on the basis not of the merit of the proposed theory but of the identity or affiliation of the theory's supporters. Although it is certainly true that one cannot reject intelligent design *solely* on the basis of who proposes it, I maintain that the latter piece of information is very relevant in the light of my contention that this is not a serious scientific debate, but rather a cultural war.

(3) ***Since most ID theorists believe in God, aren't they then covert creationists bent on proving the existence of God?*** Moody correctly points out that belief in the supernatural is not necessarily coupled with someone's scientific research and how it is conducted. After all, many scientists are Christians or theists of other persuasions. Moody also suggests that there is no intention in the ID camp to deceive and that ID cannot therefore be considered a "Trojan horse theory." Well, not exactly. For one thing, claiming that strong religious beliefs do not influence somebody's objectivity is rather naïve. Philosophers of science have long recognized that there is no such thing as a science free of ideology or immune from social pressure.[46] As we shall see in Chapter 7, what leads to emergent objectivity in science is the fact that it is an activity carried out by a variety of people subscribing to different and contrasting worldviews and ideologies. They keep each other in check, and the balance is as objective as is humanly possible. ID, on the other hand, is defended only by people subscribing to a particular ideological position, bent on defending a set of beliefs that they see as in peril and that they consider more important than any scientific enterprise. Furthermore, in reading the Wedge manifesto discussed in the next section, it is hard *not* to conjure up the image of a Trojan horse when words such as *open confrontation* and *defeat* are used, and when a programmatic strategy is being proposed with the express purpose of destroying "materialist" science. The ID program might not be as subtle as that of Odysseus, but it is a conscious attempt at deceiving the general public nonetheless.

(4) ***Wouldn't ID be the end of science?*** According to Moody, scientists would continue their work as usual after accepting the ID perspective because it has already happened in the past: A new theory replaces the old one, new questions are asked, and new avenues of research are pursued. However, in this case there are only two options, both of which are unpalatable to ID proponents. One possibility is that this *would* be the end of science simply because no further questions could possibly be meaningfully asked. This is something that has been pointed out to Dembski repeatedly in public and has always elicited very irritated reactions from him. Once science admits the supernatural into its domain of explanation, anything goes. An old cartoon shows two scientists in front of two complicated sets of equations separated by a gap. One of them has written, "God did it," in the gap. The other one observes the scene thoughtfully and comments, "You might want to work on the details of this transition here." But with ID it is not possible to work on the details. If God did it, there is no way mere mortals will be able to say anything further about how or why, by the very definition of what it means to be a god according to our conceptions. To put it in another fashion, introducing ID into science would not change the way we do science because it would not add any explanatory power to science

[46] See H. E. Longino, *Science as Social Knowledge: Values and Objectivity in Scientific Inquiry* (Princeton, NJ: Princeton University Press, 1990).

as it is today. To avoid the sudden halt caused by the "God did it" answer, scientists would still have to formulate and test hypotheses *as if* there were no supernatural designer at all. If that is the case, why bother putting ID into our science textbooks?

The second option concerns the theological implications of ID: If creationists wished to play fair and proceed in the way that science would, they would be forced to ask, "What do these designs tell us about the designer?"—just as the structures and functions of living organisms make biologists ask, "What do these tell us about the environment?" A good analogy here is with archaeology: Archaeologists, upon discovering an ancient ruin, rightly claim it as the product of intelligent design. Of course they do not stop there. Instead, they infer from the available artifacts the attributes of the designers. Where would modern archaeology be if they stopped at inferring design and didn't press on with who the agents would be and why they were doing what they were doing? But intelligent design proponents shy away from this sort of follow-up because it would open a Pandora's box of internal theological conflicts, as we have already discussed.

(5) Doesn't science deal only with natural causes? Apparently it does not, according to Moody. This is interesting because it shows what a great deal of confusion there is about science and its proper objects of investigation. Moody says—correctly—that archaeology is a science, and yet it deals with the products of intelligent agency. As I have already stressed, the problem is not with intelligence per se. If it were, archaeology, anthropology, and even the search for extraterrestrial *intelligence* would not be sciences (though clearly they are). The problem is with *supernatural* intelligence, which we have no way to detect.

(6) What could count as evidence for ID? Moody here can answer only that irreducible complexity is a falsifiable theory, and therefore scientific. The problem is that it is hard to see how it could be falsified, especially since it is a *negative* statement—that is, a prediction about what *cannot* happen. Michael Behe defines irreducible complexity in a couple of different ways. In one sense, as we have seen, it seems to be the statement that a given biological structure *cannot* possibly be produced by an incremental process over time. This is exceedingly difficult to prove, and Behe has provided no backing for his claim that irreducibly complex biological structures exist, other than to say that he cannot think of how they could be possible through natural selection. In the second sense, irreducible complexity is simply the statement that a certain biological structure—for example, the flagellum of a bacterium—will not function if enough pieces are taken away. But this is hardly controversial, and it is certainly no blow to a naturalistic interpretation of biology. Since any biological structure is both the result of hundreds of millions of years of change and the product of complex interacting developmental systems, it is no surprise that it has reached a level of sophistication such that it could not work if vital parts were removed. The fallacy is thinking that anything alive today—including bacteria—can possibly be considered "primitive." The task of evolutionary theory is to explain how currently living organisms derived from remote common ancestors, not how they derived from each other, and the same goes for any particular structure of any of these organisms.

An interesting exception is the existence of actually intelligently (humanly) designed organisms such as the genetically engineered animals, plants, and bacteria mentioned earlier (see note 39). In this case we know that they are artificially modified, but this fact could presumably be uncovered by an extraterrestrial scientist who didn't know about humans, provided that he knew enough about the natural evolution of organisms on Earth.

(7) So what do ID theorists want? Moody's answer is that they simply want to compete fair and square in the arena of scientific discussion. This sounds very logical and appeals to the sense of fairness of most people, even those who are not ideologically committed to intelligent design theory. Moody claims that the statement that ID is not science is an epistemological, not a scientific, one, and as such it is a matter of philosophy, not science. On the latter point he is certainly right which is why all of the academic discussion on ID has occurred in philosophical, not scientific, journals and why philosophers of science seem to have little doubt that intelligent design theory is not, in fact, science.[47] As for fairness, it is entirely beside the question. If ID is not science, it does not belong in science classrooms or academic departments, and it does not deserve federal funding reserved for scientific research. Complaining about its exclusion would be like complaining to the baseball commissioner that one's football team is being discriminated against in the American League. Such a complaint would, and should, be laughed off the field.

The Real Neocreationist Agenda: The Wedge Document

It is often said that the fact that a person is paranoid does not mean that the universe is not really after him. This section is about a creationist conspiracy, even though I do not think of myself as particularly paranoid. Furthermore, unlike most conspiracy theorists, I have documents to show the existence of the conspiracy. After having reviewed the basics of both standard creationism and neocreationism, I would therefore like to devote some space to the discussion of a peculiar document that shows that the ID movement has a definite ideological agenda and, in fact, makes no secret of it. It is important that not only

[47] See, for example, B. Fitelson, C. Stephens, and E. Sober, "How Not to Detect Design—Critical Notice: William A. Dembski, the Design Inference," *Philosophy of Science* 66 (1999): 472–488; also E. Sober, "The Design Argument," in W. Mann (ed.), *Blackwell Guide to Philosophy of Religion* (Oxford: Blackwell, in press); and the already cited Shanks and Joplin article (see note 42).

scientists and educators, but the general public, be aware of this agenda because it clearly shows what I have attempted to convince the reader of throughout this chapter: The evolution–creation controversy is not a scientific debate, but rather a mighty ideological struggle for the control of public education and the financing of scientific research.

The so-called Wedge strategy was first articulated in a chapter of Phillip Johnson's *Defeating Darwinism by Opening Minds*[48] as the outline of a plan to destroy secular science and replace it with a Christian-inspired version. Interestingly, Johnson opens the chapter by criticizing the pope (John Paul II) for conceding so much to the "materialist" agenda, as we have seen in considering the pontiff's letter to the Pontifical Academy of Sciences. Johnson believes that the pope is mistaken in adopting a compatibility model between science and religion, and that the reality is more like what is described by scientists such as Richard Dawkins: Science is philosophically inseparable from materialism (where this term is used to mean the philosophical view that matter is all there is, not an attachment to material objects of consumption).

Johnson rejects the theistic evolution model according to which evolution is the way God works His way into the universe because this "shallow reconciliation" leaves young people open to the dangers of materialist indoctrination. According to Johnson, if one is trained to think materialistically in the scientific realm, this then becomes an ingrained habit of thought, and people will not be able to divest themselves of it with respect to religious matters. All of this may or may not be true, of course, but it certainly sets the stage for understanding the *motivations* that push Johnson and the entire ID movement.

Interestingly, Johnson attacks also the liberal academic left's idea of deconstruction and postmodernism because they criticize materialistic evolution for the wrong reasons. Creationists don't want their position to be accepted as one of many possible and equally valid "constructions" of reality—theirs is *the* truth. In an unexpected twist, Johnson characterizes postmodernists as rationalists and hence relativists, even though few philosophers or scientists would use the label *rationalism* with respect to postmodernism, and none would subscribe to the idea that rationalism leads inevitably to relativism. But Johnson is certainly not in the business of making sophisticated philosophical arguments. His equation of materialism with relativism (the position that in ethics "anything goes" because there is no way to determine absolute truths) is based on his understanding that the materialist community rejects what has come to be known as the *naturalistic fallacy.* This was defined by Hume (in *A Treatise of Human Nature*) essentially as equating what is with what ought to be. Although the naturalistic fallacy is not accepted by every philosopher as a fallacy, we need to take a brief look at what this criticism is all about before proceeding with a discussion of the Wedge document itself.

[48] Phillip Johnson, *Defeating Darwinism by Opening Minds* (Downers Grove, IL: InterVarsity Press, 1997).

There are three basic approaches to the foundations of morality. The first one encompasses the idea that there are absolute moral truths of some kind, and that they can be discovered or otherwise known. Absolute moral truths have historically been ascribed to two sources: Either they are given by a god, or they are facts of nature not dissimilar from the existence of planets and galaxies. Obviously, Johnson and other creationists subscribe to the idea of a god-given moral code. The alternative idea that moral truths are "out there" to be discovered by the use of a more or less innate "moral sense" has been advanced by many philosophers and shows that even within the position of absolute morality, the theistic version is not the only one.

The second view, often mistakenly attributed to atheists, is that there are no moral truths at all. This is the relativistic attitude of "anything goes" that so much appalls fundamentalist Christians and triggers their opposition to secular humanism and the theory of evolution, among other things. Indeed, relativism is a very pernicious view of human morality, which can easily lead to a lack of firm ethical foundations on the basis of which to reject the atrocities of people like Hitler. The modern general framework of moral relativism is provided by the philosophical current of postmodernism, which maintains that all theories, moral or otherwise, are culturally constructed and that none is better than any other. Of course, if one takes postmodernism at its own word, one should wonder why we have to consider seriously postmodernism itself, since it too must be socially constructed (see Chapter 3). Most people (materialists or not) therefore reject relativism on the basis of its internal inconsistency and because they simply abhor some of its practical consequences.[49]

There is a third possibility: Some moral truths may have only local (situational) meaning. For example, some choices are morally neutral, as in the case of sexual relationships with other consenting adults. Others, however, such as rape, are certainly wrong for human beings. I do not wish here to enter into the merits of these positions, which would bring us too far off focus in our discussion. My point is that, contrary to what Johnson advocates, discussion about moral systems is not as simple and clear-cut as he depicts it in his writings. Johnson is so confused about philosophical issues and so upset about the impact of materialism on contemporary culture that he attacks even mainstream television shows such as the TV series *Star Trek,* nonsensically claiming that the latter is an example of *both* modernism and postmodernism!

At any rate, Johnson maintains that Christians have to take a stand because Jesus told them to "go and make disciples of all nations." The Wedge strategy is then Johnson's response to the pervasive rule of materialistic philosophy. Rather than examine the Wedge in the version published in his book, which

---

[49] I would like to note that in ethics, as opposed to science, one is absolutely justified in rejecting a theoretical framework if one does not like the consequences of that framework. Although it makes no sense to attempt to repeal the law of gravity, moral conduct can and often should be chosen and altered according to rational thought.

was very preliminary, I will use a document that was published for some time on the Discovery Institute Web page and that is significantly more detailed.[50]

The document starts out with a rhetorical statement that the idea that humans were created in the image of God is "one of the bedrock principles on which Western civilization was built" and that this idea has been under attack since Darwin, Marx, and Freud. Materialism is equated with the claim that objective moral standards do not exist, which—as we just saw—is patently false. The programmatic claim is then made that the Discovery Institute's Center for the Renewal of Science and Culture, to which essentially all of the neocreationists like Johnson, Behe, Wells, and Dembski belong, "seeks nothing less than the overthrow of materialism and its cultural legacies." How is this overthrow going to happen in practice? Through three phases of the Wedge strategy spanning a period of 20 years. The three phases are significantly labeled "research, writing and publication," "publicity and opinion-making," and "cultural confrontation and renewal."

Phase I of the Wedge is meant to provide a façade of scholarship for the ID movement. Obviously, the Wedge document does not use the word *façade,* but that is what this phase really amounts to. The whole phase was destined to last roughly five years (starting in 1999) to provide the basis for pushing the ideological and political agenda of phases II and III. The authors of the document see the Wedge program as a scientific revolution, allegedly taking place on the basis of the sloppy scholarship of people like Michael Behe and William Dembski, as if armchair speculation were all it takes to overthrow an incredibly successful scientific idea such as the neo-Darwinian theory of evolution. From these premises, one has to deduce one of two possible alternatives: Either the Wedge authors are exceedingly naïve about how science is done, or they are not interested in research at all, but simply in producing the *appearance* of solid work to back up their aim to "reform" the public education and research funding systems.

Phase II is meant to prepare the public for reception of the ideas pushed by neocreationists. Of course, given opinion polls putting disbelief in evolution among Americans solidly at about 50 percent, one would think that very little further "preparation" was actually needed. Yet the Wedge article goes on to mention the necessity of contacting and lobbying broadcast media, congressional staff, and talk show hosts. Furthermore, the Wedge document contains a direct appeal to what its authors think of as their "natural constituency" (notice the use of political campaign terminology)—that is, Christians.

Phase III will then consist of what the Wedge document calls "open confrontation" with the scientific establishment, including legal challenges to attempts to exclude ID from public school teaching. As the document puts it, "The attention, publicity, and influence of design theory should draw scien-

[50] The page has long since been removed, but it is available from the author in its entirety and original form upon request.

tific materialists into open debate with design theorists, and we will be ready." Rather than a scientific enterprise, it sounds like a declaration of war.

It is also very instructive to look at the goals stated by the Wedge document, which is an elaboration of those more sketchily published in Johnson's book. Among them we find "to defeat scientific materialism and its destructive moral, cultural and political legacies" and "to replace materialistic explanations with the theistic understanding that nature and human beings are created by God." It is notable that this "understanding" was formulated before any "research" was actually started, hardly the stuff of honest intellectual enterprises. (Compare this to the attempts of Darwin and Huxley to publicize their work, described in Chapter 1, which came *after* decades of research carried out by Darwin. There is a distinction between attempting to make one's work widely known and plans to deliberately manipulate public opinion to establish a preconceived ideology.)

Needless to say, the Wedge strategy is proceeding very well—so well that Phillip Johnson gives a regular *weekly* update of the progress on the Discovery Institute's Web site.[51] My point here is certainly not that the Discovery Institute, neocreationists, or anyone else should not engage in this sort of activity. Anybody is free to plan for whatever (peaceful) cultural revolution he wishes to realize and to let the chips fall where they may. However, it is intellectually dishonest to present such an enterprise as a program of objective research designed to overcome the stubbornness of a materialist majority bent at all costs on preventing the truth from emerging. A much fairer representation would be that this is the obstinate effort of a religious minority to silence anybody and anything that in an extreme fit of paranoia is construed as threatening one's particular concept of morality and purpose in life. I hope that at this stage few scientists and educators will still think of this issue in terms of a simple matter of science or education.

Evolution 101: What Evolution Really Is (and Is Not)

Having attempted to explain the basic tenets of the major forms of creationism, it is only fair—and most urgent, judging from my experiences at debates against creationists—to briefly explain what evolutionary theory really is, and even a few things that it is thought to be but isn't. As I try to show throughout this book, the evolution–creation debate is marred by many misunderstandings and a lot of ideological posturing, often on both sides. One major thing that creationists seem reluctant to acknowledge, however, is the distinction between what evolutionary theory actually is and what they think it is. And the difference is both huge and crucial. No matter what one's ideological position, it seems to me necessary to understand what *biologists* claim evolution to be and not to build straw men just to be able to demonize the opposition.

---

[51] Johnson's weekly update is available at http://www.arn.org/johnson/wedge.htm.

There are four key concepts that I want to cover in the following discussion. 1. Evolution is technically defined as a change in the frequencies of the genes found in natural populations. 2. A major consequence of these changes is that evolution can also be seen as gradual descent with modification linking different life forms on Earth. 3. Contrary to creationists' claims, evolution is not a theory of the origin of life. 4. Evolution is also not a theory of the origin of the universe. I will go back to the question of the origin of life in more detail in Chapter 6 to discuss the science behind it, and I will briefly (I am not a cosmologist) mention the problem of the origin of the universe in Chapter 7. For now, I simply wish to convince the reader that these topics—as interesting as they are for science and our understanding of the universe—are not really a part of the theory of biological evolution.

If one asks an evolutionary biologist—by definition the only person qualified to answer the question—she will tell you that evolution is simply a change of gene frequencies over time. This may sound rather simple and philosophically uninteresting, but it is in line with what science is all about: seeking answers to specific questions, not to questions of ultimate meaning. The theory of genetic changes in natural populations is very well understood by a branch of biology called population genetics, and modern molecular biology provides direct evidence that gene frequencies do indeed change under our very nose. Examples are abundant and are found in all classes of living organisms (humans included, of course).

Let's consider an exceedingly well understood, and therefore undeniable, current example of evolution by changes at the genetic level: the HIV virus, the causal agent of AIDS.[52] This kind of virus, which causes a disease of the immune system, has evolved for a long time using vertebrates as its vehicle. Closely related forms of the virus attack closely related types of vertebrates, as is predicted by evolutionary theory. For example, one form attacks dolphins, but it is quite distinct from the human form, which is instead much more similar to chimpanzee forms. The reason is that chimps and humans are more closely related and more similar to each other than either is to dolphins or other cetaceans. Because viruses evolve gradually (like anything else), any attacks of novel host species are restricted to organisms that are very similar to the current host.

Evolution has produced at least two major forms of HIV in humans, and a third one may be in the process of emerging. The history of one of these, HIV-1, has been particularly well characterized. We know that it is very close to a chimp virus and, according to estimates based on molecular data, it passed from chimps to humans in West Africa in the 1940s. From there it was picked up by a Norwegian sailor, who brought it to Europe, where it started spreading in the 1970s. This is a typical pattern of origin, migration, and genetic differentiation followed by many living organisms on Earth, including humans, and described mathematically by the theory of evolution.

---

[52] For a fascinating account of the details of our understanding of the evolution of HIV viruses, see S. Jones, *Darwin's Ghost: The Origin of Species Updated* (New York: Random House, 2000).

Interestingly, and again in perfect agreement with evolutionary theory, the HIV virus is changing rapidly, under our very eyes. The reason is that new mutations occur in populations of the virus, and these are selected to adapt the virus to changes in its own environment (and the virus evolves much faster than its host because of its much shorter generation time). The environment consists of the human body and our sexual behavior. Different types of HIV are now known that specialize for transmission through different sexual practices, depending on which of these are adopted by particular human populations. Unfortunately, it is this rapid evolution that makes it very difficult to find a cure for AIDS: The target literally keeps shifting in front of us as a result of mutation and selection. This incidentally makes the point that evolutionary theory is not just an academic matter, but is of vital importance for the survival and welfare of human beings.

Yet most people think of evolution in terms of *descent with modification,* to use Darwin's term, of large organisms, and in particular animals. Of course, the process is the same for humans, HIV, or anything else, but let's look at evolution seen at the level of macroscopic creatures and long timescales. This is the realm of comparative anatomy and paleontology, and the evidence for evolution of plants and animals comes from studies of their genetics, physiology, morphology, and development. Additional evidence comes from the much maligned fossil record, to which I will return in Chapter 6.

There are many great examples of morphological evolution in plants and animals, but perhaps one of the most spectacular is the evolution of modern whales, which has recently been largely elucidated by a series of paleontological findings and molecular studies. Figure 2.4 summarizes what scientists currently (as of 2002) think of the evolution of whales, a story that began with terrestrial animals hunting in the proximity of rivers several dozen million years ago, continued through a series of intermediate forms, and has arrived at the modern two major groups of whales: toothed and baleen. Notice that the animals represented in the tree are not necessarily each other's descendants, because the analysis reported here includes many contemporary forms and because evolution proceeds more like a bush with many branches than as an orderly linear sequence. This highly branched pattern of the tree of life is exactly what one would expect from a natural process that doesn't have an internal direction (as opposed to, say, a divinely inspired plan aiming at a particular end product). To confirm evolutionary theory, however, all we have to see is that some of the intermediate forms are found when and where they are expected. Finding fossils that are badly out of sequence would be one way to disprove evolution. Unfortunately for creationists, nobody has yet found a human fossil together with a dinosaur (see Chapter 7), or a dinosaur in Precambrian times.

The story of whales started about 55 million years ago, although the exact group of ancestors is currently unknown. They were terrestrial animals belonging to the Artiodactyla (the modern group that includes hippopotami,[52]

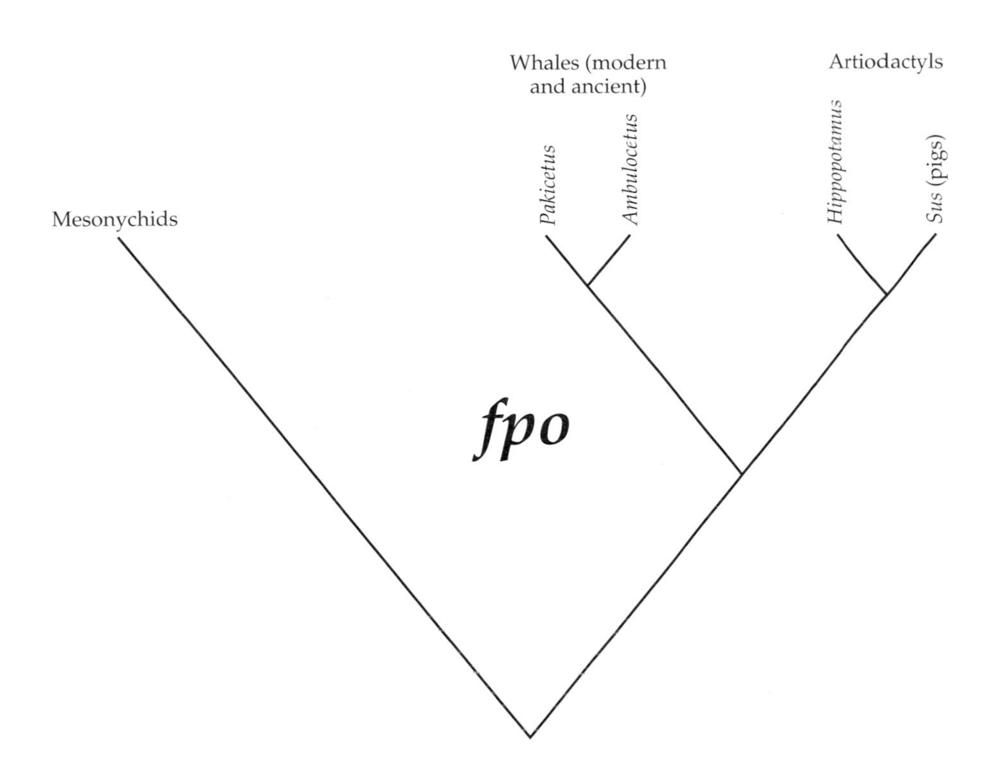

Figure 2.4 Our current knowledge of the evolution of whales. Some of these forms, such as the group of Mesonychia, or *Pakicetus* and *Ambulocetus* among the Cetacea, are extinct. The tree was obtained by the combining of molecular and paleontological evidence. (Redrawn and simplified from J. G. M. Thewissen, E. M. Williams, L. J. Roe, and S. T. Hussain, "Skeletons of Terrestrial Cetaceans and the Relationship of Whales to Artiodactyls," *Nature* 413 (2001): 217–281)

sheep, camels, and pigs). That group was itself closely related to the now extinct Mesonychia, which until recently were considered a better candidate for the direct ancestors of whales, but are now regarded as their cousins. The earliest animals belonging to the Cetacea (of which modern whales and dolphins are also members) were closely related to *Pakicetus* and *Ambulocetus* (both represented in Figure 2.4). The early relatives of these two were terrestrial artiodactyls, occupying an ecological niche similar to the one filled today by bears

[52] Interestingly, hippos are today at an intermediate evolutionary stage toward becoming fully aquatic animals, and they show many similarities to whales: Their babies swim before they walk, the mother nurses the young underwater, the male's testicles remain inside its body, they are hairless, and they do not sweat. These are all adaptations to a predominantly aquatic lifestyle and show us some of the intermediate stages that whales' ancestors probably went through (it is hard to say for sure because behaviors and soft body parts such as testicles do not easily fossilize). The next time somebody ridicules the idea of gradual evolution of marine animals from terrestrial forms, just go to the zoo and point out sea lions and hippos!

(i.e., they were probably scavengers and fish eaters), although whales and bears themselves are not closely related to each other. Between 53 and 45 million years ago (MYA) a variety of intermediate forms appeared to connect the artiodactyl ancestors to modern whales. Some of these forms have been found in the fossil record, though there probably were many more that did not survive the fossilization process. Fifty MYA the already mentioned *Pakicetus* appeared, sporting a lifestyle that included both land and water hunting (judging from its skeleton and skull). A little after that, *Ambulocetus* showed advanced features adapted to a marine life, with a skeleton very similar to that of modern otters, sea lions, and other pinnipeds that are still today in an intermediate stage of their aquatic evolution. Between 40 and 35 MYA other forms arose, in particular *Basilosaurus*. This animal was truly almost like a whale except for the still apparent limbs, which, however, were reduced enough to make it impossible for it to walk on land.

So much for what evolution *is*. Now there are a couple of important things that evolution is *not*, misleading claims by creationists notwithstanding. For example, evolution is not a theory of the origin of life, for the simple reason that evolution deals with changes in living organisms induced by a combination of random (mutation) and nonrandom (natural selection) forces. By definition, before life originated there were no mutations, and therefore there was no variation; hence, natural selection could not possibly have acted. This means that the origin of life is a (rather tough) problem for physics and chemistry to deal with, but not a proper area of inquiry for evolutionary biology. It would be like asking a geologist to explain the origin of planets: The geologist's work starts *after* planets come into existence, and it is the cosmologist who deals with the question of planetary origins.

The problem of the origin of life may not be insoluble, but it is very difficult because there are no fossils to guide us (chemical pathways, much like protowhale testicles, do not normally leave fossil evidence) and scientists still have only a very rough idea of even the conditions under which the process occurred (see Chapter 6). However, one currently promising avenue of research is based on complexity theory, a new discipline that has already made large contributions in mathematics and physics, and that deals with the property of some physical systems to self-organize—that is, to spontaneously increase in complexity. There is nothing magical or supernatural in self-organization, as it can be readily observed every day. For example, we know of the spontaneous formation of highly organized and complex convective cells in Earth's upper atmosphere, which occurs if certain environmental situations are manifested. You can replicate the phenomenon by slowly heating a thin layer of water in a pot. When the temperature is high enough (but not too high to cause boiling) you will see hexagonal cells forming spontaneously in the liquid because of slow convective movements. It could be that life originated because of self-organization of this sort under the proper environmental conditions, something that one of these days will require demonstration by a bright young scientist.

Figure 2.5 The "building blocks" of the early universe (proto-galaxies) as seen by the Hubble Space Telescope. (Courtesy of NASA/Space Telescope Science Institute)

Evolution is also most definitely *not* a theory of the origin of the universe. As interesting as this question is, it is rather the realm of physics and cosmology. Mutation and natural selection, the mechanisms of evolution, do not have anything to do with stars and galaxies. It is true that some people, even astronomers, refer to the "evolution" of the universe, but this is meant in the general sense of change through time, not the technical sense of the Darwinian theory. That the universe does "evolve" in this larger sense is clear from the fact that powerful telescopes like the Hubble can actually peer into the distant past (thanks to the fact that light travels at a finite speed) and show us firsthand what primordial galaxies looked like (Figure 2.5).

The origin of the universe, like the origin of life, is of course a perfectly valid scientific question, even though it is outside the realm of evolutionary biology. Currently, our best hope to understand it lies in what physicists call *superstring theory*,[53] which is a highly sophisticated (and entirely provisional) theory aimed at fusing the two major explanatory paradigms of modern physics into a unified theory. So far, macroscopic phenomena (such as the "evolution" of stars and galaxies) have been very successfully explained by Einstein's theory of general relativity. Similarly, quantum mechanics has been astoundingly successful in explaining the behavior of matter at the very smallest, subatomic, scale. The problem of the origin of the universe (and a few other problems, such as the behavior of black holes), however, lies in a realm in which microscopic phenomena directly affect macroscopic phenomena (since the universe

[53] For a superb introduction to superstring theory and its implication for cosmology, see B. Greene, *The Elegant Universe: Superstrings, Hidden Dimensions, and the Quest for the Ultimate Theory* (New York: Norton, 1999).

likely originated with a minute quantum fluctuation that rapidly expanded to gigantic proportions in the so-called Big Bang). Superstring theory is currently one of the most promising attempts at understanding what happens under conditions at the boundary between relativity and the quantum world. If correct, the theory may provide a sound scientific basis to explain the origin of the universe.

Is the fact that evolutionary theory can explain neither the origin of life nor the formation of the universe a "failure" of Darwinian evolution? Of course not. To apply evolutionary biology to those problems is like mixing apples and oranges, or like trying to understand a basketball play by applying the rules of baseball. Creationists often do this, but their doing so betrays either a fundamental misunderstanding of science or a good dose of intellectual dishonesty—neither of which should be condoned.

Thought Patterns in Science and Creationism

Creationists and scientists think along vastly different lines, and the rest of this book is an exploration of just how deep this divide is and of some of the common pitfalls on both sides of it. J. A. Moore[54] has nicely summarized the difference by suggesting that one frame of mind, the scientific one, relies on empirical data and uses rational thinking; the other one, the creationist, is romantic, religious, artistic. Creationist decisions are based on beliefs and preferences, not on any attempt to objectively assess the problem. As I said earlier, creationists—contrary to all principles of sound science and critical thinking—start out with a preferred conclusion and then try to find evidence to back it up. As we shall see in more detail in Chapter 4, this is exactly the modus operandi of pseudoscience and what, in the words of philosopher of science Karl Popper, distinguishes it from actual science.

Toward the end of the book I will also discuss two of the major problems that Moore identifies as perpetuating the creation–evolution controversy. On the one hand, teachers tend to ignore the fact that half of their students, and their parents as well, are deeply and emotionally involved in the controversy. It is simply not good educational strategy to shut up half of a class on the ground that a topic is off-limits. Rather, teachers should use the debate as an introduction to critical thinking. In this the creationists are right: Although I am certainly not advocating teaching creationism as science alongside evolution, what is the harm in letting the kids be exposed to the controversy? Do we have such a low opinion of the power of critical thinking that we are afraid of letting them think about creationism? If that is the case, then no amount of hiding will ameliorate the situation anyway.

The second problem is that teachers simply do not provide children with a good understanding of what science is and how it works. As we shall see in

[54] In "Thought Patterns in Science & Creationism," *Science Teacher* May 2000: 37–40.

Chapter 8, all too often science teaching, from elementary school to college, boils down to learning an assortment of disconnected factoids that make no sense and bore the student to death. We are doing an excellent job of killing the natural curiosity of young people about the world around them, thereby preparing fertile ground for creationism and other pseudoscientific fantasies.

That the situation is difficult and not easy to address is backed by data on people's belief in pseudoscience, including creationism. The results of a study conducted in 1981 at Concordia University in Canada and discussed by Moore are illuminating. At the beginning of a course on critical thinking, belief in ten kinds of paranormal phenomena was about 58 percent in the class attending the course. After the course that belief had decreased to 40 percent (which is still very high, considering that the purpose of the course was to debunk the paranormal). Alas, a year later belief in the paranormal was up to 50 percent again, showing that people need not only to be educated, but to be constantly reminded of their education.

At the time of the Concordia study, a separate study found that belief in the paranormal in the general population was as high as 80 to 90 percent, and the authors of that study (quoted in Moore's article) suggested that "the psychological mechanisms involved in [supernatural] beliefs may represent more dramatic forms of some mundane pathologies of reasoning. . . . Deficits in human inferences [are] so universal and so stubborn that they can plausibly account for many of our errant beliefs, including [supernatural] ones."

Even scientists are not immune from these pathologies. It is customary to cite the fact that Kepler believed in astrology and Newton was prone to flirting with mysticism, but one has to realize that even some highly trained contemporary scientists fall into the pseudoscience trap. I was once having a conversation about these matters with a colleague of mine whom I highly respect, when he said that he couldn't believe that all claims of paranormal experiences were unfounded. In response to my puzzled look, he reasoned that by applying the customary statistical approach to hypothesis testing (the so-called significance level of 5 percent for each test), surely in 5 percent of the cases one is bound to reject as false a paranormal phenomenon that is in fact true. He was referring to a basic rule of statistical inference about the possibility of being too conservative in assessing the significance of a measured effect. It didn't occur to him that those rules cannot establish the reality of unreal phenomena. There is no sense in which ESP is 95 percent wrong; it either is or is not. Ether either exists or it doesn't (it doesn't, it turns out), even though tests for ether may produce *false* positives a certain percentage of the time. If straight thinking on these matters can elude scientists, imagine how difficult it must be for the untrained general public.

Yet something deeper than just lack of critical thinking underlies the creationists' open war against evolution. People are not obliged to understand or share anybody else's beliefs about the world, and scientists generally don't go to preachers telling them what to say from their pulpits. So why is it that so many people feel strongly about telling scientists what to do and say? In the next chapter we shall explore the deep roots of the evolution–creation controversy: the many forms of anti-intellectualism that have pervaded American society since its very inception.

Additional Reading

Dawkins, R. 1996. *The Blind Watchmaker: Why the Evidence of Evolution Reveals a Universe Without Design.* New York: Norton. Perhaps one of the best popular explanations of the theory of evolution and the counterintuitive notion that an unsupervised natural phenomenon can build complexity and adaptation.

Futuyma, D. 1995. *Science on Trial: The Case for Evolution.* Sunderland, MA: Sinauer Associates. The forerunner of the book you are reading now. One of the first examples of a scientist taking the creationist threat seriously.

Gould, S. J. 1999. *Rocks of Ages.* New York: Ballantine. A highly debatable, but stimulating, discussion of the relationship between science and religion.

Kitcher, P. 1982. *Abusing Science: The Case against Creationism.* Cambridge, MA: MIT Press. Edited by a philosopher of science, a look at some of the best responses to specific creationist arguments.

Pennock, R. 1999. *Tower of Babel: The Evidence against the New Creationism.* Cambridge, MA: MIT Press. Also written by a philosopher of science. Pennock interestingly uses the theory of language evolution as additional ammunition against creationist criticisms of Darwinian evolution.

– 3 –

One Side of the Coin: The Dangers of Anti-Intellectualism[1]

One thing I have learned in a very long life—all our science, measured against reality, is primitive and childlike—yet it is the most precious thing we have.

– *Albert Einstein* –

The more learned and witty you bee, the more fit to act for Satan you bee.

– *John Cotton, 1642* –

---

[1] One of my readers, philosopher Elliott Sober, noted that strictly speaking, postmodernists and even intelligent design supporters are not anti-intellectuals, only "intellectuals who have the dethronement of science as one their goals . . . taking up the cause of others who might properly be called anti-intellectual." Perhaps, but although criticism (and even dethronement, if warranted) of science can be an intellectual pursuit, the destruction of an intellectual activity such as science—the declared goal of some of the people referred to here—can only be interpreted as anti-intellectual.

Is learning the most precious thing we have, or is it the work of the devil? This artificial dichotomy is at the roots of a complex cultural attitude that has pervaded different parts of the world at different times during human history: anti-intellectualism. The local chapter of the anti-intellectualist movement is not listed in any phone book because it is not an organized movement as such—although its most pernicious incarnations have raised considerable amounts of money and heavily influenced the political arena, especially in the United States.

Richard Hofstadter[2] (see Figure 3.1) so defined the problem: "Anti-intellectualism is not a single proposition but a complex of related propositions. . . . The common strain that binds together the attitudes and ideas which I call anti-intellectual is a resentment and suspicion of the life of the mind and of those who are considered to represent it; and a disposition constantly to minimize the value of that life." To intellectuals it usually comes as a shock that somebody would resent the exercise of critical thinking and discourse, but to many more people such an attitude comes naturally. This natural hostility toward intellectual pursuits is at the very basis of the creation–evolution controversy, and we cannot make sense of the latter unless we understand the nuanced complexities and history of anti-intellectualism.

Hofstadter implicitly recognized three fundamental kinds of anti-intellectualism: antirationalism, anti-elitism, and unreflective instrumentalism. To these, Rigney[3] has added what he refers to as "unreflective hedonism," and I add academic postmodernism as a fifth distinct category. Whereas the very terms with which social scientists refer to these forms of anti-intellectualism betray the tendency of intellectuals to use jargon that is perceived by most people as obscuring rather than clarifying matters, I will attempt to make clear what the technical terms mean and why they are relevant to our problem.

Figure 3.1 Richard Hofstadter, the reference source for any serious study of anti-intellectualism.

---

[2] Hofstadter is the unsurpassed analyst of anti-intellectualism in its many forms, and although his book *Anti-Intellectualism in American Life* dates back to 1963, it is a must-read on the topic. Neil Postman has touched on a subset of Hofstadter's themes from a more conservative perspective, especially in his *Amusing Ourselves to Death: Public Discourse in the Age of Show Business* (New York: Penguin, 1986).

[3] See D. Rigney, "Three Kinds of Anti-intellectualism: Rethinking Hofstadter," *Sociological Inquiry* 61(4) (1991): 434–451.

Antirationalism: Thought as Cold and Amoral

Antirationalism is the idea that rationality is not a positive good but rather something to be mistrusted. There are two fundamental manifestations of antirationalism, which are usually intertwined while reflecting distinct fears aroused by rationalism. On the one hand, there is the perception that rationality is emotionally sterile; on the other hand, there is the fear that rationalism promotes moral relativism and therefore will eventually unravel the very fabric of our society.

Both of these roots of antirationalism found expression in certain religious movements in the United States, usually the very same movements that have a problem with the science of evolution. Whereas a learned literary class dominated the early Puritan Church in the United States, the 1700s brought more emotional, and certainly less cerebral, forms of religion. The preacher came to be perceived as a popular crusader bent on evangelizing by appealing to emotions rather than reason.

The distrust of antirationalists for pure reason has some justification, as has been made clear by some recent findings in neurobiology. Antonio Damasio,[4] for example, has proposed a theory of consciousness according to which emotions are the basis of consciousness and must be intertwined with rational thinking. Research on patients affected in areas of the brain that are connected with emotions—such as the amygdala—has revealed that the ability of these individuals to relate to the world in a meaningful way is severely impaired. Studies have shown that patients with damage to the orbitofrontal cortex of the brain are profoundly affected in their personalities, developing asocial tendencies and a callous indifference to family and friends.[5] Yet their intellectual reasoning abilities are not impaired. The classic example is provided by the story of Phineas Gage, whose brain was penetrated by a metal rod during an accident while he was employed as a railroad worker. Gage became unable to fit into his previous social role, family, and occupation while otherwise demonstrating little change in his intellectual abilities. It seems that the *Star Trek* models of Mr. Spock or Data—beings devoid of emotional components and characterized by an extreme version of rationalism—would in fact produce social monsters without the ability to empathetically connect their reasoning processes with the rest of humanity.

While the "Spock specter" is exactly what antirationalists are afraid of, neurobiology (and common sense) tell us that such individuals are aberrant, sick human beings. They do not exemplify the intellectual ideal being promoted

[4] In A. R. Damasio, *The Feeling of What Happens: Body and Emotion in the Making of Consciousness* (New York: Harcourt Brace, 1999). See also V. S. Ramachandran and S. Blakeslee, *Phantoms in the Brain: Probing the Mysteries of the Human Mind* (New York: William Morrow, 1998). Ramachandran and Blakeslee provide fascinating case studies of people with neurological damage that yield profound insights into how the brain works. Damasio's work is an attempt at a general theory of consciousness, also based on recent neurobiological research.

[5] See R. J. Dolan's commentary, "On the Neurology of Morals," *Nature Neuroscience* 2(11) (1999): 927–929, and references therein.

by our colleges and universities. Furthermore, the opposite extreme—an individual dominated only by an alternation of emotional states with no rational ability whatsoever—is an equally devastating model of humanity that even the religious fundamentalists who are most averse to rationality would not embrace.[6]

Of course, antirationalism was in large part the force that brought about antievolution laws and eventually the Scopes trial in 1925. By the 1930s, in fact, religious and political absolutism were starting to merge because of the common threat of godless communism. As is well known, that tendency generated the horrors of McCarthyism, to this day one of the darkest chapters in the history of American democracy.

The typical outcome of this kind of anti-intellectualism is discussed in Bertrand Russell's famous essay "Freedom and the Colleges," originally written in 1940. Russell (see Figure 3.2)—a British philosopher and one of the intellectual giants of the twentieth century—wrote his piece immediately after an American judge, John E. McGeehan, found him unfit to teach as a professor at City College in New York City because of his philosophical and political ideas. It was one of the most public and shameful chapters of academic and judicial history in the United States, and one that is essential to bear in mind more than half a century later. Russell opens his essay by defining what he means by "academic freedom":

> *The essence of academic freedom is that teachers should be chosen for their expertness in the subject they are to teach and that the judges of this expertness should be other experts. Whether a man is a good mathematician, or physicist, or chemist, can only be judged by other mathematicians, or physicists, or chemists.*

Figure 3.2 Bertrand Russell, the British philosopher who was declared unfit to teach by a New York judge because of his views on morality.

[6] Of course, there are exceptions. Recently a religious fundamentalist sect in Tennessee went to court to challenge the forced exposure of their children by public school teachers to readings that emphasized rational argumentation. Rational thinking, according to this group, is the doing of the devil—very much in the style of the quote by John Cotton at the start of this chapter. Interestingly, however, the group employed lawyers to make an *argument* in favor of their right not to be exposed to the very idea of argumentation!

This is exactly what proponents of some radical religious agendas do not want. From William Jennings Bryan at the Scopes trial to the more recent antievolution campaign of Phillip Johnson, the "experts" are painted as a tyrannical elite bent toward undermining the morality of American youth (and therefore not to be trusted as judges of their own work).

What some religious fundamentalists and educators mean by democracy is apparently the major obstacle in this dispute. Again I refer to Russell:

> *There are two possible views as to the proper functioning of democracy. According to one view, the opinions of the majority should prevail absolutely in all fields. According to the other view, wherever a common decision is not necessary, different opinions should be represented, as nearly as possible, in proportion to their numerical frequency.*

The problem is that educators tend to subscribe to the second view, while some religious ideologues and—to some extent—political leaders much prefer the first one. In America, this tyranny of the majority is especially evident in the way some people would like public schools to operate. This idea apparently arises because the United States is not made up of citizens, but simply of taxpayers. To put it as Russell did,

> *Taxpayers think that since they pay the salaries of university teachers they have a right to decide what these men shall teach. This principle, if logically carried out, would mean that all the advantages of superior education enjoyed by university professors are to be nullified, and that their teaching is to be the same as it would be if they had no special competence.*

In other words, are we going to teach the best of what we currently know about the world (however provisional such knowledge may be), or shall we decide whether the Earth is flat or round by majority opinion? Reality has a rather nasty habit of not conforming to our wishes, no matter how majoritarian our views happen to be. Consequently, education is not a democratic process by any means, however distasteful this may sound to the American public. As Mark Lilla[7] puts it, "The danger is that one will forget that the destiny of intellect is not that of democracy in America, and that the pursuit of the first may at times require a studied indifference to the latter."

Do any negative consequences result from curtailing or entirely abolishing academic freedom? Russell offers the case of the economic collapse of Spain after the expulsion of Moors and Jews; he was prophetically hinting at a similar fate for Nazi Germany. One of the best-studied instances of the causes of the Soviet scientific collapse is the rise of Lysenkoism during the 1950s, which destroyed Russian genetics and set back the entire agricultural industry of that nation by decades (see Chapter 7). Lysenko was the embodiment of what happens when bigots take hold of academia. His theories of genetics and plant

---

[7] In M. Lilla, "Only Disconnect . . . ," *Partisan Review* 60(4) (1993): 603–608.

breeding were based not on the best science available, but on political demagoguery and ideology, no different in substance from the religious ideology currently permeating American politics and promoted by the Christian Right movement.

Russell pointed out that part of the problem is the thrill that ignorant bigots get out of dictating what people who are smarter and more educated than they are can teach or say. He speculates that if the Roman soldier who killed Archimedes was forced to take geometry in school, he must have felt a particular pleasure at repaying the man responsible for his sufferings over triangles and their hypotenuses. It would serve us well to remember that there usually are very good reasons to trust the "experts." I am not advocating here a frame of mind that rejects all criticism of the official authorities—in whatever guise they may come, including white lab coats. But I would hardly appeal to majority opinion upon entering an operating room in a hospital, or while trusting my life to the hands of an airplane pilot. Why should teachers be deprived of a similar consideration?

Again, I turn to Russell:

> *Opinions should be formed by untrammeled debate, not by allowing only one side to be heard. . . . All questions [must be] open to discussion and all opinions as open to a greater or less measure of doubt. . . . What is curious about this position [of not allowing discussions] is the belief that if impartial investigation were permitted it would lead men to the wrong conclusion, and that ignorance is, therefore, the only safeguard against error. . . . Uniformity in the opinions expressed by teachers is not only not to be sought but, if possible, to be avoided, since diversity of opinion among preceptors is essential to any sound education. . . . As soon as a censorship is imposed upon the opinions which teachers may avow, education ceases to serve [its] purpose and tends to produce, instead of a nation of men, a herd of fanatical bigots.*[8]

When I have quoted this and similar passages to creationists, the charge (made explicitly to me by Phillip Johnson during an e-mail correspondence) is that this is exactly why creationism should be taught in public schools—namely, to avoid dogmatism. But notice that Russell was not advocating the teaching of any idea just for the sake of equality. He was talking about the freedom of people who are trained in the subject matters to teach what the best scholarship has established in those subjects. To this argument Johnson, or young-Earth creationists such as Duane Gish, would reply that because evolutionists hold a monopoly on science education, obviously the only people who are going to be

---

[8] Interestingly, the concept of academic freedom was invented (chiefly by British philosopher John Locke) as a reaction against 130 years of religious wars in pre-seventeenth-century Europe. Furthermore, the principle was originally applied to defend the church against undue interference by the state. This is the same idea that is incorporated into the First Amendment of the U.S. Constitution.

trained are evolutionary biologists. This is an important point that necessitates a brief discussion.

Nobody would seriously offer as an argument against the teaching of chemistry that the only people who are allowed to teach it have to be trained in the atomic theory of the elements and not in alchemy. Analogously, few people would object to the mandatory training of physicists in quantum mechanics and general relativity. The reason is that these theories—as the theory of evolution in biology—provide the intellectual backbone of entire disciplines. This is not to say that any or all of them might not one day be proven wrong or inadequate in a fundamental aspect. Indeed, this is exactly how science proceeds. But until then, it makes sense to train people in what philosopher of science Thomas Kuhn calls the "dominant paradigm" (as well as its current, unavoidable, pitfalls, *as perceived by working scientists*). The interesting thing about the charge of ideological dogmatism leveled by creationists against evolutionary biologists is that it is demonstrably untrue. Evolutionary biologists are people of all creeds, from Christians to nonbelievers. On the contrary, the only people who have a serious problem with evolution are members of a minority of fundamentalist sects within the Judeo-Christian-Muslim tradition. If any ideological bias is evident here, it is on the side of creationism.[9]

Following the Russell episode and the McCarthy era, the "holy" alliance between right-wing radical politics and religion suffered a temporary setback during the protests of the 1960s and 1970s and the rise of the civil liberties movement. However, it came back vigorously at the onset of the 1980s with the ascent of Ronald Reagan to the American presidency. Gerard de Groot details the rise of Reagan in connection with the problem of anti-intellectualism,[10] an ascent that—ironically—started and was greatly helped by the progressive student movement in California in the 1960s, when Reagan ran for and won the governorship of that state.

As De Groot emphasizes, Reagan's aides at the time were aware of the marked anti-intellectualism of their candidate. As one of them put it, "The first question [on the campaign trail] is: 'what are you going to do about Berkeley?'—And each time the question itself . . . gets applause." Reagan promised a "code of conduct that would force [faculty] to serve as examples of good behavior and decency." The fact that he knew he could not possibly deliver on such a promise because he would have to trample on basic constitutional rights was not viewed as a problem because that is what the public wanted to hear.

---

[9] When skeptic Michael Shermer made a remark to this effect while at a conference on intelligent design organized at Baylor University in 2000, he was accused of committing the *genetic fallacy*. This is the error of attempting to invalidate someone's arguments because of that person's beliefs. But this is not the case here. The observation is brought to bear on the creationist charge of ideological bias in the evolutionist camp, not on the specific arguments pro and con evolutionary theory. These must be evaluated on their own strengths, regardless of who advances them.

[10] De Groot's account of Reagan's political history and tactics, "Reagan's Rise," was published in *History Today* September 1995: 31–36.

Reagan at the time condemned universities for "subsidizing intellectual curiosity"—obviously a heinous crime by every standard! According to De Groot, "Those most impressed were blue-collar workers without a university education—by nature Democrats—who resented the shenanigans of privileged elites on campus." This is the same voting block that later catapulted Reagan to the presidency (the so-called Reagan Democrats).

Reagan's rhetoric entered territory that most Americans would not hesitate under other circumstances to recognize as fascist, as when he said—in response to student disturbances in Santa Barbara—"If it takes a bloodbath, let's get it over with." De Groot draws a parallel with Bryan (who was certainly of much higher integrity and intellectual stature than Reagan) at the Scopes trial: "Like William Jennings Bryan, another great populist, Reagan could translate a complicated world which he barely comprehended into values he never questioned." In his inaugural speech as president, Reagan went as far as saying that "government is not part of the solution, it is the problem"—a rather peculiar statement from somebody who has just been elected to the most important post in government.

Anti-intellectualism in politics and its alliance with right-wing religion suffered a temporary hiatus with the Clinton era, but it has resumed unabashedly with the most recent American presidential election, which saw George W. Bush presenting himself as the champion of the everyday man (despite his wealth and Ivy League background) against the "pointed-head" intellectualism of Al Gore. Jonathan Chait and Todd Gitlin[11] detailed the picture emerging from the campaign trail in 2000 in light of Bush's use of the public's resentment of intellectuals. Bush actually used examples of intellectualism on Gore's part as evidence that he lacked warmth and personal appeal, the classic fallacy of antirationalism.

Unfortunately, the idea was sustained by the press, with an artificial opposition between "character" on the one hand and "expertise" on the other. The underlying assumption is that leaders are simple men of action, and that knowing something about public policy is a secondary component of the job, if not positively detrimental to it. In fact, Bush's aides have repeatedly remarked that it didn't matter if their candidate was not that smart, because once elected he could always hire smart people as advisers! The press referred to Gore as "boring," "condescending" and "the class prig," an attitude apparently shared by a large portion of Americans. In fact, some of the other candidates in the race were criticized on similar grounds. Chait quotes a review of a book written by Steve Forbes, for example, in which the author laments that the book "reveals little but his ideas . . . his column rarely mentions his family and the book index doesn't list his father, wife, or children." Why—in the context of a polit-

[11] See T. Gitlin, "The Renaissance of Anti-intellectualism," *Chronicle of Higher Education,* December 8, 2000: B6–B9; and J. Chait, "Race to the Bottom: Why, This Election Year, It's Smart to Be Dumb," *New Republic,* December 20, 1999: 26–29.

ical election—the latter kind of information is more relevant (or relevant at all), compared to "mere" ideas about how to govern, is anybody's guess.

Gitlin correctly points out that during the 2000 election, America was split between two nations—one urban and voting overwhelmingly for Gore and his "intellectualism," the other rural and going overwhelmingly for Bush on the basis of an equally strong anti-intellectualism. But this split is not simply the result of viewing Gore as stiff and unfriendly and Bush as warm and approachable. It runs deeper than that, and it brings us to the second root of anti-rationalism: the fear of moral relativism.

It is not by chance that Gore and Bush ended up representing the two sides of the moral debate as well. Gore was associated with the scandals of the Clinton presidency (and a few alleged troubles of his own), which put Bush in the position to run on "character." Chait notes that this worked in Bush's favor in 2000 because the climate was different from that of, say, the 1992 elections. In 1992 the United States was coming out of an economic recession and people wanted to hear *ideas* about how the government was going to help. But by the end of the 1990s most Americans felt comfortable and optimistic about their economic situation, which—a cynic might say—afforded them the luxury of putting morality at the forefront of the debate.[12]

To defend rationalism against the accusation of moral relativism would take a whole book, and this is certainly not my intent here. However, a sketch of this defense must be provided for the reader seriously interested in pursuing this angle of the debate. Countless philosophers have wrestled with the relationship between morality and rationality, attempting to clarify whether humans are well served by ethical rules or whether those rules simply hamper us. Would we really be better off if we all gave in to the desire to just watch out for our own interests and take the greatest advantage for ourselves whenever we could? Ayn Rand, for one, thought that the only rational behavior is egoism (which, however, she defined in a peculiar fashion), and books aiming at increasing personal wealth (presumably at the expense of someone else's wealth) regularly make the best-seller list.

Plato, Kant, and John Stuart Mill, to mention a few, have tried to show that there is more to life than selfishness. In the *Republic*, Plato has Socrates defending his philosophy against the claim that justice and fairness are only whatever rich and powerful people decide they are. But the arguments of his opponents—that we can see many examples of unjust people who have a great life and of just ones who suffer in an equally great manner—seem more convincing than the high-mindedness of the father of philosophy.

Kant rejected what he saw as the self-serving attitude of Christianity, in which a person is good now because he will get an infinite payoff later, and instead attempted to establish independent rational foundations for morality. He

---

[12] As it turned out, the economic optimism was premature, as historical events have clearly shown, and as is usually the case when a good economic turn has lasted for several years.

suggested that to decide if something is ethical or not, one has to ask what would happen if everybody were adopting the same behavior (his famous "categorical imperative"). However, Kant had a hard time explaining why his version of rational ethics is indeed rational.[13] Rand would object that adopting double standards, one for oneself and one for the rest of the universe, makes perfect sense.

Mill also tried to establish ethics on firm rational foundations, in his case improving on Jeremy Bentham's idea of utilitarianism. In Chapter 2 of his book *Utilitarianism*, Mill writes, "Actions are right in proportion as they tend to promote happiness; wrong as they tend to produce the reverse of happiness." Leaving aside the thorny question of what happiness is and the difficulty of making such calculations, one still has to answer the fundamental question of why one should care about increasing the average degree of happiness instead of just one's own.

Things got apparently worse with the advent of modern evolutionary biology. It seemed for a long time that Darwin's theory would provide the naturalistic basis for the ultimate selfish universe: Nature red in tooth and claw evokes images of "every man for himself," in pure Randian style. In fact, Herbert Spencer popularized the infamous doctrine of "Social Darwinism" (which Darwin never espoused) well before Ayn Rand wrote *Atlas Shrugged.*

Recently, however, several scientists and philosophers have been taking a second look at evolutionary theory and its relationship to ethics and are finding new ways of realizing the project of Plato, Kant, and Mill of understanding ethics from a rational viewpoint. Elliott Sober and David Sloan Wilson, in their *Unto Others: The Evolution and Psychology of Unselfish Behavior,*[14] as well as Peter Singer in *A Darwinian Left: Politics, Evolution and Cooperation,*[15] argue that human beings evolved as social animals, not as lone, self-reliant brutes. In a society, cooperative behavior (or, more precisely, a balance between cooperation and selfishness) will be selected favorably, while the individuals looking out exclusively for "number one" will be ostracized because such behavior reduces the fitness of most individuals and of the group as a whole.

All of this sounds good, but does it actually work? A recent study published in *Science* by Martin Nowak, Karen Page, and Karl Sigmund[16] provides a

---

[13] Of course, he tried. Kant's version of ethics is presented as rational on the basis of (1) what he believed was reason's intuitive apprehension of a universal moral law, and (2) the agent's conscious adoption of specific rules ("maxims") that were logically consistent with that universal moral law. This, in turn, would guarantee that ethical behavior was free from the taint of self-interest as a motive. Now the question for Kant would be, Why is motivation by an impartial moral law the only morally acceptable motive of action?

[14] E. Sober and D. S. Wilson, *Unto Others: The Evolution and Psychology of Unselfish Behavior* (Cambridge, MA: Harvard University Press, 1998).

[15] P. Singer, *A Darwinian Left: Politics, Evolution, and Cooperation* (New Haven, CT: Yale University Press, 2000).

[16] M. A. Nowak, K. M. Page, and K. Sigmund, "Fairness versus Reason in the Ultimatum Game," *Science* 289 (2000): 1773–1775.

splendid example of how mathematical evolutionary theory can be applied to ethics, and how social evolution favors fair and cooperative behavior. Nowak and his coauthors tackled the problem posed by the so-called ultimatum game. In it, two players are offered the possibility of winning a pot of money, but they have to agree on how to divide it. One of the players, the proposer, makes an offer of a split ($90 for me, $10 for you, for example) to the other player; the other player, the responder, has the option of accepting or rejecting the offer. If she rejects it, the game is over and neither of them gets any money.

It is easy to demonstrate that the best strategy is for the proposer to behave egoistically and to suggest a highly uneven split in which he takes most of the money, and for the responder to accept, since the alternative is that neither of them gets anything. However, when real human beings from a variety of cultures, using a panoply of rewards, play the game, the outcome is invariably an almost fair share of the prize. This would seem to be prima facie evidence that the human sense of fair play overwhelms mere rationality and thwarts the rationalistic prediction. On the other hand, it would also provide Ayn Rand with an argument that most humans are simply stupid, unable to appreciate the math behind the game.

Nowak and colleagues, however, simulated the evolution of the game in a situation in which several players were able to interact repeatedly. That is, they constructed the game as a social situation rather than as a series of isolated encounters. If the players have memory of previous encounters (i.e., each player builds a "reputation" in the group), then the winning strategy is to be fair because people are willing to punish dishonest proposers, thereby in turn increasing their own reputation for fairness and damaging the proposer's reputation for the next round. This means that—given the social environment—it is *rational* to be less selfish toward your neighbors.

Although we are certainly far from a satisfying mathematical and evolutionary theory of morality, it seems that science does, after all, have something to say about optimal ethical rules. And the emerging picture is one of fairness—not egoism—as the smart (and natural) choice to make. My point here is not to provide a complete rational theory of ethics—and far less to derive an *ought* from an *is*—but rather to emphasize that rationality is not incompatible with what most people would consider ethical behavior.

Furthermore, it certainly does not follow from the adoption of rational criticism of moral customs that one adopts a relaxed attitude of "anything goes" and complete moral relativism, as antirationalists seem to imply. Rationalism might reject a particular religiously based morality, but the absence of universal truths does not negate the existence of truths appropriate to certain situations and of objective values within the realm of human experience. For example, for social animals such as we are, killing our neighbors is wrong, although such behavior might be natural and justified for animals who live in solitude and direct competition with each other. Moreover, even if a moral value such as altruism *is* rooted in the evolutionary history of human behavior, its *origin* does not negate its value as a moral norm. To assert otherwise is to commit the genetic fallacy—rather like denying the value of a particular human being because she is descended from a particular group of ancestors that we dislike for some reason!

The problem may lie instead in the insecurity of antirationalists, who insist that moral principles are always absolute and universal. As Rigney pointedly puts it in the essay mentioned earlier,[17] "The absolutist fear that reason leads to relativism may mask the deeper fear that one's own views will not withstand critical scrutiny." This has certainly been my experience in conversations with creationists after public debates. Invariably, somebody comes to me very upset and demands to know why I don't go around raping and killing other people, since my commitment to evolutionary theory must make me an immoral atheist (the latter two words are, in the minds of too many people, essentially synonyms). Throughout this book my position is not the naïve response of some scientists, that science has *no* connection with philosophy and ethics.[18] Rather, the connection is more complicated than either simplistic Social Darwinism or the equation between rationalism and relativism would lead us to believe. Indeed, Rigney has acutely observed that relativism is as antirational as absolutism, as we shall see later when we discuss that peculiar form of anti-intellectualism known as postmodernism.

Anti-elitism and the Myth of a Classless Society

One of the myths of American democracy is that there is no such thing as a class system in the United States. According to most Americans, class is a concept that applies to old-fashioned Europe, and perhaps to exotic places like India, but is certainly the antithesis of the American ideal of vertical mobility. The American dream, after all, is about the possibility for *anyone* to ascend to the highest wealth and power, and American mythology is filled with examples of self-made men who went all the way to the top from very humble (and usually nonintellectual) beginnings.

Alas, this picture is far from reality, and a class structure in the United States is all too evident to any student of social sciences.[19] Nevertheless, the idea of anti-elitism is that any claim to superior knowledge within a democratic society amounts to elitism. This resonates with a current of populism in American culture, an attitude of defense of the little guy that has had a major and

---

[17] See note 3.

[18] A good example of how science and ethics are connected is the subject of race. People once believed—and some still do—that some races are naturally inferior to others in intelligence and morals. Science has decisively shown us otherwise. Knowing now that intelligence and character are not determined by race, we have from science decisive additional support for the moral argument that racism is wrong.

[19] See B. Ehrenreich, *Fear of Falling: The Inner Life of the Middle Class* (New York: Pantheon, 1989).

positive impact on American history at several crucial junctures—for example, resulting in the organization of a labor movement and in the civil rights unrest of the 1960s and 1970s.

Anti-elitism seems a strange position for three reasons. First, it is clearly possible to accept the idea that some specialists have "superior" knowledge about certain subject matters without implying that this somehow undermines the basis of a democratic system. In fact, Europeans tend to subscribe to just this kind of system, where "experts" are not regarded with the suspicion typical of the American public. There is a fundamental distinction between knowledge, which can be, and demonstrably is, unequally distributed among the citizens of any nation, and political and civil rights, which must be equally distributed for a system to be properly called a democracy.

Second, Americans themselves do not have a problem recognizing—and respecting—expertise in areas other than education. For example, few people would dream of walking up to a surgeon before undergoing an operation and arguing where and how she should conduct the procedure. That does not imply that the surgeon is "superior" in any broader sense than that she has been professionally trained to carry out certain kinds of operations on human beings. In many other areas she can be as ignorant as, or more ignorant than, the average person.

Yet it is a quintessentially American attitude that this respect for expertise does not extend to government and education. The infamous case of the Kansas Board of Education, which in 1999 excluded evolution from the topics on which Kansas high school students should be tested, is typical (the decision was repealed by a new board the following year). The board's contention was that anybody could reasonably question the teaching of a scientific theory because it is not a "fact" (I will return to this point in Chapter 5). In the United States it is common to see parents arguing with trained professional teachers about what to teach in the way of technical, usually scientific, matters. Although parents most definitely should be involved in their children's schooling and should be ready to question inappropriate behavior on the part of teachers, they are hardly qualified to decide about the content of science (or any other) curricula and should leave that decision to professionals trained for that very purpose.

The third contradiction intrinsic to anti-elitism in American culture is that Americans are among the most elitist people in the world when it comes to three areas: sports, entertainment, and finance. The greatest American heroes are businesspeople like Lee Iacocca (the former chairman of the board of Chrysler Corporation) or Bill Gates (of Microsoft), athletes like baseball player Babe Ruth or basketball legend Michael Jordan, and entertainers like Elvis Presley or Madonna. These people are not only being justly glorified for their accomplishments in their respective fields, they are also held up as all-encompassing role models for young Americans. The whole idea is extremely elitist because very few kids have any reasonable hope to become superstars of that caliber, and yet the American public seems to be entirely oblivious to this incongruity.

It seems likely to me that this astounding difference between anti-elitism in matter of intellect and unabashed elitism in sports, entertainment, and finance stems from the deeply flawed but apparently widespread view that intellectual endowment is a matter of genetics, while anybody who tries hard enough has a chance at becoming a pop star, a billionaire, or a professional quarterback. This is nonsense on stilts not only because "intelligence," whatever it is, is clearly influenced by the environment (primarily education and family), but because there is much more ground to believe that the initial playing field is unequal in the other mentioned areas: in economics because of inherited environmental effects (if you are born into a poor family, you have astronomically smaller chances to make it than if you inherit millions, glamorous exceptions notwithstanding); and in sports because there genetics really does make a large difference, which training can only sharpen. (I am not familiar with any study looking into the genetic and environmental components of Hollywood stardom.)

Populism and anti-elitism have a complex history in the United States. During the revolutionary period, when the leaders of the new nation were essentially aristocratic intellectuals, things were very different. But even then Thomas Jefferson was accused of possessing "only" theoretical learning because of his connection with France and the influence of the Enlightenment on his philosophy and policy. In the early nineteenth century, the conflict came to a head during the 1824 and 1828 elections, which saw the patrician and learned John Quincy Adams pitted against the populist Andrew Jackson. The latter's victory signaled the beginning of political populism in the United States, a current that—despite various temporary periods of dormancy—is still very much with us today.

Anti-elitism is reflected in the creation–evolution debate in the charge often raised by creationists that evolutionists are arrogant, self-styled "experts" who disrespect the will of the people. During one of my debates with young-Earth creationist Duane Gish, I explained that education is not supposed to be democratic in the broad sense of being conducted by popular consent. Gish came back angrily, accusing me of chauvinism and reiterating the argument that William Jennings Bryan used during the Scopes trial and that we saw addressed by Bertrand Russell in the essay cited earlier: The people are paying for it, so they have a right to make decisions about it. Until we succeed in changing this fundamental misconception, not only the evolution–creation debate, but also disputes on sex education, the content of history and literature curricula, and countless other controversial issues plaguing the American education system will not go away.

Unreflective Instrumentalism: If It Ain't Practical, It Ain't Good

The United States is the most capitalistic society in the world, the one that has come closest to realizing Adam Smith's vision of a pure marketplace with no governmental control. Of course, even in the United States the reality is that government still has quite a bit to say about how business is conducted, especially when it comes to issues of public health, fair treatment of workers, and, far less often, the environment. In fact, most Americans don't realize that they actually live in a social-democratic country, albeit under a social net not as extensive as those of western European nations.

The ideal of capitalism has been the engine behind the third form of anti-intellectualism outlined by Hofstadter, what Rigney calls "unreflective instrumentalism." The basic idea in this case is that education—and by extension, inquiry—is not good unless it has immediate practical value. As in the case of the other forms of anti-intellectualism, this concept has deep historical roots and directly affects the creation–evolution controversy, albeit from a different standpoint.

During nineteenth-century America, the conflict was between the cultivated gentlemen who inherited their wealth and led the country at its very beginning and the captains of industry who were making their fortunes from humble starts and had no time to do anything but create what soon became the emerging business class. The latter were obviously extremely successful without much education, a fact that seemed to reinforce the stereotype of the learned man as somebody with too much time on his hands (education and leadership for women were an entirely different issue, and one that was not at the forefront of the social debate). The attitudes of businesspeople of the time are reflected in railroad magnate Leland Stanford's comment that graduates from colleges in the East had "no definite technical knowledge of anything." Andrew Carnegie remarked that classical studies were a waste of "precious years trying to extract education from an ignorant past."[20]

During the twentieth century, the divide between "theoretical" and "practical" learning widened, with the latter gaining in importance and respect among the public. Senator William Proxmire's infamous Golden Fleece Award is a typical example. From 1975 to 1988[21] the good senator thought it appropriate to highlight examples of what he considered a waste of federal money and an insult to taxpayers. Proxmire conferred the award four times each to the National Endowments for the Arts and Humanities, seven times each to the Department of Education and to the National Aeronautics and Space Administration (NASA), and a staggering nine times to the National Science Foun-

[20] See D. Rigney's article cited in note 3.

[21] The complete story of the Golden Fleece Awards can be found at http://www.taxpayer.net/awards/goldenfleece/, which is run by an organization continuing Proxmire's crusade.

dation (NSF). Although some of the justifications for the awards are indeed funny and may even reflect actual cases of bad judgment on the part of these agencies, Proxmire gave out very few *merit* awards, and these did not get the wide publicity of the negative citations.

Similarly, during its evening news program the ABC TV network broadcasts a regular weekly series on government waste entitled "It's Your Money." No analogous series on appropriate government spending is produced, presumably because it would receive very low ratings. The point is that although lawmakers and journalists have a right—indeed a duty—to fight governmental waste and bad policy, they also have a duty to report on the positive outcomes of governmental actions and decisions. Otherwise they are guilty of generating and maintaining a cynical attitude on the part of the people toward public affairs.

One of the causes of anti-intellectualism, of course, is the lack of education about what basic science is and how it is connected to the wonders of technology and medicine from which we all benefit. There seems to be a notion that it makes sense to increase spending for applied research (e.g., the budget of the National Institutes of Health) while curtailing the "waste" that goes into basic science research (usually represented by NASA or NSF). This attitude is based on the underlying assumption that basic and applied research are entirely different activities, with essentially no connections between them. But even a cursory study of the history of science shows the contrary to be true. Basic research provides the large reservoir of ideas and information that applied scientists use to solve actual problems.

It would take a whole book to detail this connection, but a few examples will suffice to make the point here. The curiosity of physicists about the basic constituents of matter eventually led to the development of atomic energy (with the unfortunate effect of also producing the atomic bomb, given the climate during World War II). The very idea of pocket calculators became a reality because NASA had to figure out a way to equip its moon-bound astronauts with the precursors of these convenient gadgets. The modern idea of a computer was the result of basic investigations into mathematical theory, and its first practical implementation was the result of a scientist's curiosity about weather patterns.[22] An AIDS vaccine would not even be conceivable if we had not done basic research to understand the evolutionary biology of retroviruses. There are countless other examples like these.

Unfortunately, the relationship between basic and applied research is not always straightforward. Basic research often is based on a shotgun approach to understanding the world, driven by the curiosity of individual scientists and by the technical approachability of specific questions. It is nearly impossible to predict which areas of research will produce benefits for humanity years or

---

[22] See Scott McCartney, *ENIAC: The Triumphs and Tragedies of the World's First Computer* (New York: Walker, 1999).

decades down the road. While this makes for a very inefficient way of funding science, it is similar to the idea of private firms investing in research on a variety of projects, very few of which will ever come to fruition and repay the investments; the development of new drugs by pharmaceutical companies is a striking example.

There is, of course, another defense of basic research—the same one that the National Endowment for the Arts uses to defend its public role: Knowledge of the world has its own intrinsic worth. People do care for more than just the bottom line, as the crowds attending art museums or concerts, or the number of people watching science and nature shows on television, clearly attest. As Aristotle put it, "Man by nature desires to know." Regardless of any quite welcome practical application of science, basic research should be pursued because we want to know what the world is like and what our place in it really is. And the only source of relatively unbiased funding for this kind of research is the government, since corporate sponsorship naturally tends to direct funding toward areas perceived (rightly or not) to more likely lead to profitable outcomes in the short to midterm.

This is not to say that the public does not have a right to be informed about how government money is being spent and why. However, it is simply naïve to request "layperson" abstracts from professional researchers who are federally funded to be published as a justification of basic research efforts. These abstracts consist of a few sentences that convey no real information or education about the research proposals thus funded. In fact, researchers have to do back flips to emphasize an often tenuous link between the highly specific research they are conducting and general themes akin to "here is why this will—eventually—save the world." Rather, a concerted effort needs to be made by both academic researchers and governmental funding agencies to provide large-scale science education programs for the public that make clear the complex relationship between applied and basic science, as well as the intrinsic value of the latter.

There is another largely unexplored dimension to the problem raised by unreflective instrumentalism. As Rigney points out, even if we grant that a larger portion of public funding should go to applied research, this leaves completely open the broad question of who should benefit from this research. Research efforts can be carried out in many areas, and priorities have to be set among them because resources are obviously limited. It is not clear that the public interest has always been at the top of the agenda, considering the influence of powerful corporations and lobbying groups such as the health insurance industry.[23] Furthermore, even if one could somehow keep in check corporate greed for public funding for science, we would still be faced with questions of priority, such as whether it is more important to increase fund-

[23] As I am writing, for example, several pharmaceutical corporations have filed suit against the South African government to stop cheaper production of anti-AIDS drugs, on the ground that it will hurt their profits.

ing on cancer research or to find a vaccine for AIDS. I do not pretend to provide answers here, and further discussion of this issue would lead us too far from the main object of this book. But the point is that there are no simple solutions to complex questions, contrary to what some politicians and business leaders have been trumpeting in the media for quite some time now.

What does all of this have to do with the creation–evolution controversy? Although unreflective instrumentalism does not have the direct impact on our debate that other forms of anti-intellectualism do, creationists such as Jonathan Wells (in the last chapter of his *Icons of Evolution*[24]) have attacked evolutionary biology on similar grounds. Wells (more on him and his arguments in Chapter 7) has dismissed the evolutionists' contention—following a famous article by Theodosius Dobzhansky—that "nothing makes sense in biology if not in the light of evolution." Such dismissal is based precisely on unreflective instrumentalism, in that Wells says that most of biology, and especially the practical components that really matter to the American public—such as agriculture and medicine—need no input at all from evolutionary theory. Wells fails to see (or does see but neglects to acknowledge) the connections between basic and applied science in the same way that the business magnates of the nineteenth century did, although his motivation is religious rather than economic ideology. Significantly, toward the end of his book Wells calls for public protest against taxpayer-funded research in evolutionary biology, joining analogous cries by young-Earth creationists such as Kent Hovind.

Unreflective Hedonism: TV Is Better, and Not As Much Work

Rigney identifies a fourth kind of anti-intellectualism not considered by Hofstadter—what he calls unreflective hedonism, or the idea that we live in a society in which the primary goal is entertainment. This theme was perhaps best developed in Neil Postman's *Amusing Ourselves to Death*,[25] a book in which this rather conservative social critic pulls no punches on the obsession of Americans with entertainment.

That much modern ethos revolves around television, movies, and the music industry is undeniable, but what is the connection between this observation and anti-intellectualism? A good example is offered in the last chapter of Wendy Kaminer's *Sleeping with Extra-terrestrials*,[26] in which she laments that the onset of the Internet is rapidly demolishing the value of reasoned and articulated discourse. Writing after both Hofstadter and Postman, and having experienced the full power of the World Wide Web, Kaminer points out that the

---

[24] J. Wells, *Icons of Evolution: Science or Myth?: Why Much of What We Teach about Evolution Is Wrong* (Washington, DC: National Book Network, 2000).

[25] See note 2.

[26] W. Kaminer, *Sleeping with Extra-terrestrials: The Rise of Irrationalism and Perils of Piety* (New York: Vintage, 1999).

very ideas of cyber-surfing and hyperlinks are the antipodes of book learning and intellectualism.

A book slowly builds an argument while exploring the complexities of a problem and bringing to bear facts and reasoning from a variety of sources, as I am trying to do here in order to analyze and contribute to the evolution–creation debate. But clicking on hyperlinks on the Web is conducive to a style of "learning" that is fragmented, in which no discourse has the chance to unfold as the author envisioned because the "interactivity" with the user provides too much freedom to the latter. After three or four jumps in cyberspace, one is bound to have lost any meaningful contact with the starting point, dazzled by ever new and more colorful Web sites heavy on pictures and poor on text. Such Internet "surfing" probably does not help to lengthen the already short attention span of far too many students.

All of this, of course, is not a new phenomenon. Postman's indictment of television followed a similar reasoning. Channel surfing produces a fragmented experience at the end of which the viewer is hypnotized by a succession of disconnected sequences and cannot concentrate on following a story from beginning to end. Compare this to more traditional forms of entertainment, such as going to a play. Once one makes the decision of attending a particular performance, the only two choices are to stay for the whole thing or to leave and waste the admission ticket.

Biologist Richard Dawkins joined the chorus with his book *Unweaving the Rainbow*,[27] in which he attacked some "modern" ideas about teaching science (see Chapter 8 for further discussion of this point). As Dawkins put it, it seems that today's science education has to be "fun, fun, fun" to keep the students *entertained*, regardless of how much they will actually gain from such a constant level of amusement. Dawkins's point is that science—and other subject matter—does not have to be *fun* as much as *interesting*. The dichotomy between entertainment and boredom is an entirely false one that is being propagated by media conglomerates the world over. There is at least a third option, which is that some things are fascinating without necessarily leading to laughter or amusement. Human beings probably don't need boredom (which also in other primates leads to depression), but they do need *both* interesting and entertaining activities (not necessarily at the same time).

Kaminer also points out that a major problem with the Internet is that there is a very low level of quality control. True, one can find tons of information about almost any subject on the World Wide Web, but *more* information is not what most of us need. It is easy to drown in information without having found what we actually wanted. Surfing the Net can generate the frustrating sensation of being stranded on the ocean: We are thirsty and we are surrounded by water that won't do us any good. I have had repeated experiences with stu-

[27] R. Dawkins, *Unweaving the Rainbow: Science, Delusion, and the Appetite for Wonder* (Boston: Houghton Mifflin, 1998).

dents who come to class with an assignment based on Internet resources as major references. It seems hard to convey the idea that because *anybody* can put up a colorful Web page, one often knows very little about where the contents come from and how reliable they are (this, of course, does not negate the fact that there are plenty of very valuable Web sites out there).

Obviously the very idea of quality "control" over information is what terrifies enthusiasts of the new medium and why they hail the Web as the ultimate marketplace of ideas, the long-awaited refuge from the thought police and the much-dreaded "experts." That may be true—such openness may have its value—but most of us simply do not have the time to wade through a pile of garbage to get to the few relevant and reliable pieces of information. To give the reader a small example of what I am talking about, I just typed into a search engine the word *anti-intellectualism,* which ought to be a poorly discussed topic on the Web. Yet I was immediately presented with 2,020 documents with titles ranging from "Garlic, Vodka, and the Politics of Gender" to "The Cult of Anti-intellectualism amongst Blacks" to "Promise Keepers, Evangelicalism, and Anti-intellectualism" and even a piece titled "The Aesthetic Pedagogy of Francis of Assisi." In this case, most of the first 60 hits were actually relevant to the topic at hand, although I have no idea of the reliability of most of the sources for these articles. A search on *creationism* returned 31,300 hits; on *evolution,* 1,410,000! Try wading through *that* pile for a school assignment due the following morning.

Yes, my librarian has explained to me how to narrow searches in cases like these, but the point is that nobody has the time to sift through all of this "information" to learn what we want to learn. Similarly, nobody has the time, inclination, or money to see every movie ever produced. That is why we rely not just on catalogues of movies arranged by genre or director, but on *experts* to tell us if it's worth paying the full price for a given selection, or if it is better to go to a matinée or even wait until the video comes out. Of course, using somebody's advice limits our choices and biases them in certain directions. But we have no acceptable alternative, and this is why it is so important to make sure that the "experts" are well trained (or, in the case of movies, that you pick one you often agree with).

In the specific case of the creation–evolution affair, unreflective hedonism is particularly evident when one considers how many people attend sound bite–style public debates on the topic, as opposed to how many bother to actually read at least one book from each side of the issue. I do engage in debates on a regular basis, but this is not because I think I am educating or "convincing" anybody. My goal is simply to plant seeds of doubt that just might lead some people to escape their unreflective hedonism, go to their public library, and look up what the creation–evolution issue is all about. As I discuss in Chapter 8, it works, at least in some cases, as is testified to by the encouraging e-mails and letters I get weeks or months after the debate (often from fence-sitters and even those who were initially on the "other side"). But the dynamic of the debate itself is fascinating to observe as an example of unreflective hedonism at work.

In some sense, a debate is the MTV equivalent of rational discourse. Typically, the speakers have a limited amount of time to present their views on complex and often technical subject matters. They have even less time to rebut their opponent's positions, after which the debate turns into a series of questions and answers for each of which one- or two-minute slots are allocated. If the question is, for example, "What is the evidence in favor of the evolution of humans from other primates?" how on earth is one supposed to give a meaningful answer in one minute? That is why I usually simply give brief examples or pointers and then encourage the questioner to e-mail me or visit my Web site for more information.

The worst aspect of a debate is not the limited amount of time available to construct a reasonable line of argument, but rather the tendency of creationists to play on the audience's emotions or to mischaracterize evolutionists' views. Two examples will suffice. During every debate I have had against young-Earth creationist Duane Gish, he has managed to do two things that are meant exclusively to appeal to the emotions of the audience. One is to slip in the fact that I am an atheist; the second is to invariably conclude his last rebuttal with words such as these: "And so I maintain that still the best answer to the origin question is 'God created the Heavens and the Earth.'" This biblical quotation is followed by a predictable chorus of "Amen." Because I know what to expect, I usually take the precaution of warning the audience that both things will happen (I know that Gish does not bother to—or cannot—deviate from his prepared speech), which at least generates chuckles when my predictions infallibly come true. I say that the fact that I am an atheist (by which I mean somebody without a belief in the supernatural, not somebody who *knows* that there is no god) is irrelevant to what is allegedly a "scientific" debate. More to the point, I add that plenty of evolutionists are theists and Christians, which should take the wind out of any argument based on my personal beliefs. I don't know how many people are prompted to pause and reflect on these points, but I do realize that many attendees are there only to see their "champion" smash the devil, regardless of the reasonableness of the latter.

The mischaracterization of one's own arguments is probably even more infuriating, especially because it is never clear whether this is done in a conscious and malicious way or just out of plain stupidity.[28] Perhaps my favorite example of this occurred during several debates with arch-young-Earther Kent Hovind (who calls himself "Dr. Dino," even though he does not hold a recognized degree, and certainly not in paleontology). Hovind repeatedly—and despite

[28] For an interesting discussion about creationists as people who don't understand science versus the hypothesis that they do understand and lie anyway, see J. Arthur, "Creationism: Bad Science or Immoral Pseudoscience?" *Skeptic* 4(4) (1996): 88–107, as well as the more comprehensive book by Ian Plimer, *Telling Lies for God: Reason vs Creationism* (Milsons Point, New South Wales: Random House Australia,1994).

my warnings—tried to convince the audience that evolutionists believe that humans came from rocks.[29] Furthermore, he challenged me to explain how a human being can come from a banana. Exasperated, I subtly pointed out that he might have a psychological problem with bananas that would keep a Freudian psychoanalyst very busy, at least getting some laughs out of the audience. The point is, however, that it is very easy for creationists to introduce sound bites similar to those used by Gish and Hovind, and exceedingly difficult to come up with reasoned responses in an equal amount of time.

Should we therefore *not* debate creationists? Some of my esteemed colleagues—for example, Eugenie Scott of the National Center for Science Education—most definitely think we should not. I am a pluralist in these matters and think that many strategies can work, depending on what the goals are and how they are carried out (on science education, see Chapter 8). Although I would certainly not limit our efforts at public education to debates, it is also true that these occasions are often the only times many people get to see a scientist, particularly an evolutionary biologist, in action. It is very encouraging to be approached after such a debate by people who compliment me for having presented good arguments, or even ask me for references for further readings. People need to see scientists in the public arena much more often than is the case now, and we should use even MTV-like occasions as a way to draw the audience toward better sources of information and more intellectual engagements. Carl Sagan did not scorn writing for *Parade* magazine (though he was scorned by his fellow academicians for doing so and even denied admittance into the National Academy of Sciences), and I know many people whose understanding of science has been improved by those short essays. Given the rampant level of unreflective hedonism in our society, it is simply unrealistic to expect that most people will flock to public and university libraries to read the latest on evolutionary theory.

But is the situation really that bad? It is easy to shrug off Postman's, Kaminer's, Dawkins's, and my own warnings as the last gasps of a dying elite of curmudgeons who cannot adapt to a novel environment. After all, were people not predicting the end of the world when the printing press was invented? Were not similar dire pronouncements being made when the idea of public education was being implemented? (In fact, Postman himself considers these two milestones of modern civilization rather suspiciously in his *Technopoly.*[30]) I do have a strong belief in the possibility that humanity will survive, indeed even improve, over the long run. But what concerns us here is the all-important short time of the immediate future. I, like these other writers, am concerned about the waste of human potential when tens of millions of young people of

---

[29] It is ironic that Hovind makes fun of evolutionary biologists for claiming that organisms come from rocks, given that from a literal reading of the Bible one learns that humans were created from "the dust of the earth." This seems but a small erosional process away from having us being created from rocks.

[30] N. Postman, *Technopoly: The Surrender of Culture to Technology* (New York: Knopf, 1992).

this generation, including my own daughter, will not be able to partake of some of humanity's greatest achievements because they would rather spend their days in front of a screen—be that a television or an Internet terminal.

For example, it is astonishing how many millions of children and adolescents are now being diagnosed with "attention deficit disorder" and actually put under the influence of calming drugs. Surely there are some cases of truly pathologically short attention span. But the root cause of this phenomenon may very well be that people are just not used to concentrating for long periods of time, conditioned as they are by television programs (including "educational" ones for kids, such as *Sesame Street*) that consist of very short sequences and flashing colorful images. It is our responsibility as parents and educators to avoid allowing our children to go on Ritalin or Prozac if they do not have serious neurological disorders, and the spread of unreflective hedonism is entirely the result of our failure to take this responsibility seriously.

Attack from Within: Postmodernism as Anti-intellectualism

Perhaps the most astounding version of anti-intellectualism is presented by postmodernism and related "isms," such as deconstructionism and poststructuralism. This is an *intellectual* movement that originated in France and was exported to the United States, where it has quickly become much more radical than in its birthplace. As Lilla puts it in the essay cited earlier,[31] "The French thinkers who inspired the various schools of 'post-structuralism' never drew the leveling, democratic [not in the good sense of the term] conclusions regarding culture which their American disciples have."

What are these leveling conclusions? Essentially, that science is *entirely* a social construct,[32] and that as such it does not have any more claim to truth than other "stories" told within a number of cultural traditions—including creationism. This is an all-out attack on the notion of rationality itself, and it is being carried out from within academia, the bastion of intellectualism and rationality throughout the ages.

Postmodernism in America is a hodgepodge of currents and positions, rather than a coherent movement. It attracts many exponents of the academic left who subscribe to radical environmentalism, afrocentrism, and radical feminism, among others. As Norman Levitt sharply put it,[33] "The left, already des-

[31] See note 7.

[32] Of course, science is indeed affected by social forces and cultural prejudices. Many philosophers of science have made this point and explored it in depth. This view, however, does not have to imply the belief that science does not provide us with the best means to understand the natural world. For a detailed treatment of this point, see H. E. Longino, *Science as Social Knowledge: Values and Objectivity in Scientific Inquiry* (Princeton, NJ: Princeton University Press, 1990); and P. R. Gross and N. Levitt, *Higher Superstition: The Academic Left and Its Quarrels with Science* (Baltimore, MD: Johns Hopkins University Press, 1998). This discussion is resumed in Chapter 7 of this book.

[33] In N. Levitt, "Why Professors Believe Weird Things: Sex, Race, and the Trials of the New Left," *Skeptic* 6(3) (1998): 28–35.

perately marginalized, is likely to disintegrate altogether as it is morphed into a species of academic silliness." A few examples of such silliness are necessary to appreciate the dangers of the postmodernist program and how it plays into the creation–evolution affair.

One of Levitt's favorite instances of silliness is a book by Jodi Dean entitled *Aliens in America*,[34] in which she states that it is simply inadmissible to label anything as delusional, including alleged sexual experiences with extraterrestrial beings (a biological impossibility, by the way).[35] The reason—apparently—is that truth is everywhere and nowhere, and science is just a corrupted stew of racism and violence. To quote from Dean's book: "The early ufologists fought against essentialist understandings of truth that would inscribe truth in objects (and relations between objects) in the world. Rejecting this idea, they relied on an understanding of truth as consensual. If our living in the world is an outcome of a consensus on reality, then stop and notice that not everyone is consenting to the view of reality espoused by science and government." Although the latter statement is certainly true (albeit inconsequential when it comes time to decide what is true and what is not), her characterization of ufologists is naïve to say the least; in fact, it is downright wrong. I have read plenty of literature on UFOs and alien abductions, and I have met with both ufologists and people who claim to have observed flying saucers in action. They are most definitely *not* claiming that truth is consensual. They think of UFOs and aliens as real entities, objectively interacting with the world. They simply have a much lower standard of proof for believing these claims than a scientist would. The same is true of creationists, who are not happy about the help they are getting from postmodernists: They don't want their truth to be one among many; their God is a jealous God, and He wants a monopoly.

As Levitt also reports, a more direct attack on science is presented in a chapter written by Andrew Ross for a collection of essays in defense of O. J. Simpson,[36] edited by novelist Toni Morrison. Ross goes so far as to claim that admitting DNA evidence at trials is equivalent to accepting the racial theories of Herrnstein and Murray, the authors of the infamous *Bell Curve*,[37] a book on the alleged absolute genetic determination of intelligence. But this is obviously nonsense, an egregious non sequitur based on the idea that if one does not like the consequences of the scientific enterprise, then that enterprise is "bad" (as we shall see in Chapter 5, this is exactly one of the creationists' most common fallacies). Ross states that "if complex scientific testimony increasingly becomes

---

[34] J. Dean, *Aliens in America: Conspiracy Cultures from Outerspace to Cyberspace* (Ithaca, NY: Cornell University Press, 1998).

[35] For a similar amount of nonsense from Harvard psychiatrist John Mack, see his *Abduction: Human Encounters with Aliens* (New York: Scribner's, 1994).

[36] The ex-football star who was accused of slaughtering his ex-wife and her friend in one of several "trials of the century" in late-twentieth-century America.

[37] R. J. Herrnstein and C. Murray, *The Bell Curve: Intelligence and Class Structure in American Life* (New York: Free Press, 1994).

a customary presence in the courts, what are the prospects for continued respect for lay judgment?" In other words, forget the technicalities; if the jury does not understand the evidence, we should still rely on their common sense and intuition. Indeed, it has now become common for trial lawyers to excuse prospective members of a jury on the ground that they are "biased" if they know anything about DNA evidence. In what sense of the word *bias* are education and knowledge condemnable?

Sheila Jasanoff in her *Science at the Bar*[38] takes a position similar to Ross's, unashamedly proposing that science should be forced to battle it out with its alternatives, including all sorts of pseudoscience (something also seriously proposed by radical philosopher of science Paul Feyerabend back in the 1960s). Referring to a specific class of litigation, she says that "allowing orthodox scientific practice systematically to dominate over other types of meaningful knowledge production may not be the best way to bring closure to such controversies. Instead, we need to search for mechanisms that strike a better balance between scientific and subjective knowledge in toxic tort litigation." In what sense nontechnical knowledge of "toxic tort" is "meaningful" is—of course—not explained. A creationist demanding equal time for creation "science" in public schools couldn't come up with better reasoning. Indeed, Jasanoff maintains that in communities with preponderantly fundamentalist views, creationists should be allowed to have their way. Their "locally constructed scientific agreement" trumps "more universalizing notions of science." That is, opinion should overcome reality as a guiding principle in teaching our children about the world.

A direct postmodernist attack against Darwinism is contained in another book criticized by Levitt: Marilynne Robinson's *The Death of Adam*.[39] According to Robinson, Darwinism is a cynical ideology used to preserve an unjust social order: "Where does this theory get its seemingly unlimited power over our moral imaginations, when it can rationalize stealing candy from babies—or, a more contemporary illustration, stealing medical care or schooling from babies? . . . Why does it have the stature of science . . . when it is . . . only mythical, respectabilized resentment, with a long, dark history behind it?" I am wondering which "Darwinism" Robinson is thinking about. Certainly not the modern theory of evolution, since that has little or nothing to do with ideological positions on candies and health care. And as I have already pointed out, people like Singer, and Sober and Wilson, have now broken the right-wing monopoly over social interpretations of Darwinism.

Of course, the best argument against postmodernism is postmodernism itself. If indeed all truths are just-so stories and there is no reason to prefer one to the other, why on earth should we think that postmodernism has got it

---

[38] S. Jasanoff, *Science at the Bar: Law, Science, and Technology in America* (Cambridge, MA: Harvard University Press, 1995).

[39] M. Robinson, *The Death of Adam: Essays on Modern Thought* (Boston: Houghton Mifflin, 1998).

right? As Paul Hollander writes,[40] "It is significant and paradoxical that the postmodernists who claim to believe that everything is a matter of opinion, are among the fiercest, most virulent, and radical critics of Western culture." Gertrude Himmelfarb put it this way in *On Looking into the Abyss:*[41] "The presumption of postmodernism is that . . . because there is no absolute, total truth there can be no partial, contingent truths." Gross and Levitt, in *Higher Superstition,*[42] echo that postmodernism leaves "no ground whatsoever for distinguishing reliable knowledge from superstition." It would seem that for a postmodernist the only coherent thing to do is simply to shut up.

For the best example of how true (not in a relative sense) these accusations of postmodernism are, one need only consider the famous case of the so-called Sokal hoax.[43] A. D. Sokal is a physicist who wrote an essay entitled "Toward a Transformative Hermeneutics of Quantum Gravity" and sent it to the editors of the peer-reviewed postmodernist journal *Social Text* (edited by the same Andrew Ross mentioned earlier). The paper presented a postmodernist critique and re-evaluation of some cutting-edge aspects of research in subatomic physics and was immediately accepted for publication. Afterward, Sokal revealed to the magazine *Lingua Franca* that the article was a hoax. The entire text—some 35 pages long—was a string of nonsense made to sound scientific and peppered with stereotypical postmodernist phrases.[44] Needless to say, the postmodernist elite did not take kindly to Sokal's joke, but the damage inflicted to postmodernism by this exercise in critical thinking (or lack thereof) is incalculable.

The postmodernist form of anti-intellectualism is directly related to the evolution–creation controversy. M. Cartmill outlined this link very effectively in an article comparing fundamentalist creationist and postmodernist attacks on evolutionary theory.[45] Cartmill hit on the commonality between these otherwise antipodal movements when he characterized them as "oppressed" by evolution. Hollander went a step further and traced the multifaceted reaction against science in general to a resentment of modernism. Indeed, that was the theme underlying the first outburst against science and evolution in the Unit-

---

[40] In P. Hollander, "The attack on science and reason," *Orbis* 38(4) (1994): 673–677.

[41] G. Himmelfarb, *On Looking into the Abyss: Untimely Thoughts on Culture and Society* (New York: Knopf, 1994).

[42] See note 32.

[43] See *The Sokal Hoax: The Sham That Shook the Academy*, edited by the editors of *Lingua Franca* (Lincoln: University of Nebraska Press, 2000).

[44] Here are a couple of examples: "In mathematical terms, [deconstructionist philosopher] Derrida's observation relates to the invariance of the Einstein field equation $G_{\mu\nu} = 8\pi G T_{\mu\nu}$ under nonlinear space-time diffeomorphisms (self-mapping of the space-time manifold that are infinitely differentiable but not necessarily analytic)." Or: "The content and methodology of postmodern science thus provide powerful intellectual support for the progressive political project, understood in its broadest sense: the transgressing of boundaries, the breaking down of barriers, the radical democratization of all aspects of social, economic, political, and cultural life." And so on; you get the drift.

[45] See M. Cartmill, "Oppressed by Evolution," *Discover* 19(3) (1998): 78–83.

ed States at the beginning of the twentieth century, the Scopes "monkey" trial in Dayton, Tennessee (see Chapter 1). We have therefore come full circle, connecting postmodernism with the very first kind of anti-intellectualism we have encountered: antirationalism and its fear of moral relativism. Ironically, postmodernism and antirationalism are perceived as threats also within the Christian religion itself, as I will discuss in the next section.

Trouble in the House of the Lord: Relativism and Christian Evangelicalism

The influence of postmodernism is being increasingly felt within the Christian evangelical movement itself, generating a fascinating debate inside Christianity that reflects the themes I have been discussing so far, from antirationalism to the fear of moral relativism. That such fear is very real is highlighted by a survey quoted in *Christian Apologetics in the Postmodern World* (edited by T. Phillips and D. Okholm),[46] which cites that "only 28% of Americans have a strong belief in 'absolute truth,' and . . . a corresponding relativism is on the rise." According to Charlotte Allen,[47] the result of the survey has raised the paradoxical fear that evangelical colleges are becoming too liberal and relativist.

At this point the reader should keep in mind an important distinction between fundamentalism and evangelicalism in the United States. Fundamentalism is a movement that originated during the nineteenth century from the millenarian movement of the 1830s and 1840s (so called in reference to the expectation of the Second Coming of Jesus Christ and the ensuing thousand years of his reign). Fundamentalists advocate a return to the fundamentals—that is, to a literal interpretation of the Bible—and believe in the imminent coming of Jesus, the Virgin Birth, the Resurrection, and the Atonement. In 1902 the American Bible League was established with the production of a series of 12 pamphlets entitled *The Fundamentals,* which are a defense of the Bible in response to biblical criticism and modernism.

Although several common threads link the histories of fundamentalism and evangelicalism, the latter is distinct in its emphasis on personal conversion, witnessing for the faith, and active preaching in order to convert to Christianity. The roots of evangelicalism in America go as far back as the Great Awakening between the 1720s and 1740s, itself an offshoot of a religious movement that swept Europe around that time and that is referred to as *pietism* or *quietism* (evangelicalism itself goes all the way back to Martin Luther and his followers, who called themselves evangelicals). In the 1790s a Second Great Awakening began in New England, which led to the establishment of colleges and seminaries and to the beginnings of camp-meeting revivals in Kentucky. Although fundamentalists tend to be evan-

[46] T. R. Phillips and D. L. Okholm, *Christian Apologetics in the Postmodern World* (Downers Grove, IL: InterVarsity Press, 1995).

[47] In C. Allen, "The Postmodern Mission," *Lingua Franca* December/January 2000: 46–59.

gelical and evangelicalists tend be fundamentalist, there are exceptions, so the two terms not only have different historical roots but cannot be used as synonyms.

The problem within the Christian community seems to be that an effort is being made by several Christian (mostly evangelical) colleges to adapt to postmodernism, an adaptation that entails a profound contradiction for a faith that has always fought hard against the notion that every story is just as good as any other. Allen's article points out a closely related piece of irony. The conservative Catholic philosopher Alasdair MacIntyre believes that American Protestantism has been harmed by buying into the Enlightenment notion that there are facts about the world, a notion that goes well with the Protestant "heresy" (from a Catholic perspective) that individuals can read the Bible without the mediation of the church. Indeed, Philip Kenneson for one (also quoted by Allen) goes as far as saying that postmodern evangelicalism should do away with the problem of truth altogether and that Christians should point instead to their behavior as the living truth of their convictions. Allen perceptively observes that "in a sense one might say that postmodernism allows evangelicals to have their Scripture and eat it too."

This attitude has of course generated reactions against evangelical colleges. Some of them have even gotten into legal trouble because of their pledges. In 1996, Andrea Sisam sued Bethel College (St. Paul, Minnesota) for violating a contract to provide her with a Christian education. Why? Because she was forced to watch the 1979 German film version of Günter Grass's *The Tin Drum,* which included a scene depicting oral sex. She was also forced to see Spike Lee's *Do the Right Thing* and to read "objectionable" excerpts from Ralph Ellison's *Invisible Man.* The suit was dismissed, but the episode is a good indication of how complex the anti-intellectualist movement really is.

Another window into the same controversy is provided by the spread of colleges that cater to home-schooled children and that are fiercely competing for the education of young Christians against not only public schools, but classic evangelical schools as well. In the year 2000 alone a staggering 1.5 million children were home-schooled in America, and as a brochure from Patrick Henry College in Virginia states, "The value of sending fifty young people each year to Capitol Hill with excellent skills and unshakable convictions cannot be overstated." Indeed. But what kind of people are we sending to Washington to aid our elected leaders? At Patrick Henry, dating is prohibited because it is considered "serial infidelity." Michael Farris, chairman of the Home School Legal Defense Association, when interviewed by Allen, said that he considers the public school system "a godless monstrosity." Farris is also a vigorous attacker of homosexual rights and an advocate of women's "duty to be a loving and submissive aid to their husbands."

The problem of anti-intellectualism is acutely sensed and discussed among Christian evangelicals. *Christianity Today*[48] asked a group of Christian intellec-

[48] See "Scandal? A Forum on the Evangelical Mind," *Christianity Today* August 14, 1995: 21–27.

tuals about their thoughts on this matter, and the responses yield insights into the contradictions of the dialogue between reason and faith. Mark Noll, a professor of Christian thought at Wheaton College (Wheaton, Illinois), decried anti-intellectualism within Christianity, expressing his concern about six-day creationism and the hijacking of Christian politics by right-wing extremists. Indeed, Alister McGrath (theology professor at Oxford) observes that when he tries to explain the problem with the idea of a six-day creation, he is accused of being "soft on [biblical] inerrantism." Ironically, Noll observes that there is a disconnection between Christian intellectuals and the majority of Christian believers, just the situation that secular intellectuals complain about. Might there be a common problem, and perhaps even a common solution? Interestingly, Noll admits that some parts of the Christian faith *have* to be anti-intellectual, which at least is a problem that secular intellectuals don't have—unless they happen to be postmodernists.

Darrell Bock of Dallas Theological Seminary has provided a historical perspective on the problem by pointing out that the Christian Church was "bleeding" in the face of modernism at the beginning of the twentieth century, resulting in an attempt to survive that eventually led to the excesses of anti-intellectualism. In fact, according to Richard Mouw from Fuller Theological Seminary, Christianity experienced two "glorious" moments of anti-intellectualism (which he euphemistically calls "bold suspicion against the mind"): the first one in the seventeenth century, when pietists reacted against what he defines as rational orthodoxy, the second in the eighteenth century against the Enlightenment's idea that human consciousness is the highest standard of truth. (I doubt that the latter is a fair characterization of the Enlightenment, but certainly the emphasis on rationality as the only criterion for uncovering truth is a reasonable "charge," if one wishes to characterize it as such.) Noll adds that "the evangelical, fundamentalist, dispensational protest against naturalistic science has been a glorious protest. . . . A humble empiricism is a Christian (and biblical) way of approaching the natural world." Although I am baffled by what one could possibly mean by "humble empiricism" as a Christian value, the internal struggle of these intellectuals to reconcile their reason and faith is clear and instructive and can be traced back all the way to Augustine, one of the early fathers of the Christian Church.

For example, Mouw admits that "in America, there is a populist, anti-intellectual Christian remnant that feeds on overstatement, rhetorical overkill, proof-texting, and sloganeering, which grows out of distorted, grassroots pietism." In attacking two of the most vocal right-wing Christian anti-intellectuals, Mouw remarked, "Jerry Falwell and Pat Robertson . . . haven't thought these issues through theologically. The result is that the theological basis for what their political followers have advocated has been at best minimal and at worst perverse."

Yet McGrath told *Christianity Today,* "I'd still like to maintain the emphasis that what we are defending [the content of Scripture] is actually far more important than the means of defense we bring," implying that anti-intellectualism might be a necessary evil to ensure the good end of the affirmation of the Bible. One can hardly find a more overtly anti-intellectual posture in a self-styled intellectual such as McGrath.

The problem is that evangelicalism needs populism, as Noll puts it. He recognizes that "nonthinking" comes at the transition between academic and populist evangelicalism, but he does not seem to have any idea of how to solve the problem. In fact, in a clearly condescending manner, he characterizes Christian populism as "immature intellectually" and yet "necessary and a delight." Few, if any, secular intellectuals would dare to publicly display this level of paternalism.

Given this degree of internal turmoil within Christian evangelicalism, it seems fitting to remind the reader of the biblical passage from which the title of the play and movie *Inherit the Wind* (originally intended as an indictment of McCarthyism) came. Proverbs 11:29 says, "He who brings trouble on his family will inherit only wind, and the fool will be servant to the wise." It remains a matter of heated discussion who the fools and the wise are in this case.

So What Do We Do about It?

From what we have seen, anti-intellectualism is a complex problem that is the cause not only of the creation–evolution controversy, but also of much political and social wrangling in modern America. Clearly, a problem of this magnitude and intricacy cannot be solved with a couple of silver bullets. However, I shall provide my thoughts on what to do about the whole shebang in Chapter 8. For now, let me briefly discuss the thoughts and solutions of others, based mostly on Rigney's article cited earlier[49] and on Alvin Gouldner's *The Future of Intellectuals and the Rise of the New Class.*[50]

Gouldner—rather optimistically, I think—suggests that the emergence of a new class of knowledge workers during the information age (which was barely beginning when he wrote his book) will eventually cause the retreat of antirationalism. As he puts it, the new class will be "subtly drawing even the enemies of reason into the web of critical discourse" because of a shared ideology of a "culture of critical discourse." Perhaps, but more than 20 years later antirationalism is still strong, and fundamentalists have simply learned to use the new media to their advantage, with the result that the Internet is even more polluted by nonsense than it would have been otherwise. The human mind seems to have a remarkable capacity to adapt to new technologies while desperately clinging to old ideas. As we shall see in Chapter 8, this ability to accept the paradox is likely due to the high threshold for cognitive dissonance

---

[49] See note 3.

[50] A. Gouldner, *The Future of Intellectuals and the Rise of the New Class* (New York: Seabury Press, 1979).

that we all have and that—ironically—probably arose as a result of natural selection.

About anti-elitism, Rigney says that "a dialogical strategy . . . lies not in abandoning high standards of critical discourse on the grounds that they are anti-democratic, but in opening the conversation up to those who have been systematically excluded from it." This is most certainly the way to go, but it is also exceedingly difficult and frustrating. What is necessary is a commitment to long-term involvement in education and a willingness by intellectuals to descend from the ivory tower and mingle with mere mortals (again, see Chapter 8).

Rigney also suggests that the appropriate response to unreflective instrumentalism is serious debate and reflection on society's goals, a task that should be the primary concern of education. Yes, but how are we to start and sustain such debate? In some states in America (e.g., Kentucky) schools have adopted rules to the effect that certain subject matters deemed "controversial" can be excised from curricula because they would upset the students. Such a policy is paradoxical, since one could argue that the whole purpose of a good education is indeed to upset people at least once a week and possibly more often. Controversial subject matters, from evolution to gay rights, from abortion to the separation of state and church, are precisely those topics on which we need to focus the discussion as soon as possible, so that young people can be informed and will have given some critical thought to these topics *before* they enter the voting booth.

The current situation can change only if one or the other of two paths is followed: Either an enlightened leader will come along and build a coalition strong enough to implement dramatic reforms of the public school system, or a grassroots movement will originate after enough people have gained access to higher education to realize that this is a vital matter for the future of humanity. The first option is exceedingly unlikely (and would probably not be the best way for a society to mature because the solution would be imposed from above); the second option will not materialize for probably several decades, if not centuries. It is difficult to work hard for a goal that not even one's own grandchildren are likely to see, but this is the way all great cultural changes have occurred so far, from the fight against slavery to the emancipation of women. So, let us roll up our sleeves, stop whining, and start doing something about this mess.

Additional Reading

Dawkins, R. 1998. *Unweaving the Rainbow: Science, Delusion, and the Appetite for Wonder.* Boston, MA: Houghton Mifflin. A delightful book rebuffing the charge that science spoils our poetic view of the universe. The title comes from the poet John Keats's rallying cry against Isaac Newton, who, by "unweaving" the rainbow to study its nature with his prism, allegedly robbed us of the poetry of that celestial spectacle.

Hofstadter, R. 1963. *Anti-intellectualism in American Life.* New York: Knopf. The classic book on anti-intellectualism and its many facets.

Kaminer, W. 1999. *Sleeping with Extra-terrestrials: The Rise of Irrationalism and Perils of Piety.* New York, NY: Vintage. An easy-to-read book on the basic principles of skepticism.

Mill, J. S. [1861] 1979. *Utilitarianism.* Indianapolis, IN: Hackett. The classic rendition of the utilitarian doctrine in ethics.

Pigliucci, M. 2000. *Tales of the Rational: Skeptical Essays about Nature and Science.* Atlanta, GA: Freethought Press. A collection of essays by this author on topics including creationism, skepticism, and the nature of science.

Postman, N. 1985. *Amusing Ourselves to Death: Public Discourse in the Age of Show Business.* New York: Penguin. An amusing and cunning critique of modern American society and its love for mindless entertainment.

Sagan, C. 1995. *The Demon-Haunted World: Science as a Candle in the Dark.* New York: Random House. A classic on science and skepticism, featuring the famous—and very useful—"baloney detector kit."

Singer, P. 2000. *A Darwinian Left: Politics, Evolution, and Cooperation.* New Haven, CT: Yale University Press. A philosopher's response to Social Darwinism. Singer explains why the findings of evolutionary biology can be used to confirm some aspects of progressive social theories.

Sober, E., and D. S. Wilson. 1998. *Unto Others: The Evolution and Psychology of Unselfish Behavior.* Cambridge, MA: Harvard University Press. A wonderful book on the problem of the evolution of altruism, attacked from the points of view of philosophy, biology, and psychology.

– 4 –

Scientific Fundamentalism and the True Nature of Science

The saddest aspect of life right now is that science gathers knowledge faster than society gathers wisdom.

– *Isaac Asimov* –

When one tries to rise above Nature one is liable to fall below it.

– *Sherlock Holmes, in Arthur Conan Doyle's "The Adventure of the Creeping Man"* –

Edward Teller is considered the father of the hydrogen bomb. Although there has been some question about how much of his contribution to the development of that weapon was original or crucial, the fact remains that his entire scientific career has been associated with that particular enterprise. Given that we are talking of a means of mass destruction that could conceivably end life as we know it on planet Earth, you might have hoped that Teller would have distanced himself from the whole episode, perhaps even to have actively promoted critical discussions of the ethical role of science and scientists in modern society, as Einstein and Oppenheimer did. You would be sorely disappointed. Not only is Teller not in the least apologetic about having fathered the bomb, but he is mighty proud of it.

Indeed, he is so convinced that science can do no evil that he has proposed over the years a string of ludicrous "peaceful" applications of the H-bomb, from finding out the chemical composition of the moon by exploding one on its surface to paving mountains to make road construction easier, from dredging harbors and canals to testing the theory of relativity by exploding a bomb on the far side of the sun. Perhaps Teller's suggestions can charitably be interpreted as the desperate attempt of a man to justify his life's work in the face of the obvious dark consequences that it could carry for humanity. Nevertheless, it is difficult to find a more clear and disturbing example of scientism, the fundamentalist belief that science can do no wrong and will ultimately answer any question worth answering while in the process saving humankind as a bonus.

Milder but illustrative examples of scientism are writings by first-rate scientists such as biologist E. O. Wilson and physicist Steven Weinberg.[1] Curiously, both Wilson and Weinberg address what they consider the inutility of philosophy, albeit from different perspectives. For Weinberg,[2] philosophical inquiry is a waste of time and is indeed detrimental to true science. His argument is that not only do philosophers never accomplish anything—most especially they have never helped science to solve any problem—but in at least a few cases philosophical positions held by scientists have positively hindered progress. For example, Weinberg maintains that a philosophical school known as logical positivism, which was particularly popular at the beginning of the twentieth century, significantly slowed the acceptance of quantum mechanics because of its rejection of "unobservables" (i.e., quantities that cannot be measured directly) as scientifically valid constructs. Quantum mechanics is arguably the scientific theory relying on the highest number of unobservables, and yet

---

[1] I should preface what follows by saying that I consider both Weinberg and Wilson brilliant scientists and critical thinkers. I am just focusing here on a particular portion of their work I happen to disagree with. One of the crucial differences between science and religion is that in science one gets to criticize (constructively) anybody who deserves criticism, while maintaining intact one's respect for the work of one's critics.

[2] See S. Weinberg, "Against Philosophy," pp. 166–190 in *Dreams of a Final Theory* (New York: Pantheon, 1992).

it has now been demonstrated to be so accurate in predicting and explaining a variety of natural phenomena that few seriously doubt its validity.

Weinberg may have a point in the case of logical positivism (a position, incidentally, that few if any philosophers espouse today, just as it is hard to find "Newtonians" among physicists or "Lamarckians" among biologists). Where he is completely off track and shows his scientism is in his misunderstanding of the role of philosophy. Philosophy is not meant to "solve" problems the way science does, and most certainly not to solve *scientific* problems. The role of the philosopher—at least in the modern understanding of the word—is to be a metathinker, to think about how we reach certain conclusions and by what means we proceed in our inquiries (epistemology), as well as to elaborate on the relationship between what we do and what we should do (ethics) and on the big picture of reality (metaphysics). Although all these philosophical activities must be informed by science (or the philosopher will condemn herself to a rather sterile exercise in logic decoupled from the real world), they are not scientific and their effectiveness cannot be judged by scientific standards. It is precisely this obsession with applying scientific standards to everything else that characterizes scientism.

Wilson's scientism is different from Weinberg's but, I think, equally misguided. In his book *Consilience: The Unity of Knowledge,*[3] Wilson sets an ambitious agenda for science. He thinks that there is only one kind of knowledge and that eventually all fields, from science to philosophy to social sciences to the arts and even religion, will converge on the same answers. This sounds much more inclusive than Weinberg's scientification of everything, but the difference is only superficial. Early in Wilson's book it clearly appears that what he means by unification of knowledge is in fact a reduction of everything else to science. He sees a future in which social sciences, religion, art, and philosophy will all be explained by, and therefore reduced to, science. This is a form of hubris that the ancient Greek philosophers had already warned us against 25 centuries ago.

Of course, nobody thinks of himself as espousing scientism (in fact, the term does not come with an associated word to describe such a person: Scientists are those who do science, not those who endorse the ideology of scientism). Indeed, in philosophy it has become a widespread sport to accuse your opponent of scientism to ridicule his "faith" in science and the consequent attempt to reduce philosophical disputes to scientific ones. Creationists, too, profess to respect science, asserting that it is scientism that they loathe, although they define as "true science" only whatever does not contradict their interpretation of the Bible, and as "scientistic," any position that disagrees with theirs. Both scientism and antiscience sentiments are sides of the same coin—a blind commitment to a particular ideology, come what may. It will be useful to examine one case of antiscientific ideology before we attempt to clear the path and try to understand what science actually is—regardless of what its detractors and ideologues claim.

---

[3] E. O. Wilson, *Consilience: The Unity of Knowlege* (New York: Knopf, 1998).

Huston Smith and the Place of Science

"Science bumps the ceiling of the corporeal plane. . . . From the metaphysical point of view its arms, lifted toward a zone of freedom that transcends coagulation, form the homing arc of the 'love loop.' They are science responding to Eternity's love for the productions of time." This grandiose bit of poetical nonsense concludes a chapter of Huston Smith's *Forgotten Truth*[4], which is dedicated to putting science in its place. Smith is one of the world's foremost authorities on religions, and his aim is to demonstrate that science is not an omnipotent force that can answer all questions posed by humanity. That is, science needs to be put in its place.

As I have said, I certainly do not disagree with the general spirit of Smith's criticism, but what is astounding in Smith's essay is his attempt to develop a parallel between science and mysticism in order to demonstrate that the world's great religions are capable of insights at least as powerful as those of science because they use similar tools. In some sense, Smith—like "scientific" creationists—wants to have his cake and eat it too: On the one hand, he would have us take a more modest view of science than scientists allegedly desire, yet on the other he claims that mysticism is as powerful as science precisely because it uses a similar approach at a deep level of understanding. I maintain that this is a perverse version of scientism, and I will briefly examine this alleged parallelism and in the process try to understand what the place of both science and religion ought to be.

Smith's first insight is that science and religion both claim that things are not as they seem. For example, one has the perception that the chair on which one is sitting is solid, but modern physics says that it is made of mostly empty space. This, apparently, is analogous to the following bit from C. S. Lewis: "Christianity claims to be telling us about another world, about something behind the world we can touch and hear and see." Never mind, of course, that physicists can provide sophisticated empirical evidence to support their claim about the emptiness of space, while Christianity is made up of a series of partially contradictory stories backed by little evidence.

Second, according to Smith, both science and religion claim not only that the world is different from what we perceive, but that there is "more" than we can see, and that the additional part is "stupendous." Of course, electrons, quarks, and neutrinos are more than we can see, although they probably seem stupendous to only those few scientists who spend their lives working on them. Smith's idea here is allegedly the same as Shankara's "notion of the extravagance of his vision of the *summum bonum* when he says that it cannot be obtained except through the merits of 100 billion well-lived incarnations," a cornerstone of an Indian sacred text (of course, Smith omits telling the reader who Shankara is, and I shall leave that to my readers as an exercise; the point is that

[4] H. Smith, *Forgotten Truth: The Common Vision of the World's Religions* (San Francisco, CA: HarperSanFrancisco, 1993).

name dropping is helpful in pseudoscientific statements, but not acceptable in serious discourse). What Smith is doing is reasoning by (superficial) analogy, making the parallel between science and religion on the basis of a very general reading of only the broad pronouncements typical of the two enterprises.

Smith proceeds to the third point of his analogy by stating that the two quests for truth also share the quality that this "more" that they seek to explore cannot be known in ordinary ways (otherwise, presumably, one would need neither science nor religion to get there). The ways of science lead to apparent contradictions, such as in the case of some aspects of quantum mechanical theory. To this statement, Smith juxtaposes some gems from the Christian literature that he says uncannily resemble modern notions of quantum physics. For example, did not Nicholas of Cusa (in his *De Visione Dei*, "Of Seeing God") write that "the wall of the Paradise in which Thou, Lord, dwellest is built of contradictories," a statement that pretty well characterizes the dual particle-wave nature of light? And did not Dionysius the Areopagite (in "The Divine Names") say, "He is both at rest and in motion, and yet is in neither state," thus anticipating Heisenberg's indeterminacy principle? I am not making these examples up; these are Smith's very own.[5]

Fourth, both science and religion have found other ways of knowing this "more" that cannot be accessed by our ordinary senses. The language through which science accomplishes this is mathematics; the language of religion is, of course, mysticism, which Smith describes as a "comparably specialized way of knowing reality's highest transcorporeal reaches" (I honestly cannot figure out what that means). This, according to Smith, is "not a state to be achieved but a condition to be recognized, for God has united his divine essence with our inmost being. *Tat tvan asi;* That thou art. Atman *is* Brahman; samsara, Nirvana." Well, of course!

The fifth parallel asserted by Smith is that in both science and religion these alternative ways of knowing need to be properly cultivated. A scientist needs to dedicate a lifetime to her education and research if she wants to make a significant contribution. This is apparently similar to the asceticism of saints because, as Bayezid, of the Ottoman Empire, "correctly" pointed out, "The knowledge of God cannot be attained by seeking, but only those who seek it find it." (This statement, incidentally, is a plain logical contradiction.[6])

---

[5] My experience is that at this point readers tend to be divided along two lines: One group clearly sees the superficiality of Smith's parallels with science; the other starts being intrigued by the depth of the similarities. If you do not fall into the former category I suggest you try to *explain* (not simply reiterate) the parallels to a third party. If you succeed in an explanation, please pass it along to me!

[6] It is also an indication of a major difference between science and mystical religion: Science has both a methodology (naturalism) and an epistemology (empiricism). Mystical religion has neither. Mysticism is a self-induced state of experience, very far from any respectable epistemological standpoint. What Smith is saying reminds me of what one of the twentieth-century philosophical naturalists, Sidney Hook, said—namely, that science is a way of knowing the world, while religion and art are ways of experiencing the world. There is an important epistemological difference.

Sixth, and finally, according to Smith, in both science and religion profound knowing requires instruments. In science, these are microscopes, telescopes, and particle accelerators. In religion, the equivalent is provided by the Revealed Texts, "Palomar telescopes that disclose the heavens that declare God's glory." If gods who dictate texts are not palatable to you, there is an alternative: "Spirit (the divine in man) and the Infinite (the divine in its transpersonal finality) are identical—man's deepest unconscious is the mountain at the bottom of the lake." Get it? Good, because I don't.

I would not have bothered the reader with this mountain of nonsense if it came from the local televangelist screaming bloody murder against the humanists' corruption of the world. But this is Huston Smith, one of the most respected intellectual exponents of modern religion, one who is hailed as offering the deepest insights that not just one but *all* of the world's religions can offer! This is a maddening example of what Richard Dawkins calls "bad poetry."[7] Metaphors make much of the world's literature a pleasure to read, but they can also be exceedingly misleading. One can practice science or religion or both, but to pretend that they yield common insights into the nature of the world is an intellectual travesty. To go further, as Smith and so many religionists do, and assert that science is arrogant because it claims to provide the best answers to a circumscribed set of questions is astonishing, especially when the proposed alternative is so obviously the result of Pindaric flights of imagination. Smith, it seems to me, is making the same sort of mistake that Weinberg and Wilson make: His hubris does not allow him to see that the two extremes of scientism and antiscience are equally flawed and sterile positions.

A Brief History of Science

After having encountered scientism and antiscience, it seems appropriate to ask ourselves what "true" science is actually all about. Often the best way to understand a concept is to trace the history of its development, an approach that is sadly becoming less and less common in science itself because of the obsession of teachers to cover all the "facts," which leads them to neglect the all-important study of how we came to know these facts (see Chapter 8 for the consequences that this misguided focus has on science education). I cannot do justice to the history of science in this essay, so I recommend D. C. Lindberg's *The Beginnings of Western Science*[8] to the interested reader. The events and personalities chronicled in that book formed the backbone of the modern scientific method and help me now to illustrate what science is all about.

---

[7] R. Dawkins, *Unweaving the Rainbow: Science, Delusion, and the Appetite for Wonder* (Boston: Houghton Mifflin, 1998).

[8] D. C. Lindberg, *The Beginnings of Western Science: The European Scientific Tradition in Philosophical, Religious, and Institutional Context, 600 B.C. to A.D. 1450* (Chicago: University of Chicago Press, 1992).

The English philosopher John Locke (1632–1704) was the first thinker to use the term *science* in a context resembling the modern meaning of the word. But the historical foundations of the scientific method go as far back in time as ancient Greece, and in particular to the remarkable civilization flourishing in Athens during the decades culminating around 450 B.C.E., the Athenian golden age. In fact, the Greeks distinguished between two fundamental concepts that even today separate rational from irrational attitudes toward the world. They used the word *mythos* to indicate an approach based on supernatural explanations, and the word *logos* to imply the use of rationality. Mythology and logic are to this day diametrically opposite tools to make sense of the world.

Perhaps the greatest Greek contributor to science was Aristotle (384–322 B.C.E.). His many books on natural history influenced thinking in Western society until the end of the Middle Ages. Yet Aristotle did something far more important and lasting for science: He formalized the logic of *deduction,* which enables us to test logically our assumptions about the world.[9] Today deduction is an important component of the scientific method, even though almost all of Aristotle's specific conclusions and hypotheses about the nature of the world have been rejected by modern science because his assumptions were not sufficiently empirical. This, once again, dramatically emphasizes that science is a *method,* not simply a particular body of knowledge: The knowledge can be superseded by better knowledge, but this process is accomplished through the use of a set of methods based on rational inquiry and empirical evidence.

Deduction is the process of deriving certain conclusions from the logical analysis of a set of premises.[10] If the premises are correct and the reasoning is not flawed, the conclusions *must* follow. Deduction does not enable us to acquire new knowledge of the world; only inductive reasoning (discussed later) does that. However, deduction is useful in two ways: (1) It shows us the conclusions to which we are logically committed on the basis of our assumptions, and (2) it thereby enables us to clarify, or test logically, what is implicit in these assumptions. It can point the way very fruitfully to the kind of empirical evi-

[9] Aristotle did not of course invent the method of deduction itself. It is part of the way humans naturally reason; that is, we tend to derive conclusions from assumptions that we consciously or unconsciously see as indicating large categories within which to place things. Aristotle invented the syllogism, which enables us to illustrate many of these ways of reasoning and to evaluate them. One of the things he was trying to do was to show that deduction enables us to see natural relationships between individual things and categories of things, or among categories themselves. For example, we can see clearly the relationship between Socrates and human mortality when we know that (1) all humans are mortal and (2) Socrates is a human. We can see that Socrates is a member of two other, overlapping (but not completely coextensive) classes of things—namely, humans and other mortal beings.

[10] As an aside, it is interesting to note that popular culture associates the use of deduction with the fictional character Sherlock Holmes. However, this association reflects a common misconception about Holmesian reasoning in Sir Arthur Conan Doyle's work. The author calls it deduction, but it is actually a form of induction called "inference to the best explanation." A good example of this is found in Chapter 2 of "A Study in Scarlet." I owe this clarification to my colleague Barbara Forrest.

dence a scientist must look for, and it can indicate when a hypothesis is conclusively disproved.

For example, in the early stages of the formulation of his evolutionary hypothesis, Darwin had to specify the logical commitments involved. In particular, the hypothesis that (1) current life forms had evolved over time from earlier ones, combined with the premises that (2) Earth is geologically very old, (3) the ancient age of Earth is recorded in the geological strata, and (4) the life forms of each geological epoch would be fossilized in their respective strata, would lead to the logically necessary (deductive) conclusion that the older the organism is, the deeper in the geological strata its fossils should be found (assuming the successful formation of fossils, of course). One would also have to conclude that no fossils from different geological epochs should be found together in the same strata. Thus, deductive reasoning pointed to the type of empirical evidence Darwin needed to strengthen his hypothesis: Finding fossils in their proper geological strata would strengthen the probability of evolution as change over time; finding fossils from different geological epochs in the same stratum would conclusively disprove it.

Aristotle gave us the method by which to illustrate and evaluate the various forms of deductive reasoning: his famous syllogisms. One kind of syllogism is made of a general premise, a specific premise, and a necessarily true conclusion (which follows logically from the truth of the premises). Here is an example:

1. (All) biblical fundamentalists believe the Earth is 6,000 years old.
2. Dr. Duane Gish is a biblical fundamentalist.
3. Therefore, Dr. Gish believes that Earth is 6,000 years old.

The first sentence is based on the wide-ranging (in deductive logic, a *universal*) premise (based on common observation) that biblical fundamentalists believe Earth to be young in geological terms (because they can count the generations back to Adam and Eve). The second sentence specifies the more limited (in deduction, a *particular*) fact that Dr. Gish is a biblical fundamentalist. (I had to verify this in person by asking him because ascertaining the truth value of one's premises in a deductive argument is not itself a logical process unless the premises are true by definition.) The third sentence shows the power of deduction: From the truth of the two premises, we can derive analytically the already implicit fact that Dr. Gish also believes that Earth is 6,000 years old. (I also verified this in person, although in deductive reasoning, if one is sure of the truth of one's premises, one need not verify independently the truth of the conclusion because true deductive premises preserve truth; that is, they *guarantee* in advance the truth of the conclusion.)

Although deduction is a very powerful logical tool, its limitations are also clear from the example just given. What if one or the other of the premises were not true, or not always true? It could be that *most,* but not all, biblical fundamentalists believe in a young Earth. In that case, our conclusion regarding Gish would *probably* be correct, but not certainly[11] (more on the probability of our beliefs later). That is why the specific body of knowledge that Aristotle left us has in some ways actually hindered scientific progress, not helped. For a long time people thought that deduction was an infallible method of reasoning about the world such that whatever Aristotle was able to deduce about the world must be true, regardless of the evidence. But deduction is an infallible method only of *clarifying* what follows logically from what one already assumes to be true.

Medieval scholars' rigid adherence to scholastic—that is, deductive—reasoning, based on nonempirical premises, actually impeded human knowledge of the world rather than advancing it. Their insufficient understanding of the function and importance of inductive reasoning and empirical observation slowed the progress of science and was directly responsible for the executions of several courageous dissidents who were burned at the stake as "heretics." (In fairness to the medievals, inductive logic was not recognized and seriously studied as a branch of logic in its own right until the nineteenth century.) Clearly the methods of reasoning needed some improvement.

Several improvements came in a relatively brief period of time after the earliest universities throughout Europe were established (Figure 4.1). The first to formulate a modern version of the scientific method was the Frenchman René Descartes (1596–1650). In his *Discourse on Method* (1637), Descartes summarized his way of approaching questions about the world in the following manner:

- *The first [rule] was never to accept anything for true which I did not clearly know to be such; that is to say, carefully to avoid precipitancy and prejudice . . .*
- *The second, to divide each of the difficulties under examination into as many parts as possible, and as might be necessary for its adequate solution.*
- *The third, to conduct my thoughts in such order that, by commencing with objects the simplest and easiest to know, I might ascend by little and little, and, as it were, step by step, to the knowledge of the more complex . . .*
- *And the last, in every case to make enumerations so complete, and reviews so general, that I might be assured that nothing was omitted.*

Descartes did not formulate his method for use in tandem with an empirical approach to the acquisition of knowledge (he distrusted sense experience). Rather, he intended it to be applied to the analysis of rationally apprehended ideas in order to determine the correctness of those ideas as one's initial assumptions, which, if correct, could then be the starting points of a chain of deductive reasoning, as in a geometrical proof. Notice, however, Descartes's ingenious, if unintended, introduction of a technique that has made possible many successes in modern empirical science: dividing large and complex problems into smaller, more manageable units, a process that lends itself equally

[11] We would have a flawed deductive argument, but a pretty reliable inductive one.

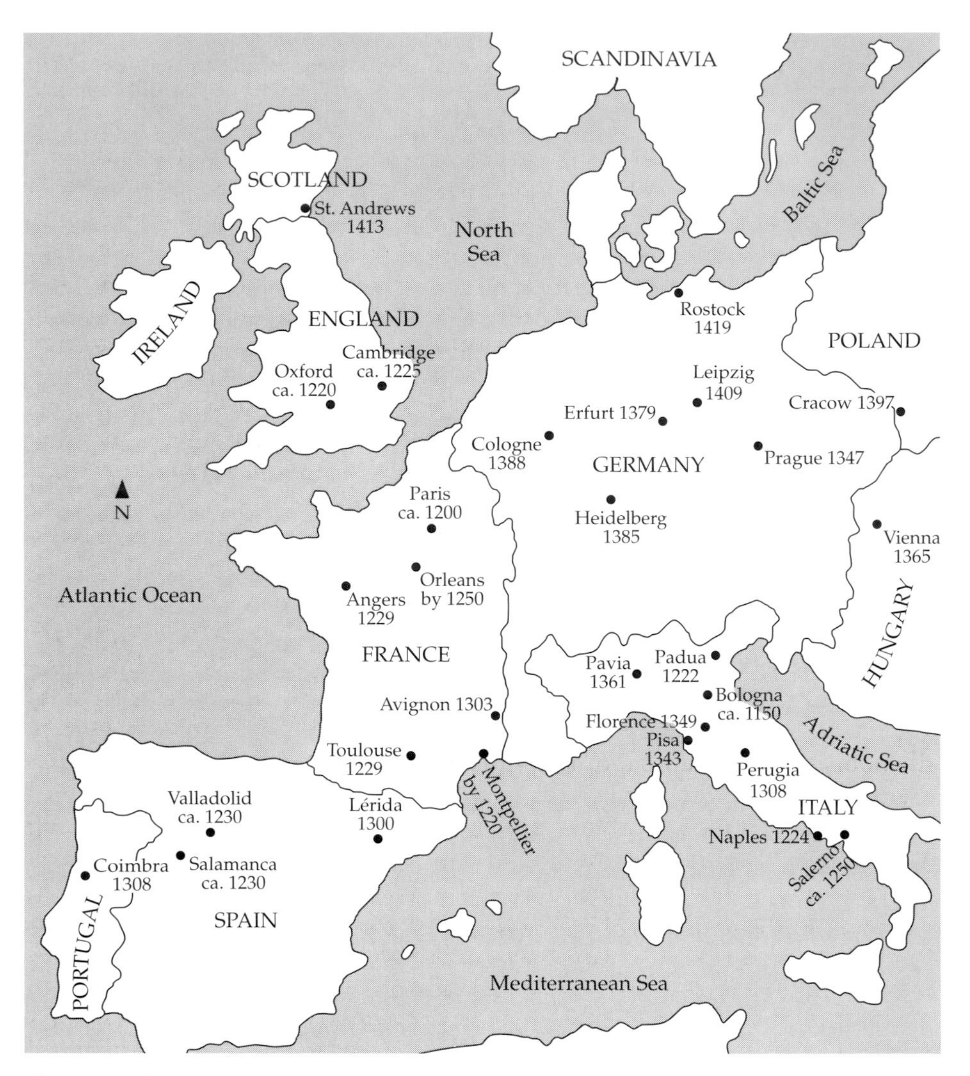

Figure 4.1 Map showing the dates when the earliest universities throughout Europe were established. The invention of the modern version of the scientific method was fostered by the widespread phenomenon of academic teaching and research at the beginning of the modern era.

well to empirically based, scientific reasoning. When enough of these units have been understood, they can be combined to yield broader generalizations. This is what has been called the *reductionist* approach. Once you have put enough pieces into position, the overall picture will emerge and yield itself to be understood in its entirety.

The reductionist approach has gotten a bad reputation of late, being countered by the *holistic* approach. The holistic argument is that one must consider problems in their entirety, in a gestaltlike fashion, because focusing on the details leads one to miss the Big Picture. I have come to the conclusion that a better way to think about the problem is in terms of reasonable or moderate (as opposed to fundamentalist) reductionism: A reasonable reductionist approach to tackling a problem *does not* ignore the Big Picture. *Au contraire,* the goal is exactly to reconstruct the whole picture, but only an analysis of the individual elements will enable one to construct a comprehensive explanation. One may appreciate the functioning of the V-8 engine that enables rapid movement from home to work, but one can *understand* its functioning only by understanding the functioning of each of the engine's parts (considered in the context of the whole). One could certainly never on one's own solve the problem of getting to work if the engine broke down *without* an understanding of the function of its various parts.

Part of the reason why moderate reductionism works (and both extreme reductionism and holism do not) is pragmatic: The human mind has a hard time considering the whole of a complex data set at once. The intrinsic limitations of our brains give us really no choice. It may well be that a holistic approach is more comprehensive and intellectually satisfying, but if our brains cannot implement it, we'll have to make do with what we have. As in the case of science versus other approaches to explaining the world, the dispute between reductionism and holism is ultimately to be decided in the field, not by armchair discussions. And it is quite clear that (reasonable) reductionism wins hands down. Reductionist science has been able to split the atom (with the debatable benefits and evils that humans have derived from such an accomplishment), to uncover the structure of DNA, and to discover galaxies and planets. Holism has simply been stuck at the starting line, pontificating that things are "more complex than they seem," but without providing specific (or general, for that matter) solution to any problem.

Another important piece of the scientific method is implied in Descartes's third point (to answer questions by beginning with the simplest objects to know). Descartes believed that the simplest things to know were the "clear and distinct" truths of rational intuition, not sense experience. He never fully appreciated the importance of empirical observation and inductive reasoning. The need for the latter was clearly stated by the English philosopher Francis Bacon (1561–1626). Induction performs a very different task from that of deduction. Instead of showing us the conclusions to which our assumptions logically (i.e., necessarily) commit us, induction starts from empirical (observation-based) premises and enables us to draw conclusions that do provide additional knowledge. For example, having observed the sun rising hundreds of times in the past, by inductive reasoning I can be relatively confident that it will do so tomorrow. Notice that, unlike deduction, induction does not make any claim about the causality of the observed phenomena; it simply generalizes from a number of observations. Another way to think about it is to regard deduction as a logically necessary, analytical derivation of predictions from an initial set of assumptions (after all, in a syllogism one has to be reasonably sure that one's premises are correct). Induction, on the other hand, is the statistical, inferential derivation of *probable* predictions based on a sampling of phenomena, *regardless* of what causes the phenomena themselves.[12]

Let us consider an illustrative example. The problem of predicting the weather is, as everybody knows, one of the most arduous that confronts modern science. Despite very sophisticated and expensive weather satellites and dedicated computers, meteorology is still far from being an exact science.[13] Both deductive and inductive methods are used in meteorology (as they are in all other science)—one for short-term, the other for long-term predictions. Short-term predictions rely on the analysis of up-to-date weather patterns and on generalizations about weather patterns (general principles of meteorology—i.e., generalizations based on a great deal of cumulative weather data). Current observations, combined with these meteorological principles, indicate how these patterns should evolve in the near future, given current conditions. This is an example of deduction based on a theory describing the causal mechanisms driving the atmosphere (general premise) and the actual observations pertinent to recent events (particular premise). The development of weather changes in the near future is deduced as a consequence of such premises.

However, the general principles of meteorology are themselves based on observations, which are by their nature extremely time and location specific. Therefore, when it comes to predicting the average weather conditions in a certain area during a certain season, the deductive method fails completely because no sufficiently stable theory has yet been developed for long-term forecasts. That is where we resort to induction. We gather as much information on the weather in the region and season of interest for as many years as possible for which we have reliable data. We then use statistical analyses to infer the *probability* of, say, a frost or so many inches of snow to occur next year in a particular place and season. This is induction in that the prediction does not rely at all on an actual understanding of the underlying causal dynamics, but is simply a statistical projection of what might happen *if* the underlying causes have remained essentially the same. Although the inductive method works

[12] An argument can be made that ultimately all scientific knowledge, and in fact all knowledge in general, relies on induction, since the premises of deductive reasoning are themselves derived by inductive methods. The problem, then, becomes to establish how we know that induction works. If one answers, "Because it has worked in past," one is still using inductive reasoning, and it does not make sense to use induction to justify induction because we fall into circularity or an infinite regress. This is known in philosophy as "the problem of induction." As far as we know, it has no theoretical solution, which hasn't stopped science from making progress in a pragmatic sense. And the pragmatic success of induction provides a great measure of justification for its use. As philosopher Ludwig Wittgenstein said, there comes a point where one has to stop asking why and just accept that that is the way things are.

[13] And there are very good reasons, based on chaos theory, to think that it will never be. See James Gleick's *Chaos* (New York: Penguin, 1988).

for long-term predictions, it does not work for short-term predictions because in those cases the amount of data available (from the previous few days) is simply insufficient to derive a reliable statistical inference. The two approaches, then, are mutually beneficial.

Induction, even though formally introduced in science by Bacon, is probably much more ancient than deduction. In fact, most animals make their way in the world by inductive reasoning. Fascinating experiments (discussed, for example, by Carl Sagan and Ann Druyan in *Shadows of Forgotten Ancestors*[14]) show that animals such as rats can be as "superstitious" as humans because they use induction to make causal associations, and to predict future events on the basis of these associations. For example, if a rat is given a reward (typically food) after it performs a particular (and random) sequence of movements (e.g., two steps to the right, one to the left), it will quickly associate the sequence of steps with the reward. Now suppose the experimenter gives the reward from time to time even if the rat has *not* performed the "correct" steps. The rat will keep attempting to get the reward by repeating that sequence, as long as occasionally such behavior is rewarded. In a sense, the rat's brain has been hardwired for finding causal associations between phenomena in the world. The reason is that it is clearly advantageous for the rat to be able to predict, even if only partially successfully, what is going to happen. The rat perceives a causal connection between two events (without actually understanding the causal basis, if any, of such correlations) and attempts to use that knowledge to manipulate the appearance of the reward.

A strikingly similar tendency to make causal connections occurs in humans and represents the basis for the commonly observed phenomenon of superstition, as well as for better-founded expectations. If you get a good grade the day you wear a certain pair of shoes, you will be more likely to wear the same "lucky" shoes on several other occasions, even though your scores will be high only some of the time because you have associated phenomena that are not truly causally related. (One would be better advised to associate getting good grades with the causally relevant activity of studying!) However, although it is irrational (meaning, literally, not based on a reasoned understanding of what is going on), such thinking probably paid off often enough during the history of humankind to be ingrained in our brains. Superstition, in other words, saved enough lives in the past (by chance) that even the current availability to the human race of more powerful tools (such as logical thinking) does not completely supersede it. It is ironic in the highest degree that part of the reason for widespread irrationality in modern human beings is likely the result of natural selection for the ability to identify probably true causal connections among natural phenomena.[15]

[14] C. Sagan and A. Druyan, *Shadows of Forgotten Ancestors: A Search for Who We Are* (New York: Random House, 1992).

[15] This may have to do with the low cost of incorrect hits (thinking you heard a lion and running when it wasn't true) compared to the high cost of unperceived misses (not detecting a lion that then ate you).

Lack of understanding of causal mechanisms, however, is also one of the fatal flaws of induction and is the reason it is used only as a preliminary investigative tool in science. The limits of induction were best explained by philosopher Bertrand Russell, with his example of the inductivist turkey. Russell's imaginary turkey has just been brought onto a farm, and he notices that he is fed in the morning, say around 7:00 A.M. This feeding pattern goes on for several weeks, but—being a good inductivist—the turkey does not risk a prediction of what will happen the following morning because he feels that the data base is not sufficient just yet. After several months, the turkey's data sheet is quite long, and he still is fed around 7:00 A.M. every morning. But he insists on collecting more data. Finally, after 364 days worth of reliable data, the turkey feels confident enough to make one small prediction: that tomorrow morning around 7:00 A.M. he will be fed. Of course, the following day is Thanksgiving, and the turkey is slaughtered for the annual feast. Such can be the tragic limitation of a purely statistical analysis. The turkey's prediction is reasonable as far as it goes, but it does not take into consideration the other possible conclusions that are compatible with his own because he is unaware of other factors that are causally related to his well-being (or, in his case, the lack thereof).

The next fundamental historical and conceptual step in shaping the scientific method was introduced by the Italian Galileo Galilei (1564–1642), although it had been foreseen by Bacon: the use of controlled experiments. Galilei was the first to realize (and to implement) the idea that one could test hypotheses and gain knowledge about causal relationships by carefully controlling and measuring what goes on during the unfolding of a given phenomenon. Of course, experimentation is limited to things that one can experiment upon. It cannot be carried out, for example, in many fields of astronomy because one cannot manipulate stars. As I will discuss in the next section, therefore, experimentation is a very useful—but by no means essential—part of the scientific method. One can think of it as a powerful tool that should be used anytime it can be.

The last great addition to the methods of science came from the Englishman Sir Isaac Newton (1642–1727), though it had been foreseen by Galilei. Newton realized that mathematics is a very powerful intellectual tool for tackling all sorts of scientific problems. Indeed, mathematics (and its stepsister statistics) is now almost universally employed by physicists, chemists, biologists, geologists, and social scientists the world over. It is important to realize, however, that mathematics is simply a very convenient, yet arbitrary, language. It is a toolbox invented by humans to think about complex problems better than they can with their "natural" language. Statements to the effect that "the book of nature is written in mathematical symbols" (as Einstein said) are just poetic license; they become utter nonsense if taken literally. Nature, as far as we know, does not have a language because it doesn't have anybody to communicate with. It just is. But, as George Orwell made dramatically clear in his novel *1984*, language directly shapes the ability of humans to think.[16] Mathematics, by adding incredibly useful devices to our ability to think, has made it possible for us to grasp some of the most fundamental aspects of the universe in which we live.

Overall, then, what distinguishes the scientific method from other ways of thinking about and investigating reality is a combination of the pieces that Aristotle, Descartes, Bacon, Galilei, and Newton—among others—have put together over the span of two and a half millennia. Modern science is both a very recent activity in human history (becoming a widespread form of systematic inquiry only during the last couple of centuries), and one that has been characterized by a long period of development spanning most of what we refer to as human civilization. The fundamental point to understand is that science is pragmatic, even when it approaches such esoteric problems as the fundamental makeup of matter. Its methods incorporate anything that works—induction, deduction, experimentation, mathematics—and exclude everything that doesn't. The result is a continuous process of refinement of the method itself, which extends back for centuries and adopts the best strategies that the most brilliant human minds have been able to produce, while discarding the rest. This historical development is why, when somebody sits down and tries to summarize *the* scientific method, the result ends up sounding like a rich and convoluted family tradition passed down orally from grandmothers to granddaughters. This is indeed not a bad description of what has actually happened.

A Closer Look at This Thing Called Science

Now that we have some historical perspective on how science developed, it is instructive to understand what modern science is and how it works. We can start by dispelling a common myth. Science is *not* a body of knowledge. The body of knowledge that we refer to as "scientific" is a *product* of science, but it does not define it. For example, the theory of general relativity is now almost universally accepted as the best explanation of many (albeit not all) of the basic physical properties of the universe. It has entirely supplanted Newtonian (*classical*) physics as a model for understanding the world. Does that mean that Newtonian mechanics *was* science, but it isn't any more because of Albert Einstein? Of course not. Both are theories developed within the realm of science using scientific methods. Both are valid within certain limits, but Einstein's theory is of much broader applicability, and it makes more precise predictions. Therefore, it is a more satisfactory explanation of what is out there. That (and the fact that Newtonian mechanics is conceptually simpler) is why physics stu-

[16] In Orwell's novel, the omnipresent government known as "Big Brother" engages in a slow but relentless reduction of the available vocabulary, by printing new editions of the official dictionary that include fewer and fewer terms. The idea is that if a word becomes lost from common usage, the very concept to which that word corresponds will be lost as well from human consciousness.

dents are taught classical mechanics at the beginning of their academic career and then are told to forget about it when they are initiated into the wonders of relativity.

So science is a *method* used to uncover and provisionally explain observations about the world, as well as to predict future observations. I will explain shortly the fundamental—and often neglected—meaning of the word *provisional* inserted in this definition. For now, let us be content with the distinction between science as a method and the scientific body of knowledge that is the result of the application of this method. But what *is* the method by which science operates? Just saying that scientists uncover things and explain phenomena in a scientific way does not amount to much more than putting a label on a mysterious box. When the box is opened, however, it is plain for everybody to see that science is—at its core—a very simple and straightforward approach to reality, as much as some of its conclusions turn out to be completely counterintuitive.

As I mentioned earlier, philosophers of science have found it difficult to define the scientific method. Indeed, it is whispered that even scientists themselves enter a difficult territory when asked to define how science actually works. There is a simple explanation for these perceived problems: Some things we learn best only by doing. Philosophers do not actually *do* science, so everything they conclude about it is through indirect methods of observation, discussion, and reading about how science is done. This is like trying to understand a car mechanic's methodology by reading a repair manual. Scientists often find it difficult to precisely articulate something they do automatically as the result of years of training and experience. Think of how difficult it is to explain in plain words simple things such as how to ride a bicycle. It is far more effective to simply *show* somebody how to do it. That is how scientific training is done in practice. When a new graduate student joins my laboratory, we may occasionally have philosophical discussions about the nature of science (usually over a beer). But she learns how to design a research methodology by daily interactions with people who do science. Even the gradual refining of her critical thinking skills over time does not happen after theoretical lessons on how to think, but through a long and painstaking process of discussing research papers published in the literature, as well as composing her own papers and responding to criticisms. In other words, science is learned in very much the same way that many other complex human activities are learned, such as music, the visual arts, or business administration, by apprenticeship.

The goal of science is an understanding of the general principles underlying the functioning of the universe. Such understanding is achieved by a combination of four approaches:

1. Observation of specific facts or phenomena
2. Formulation of generalizations about such phenomena
3. Production of causal hypotheses relating different phenomena

4. Testing of the causal hypotheses by means of further observation and experimentation

The scientific method is based on two fundamental, well-founded assumptions:

1. A naturalistic explanation is sufficient to account for the functioning of the universe.
2. The universe can be understood through logic and rational thinking.

Practicing scientists accept the following two types of evidence:

1. Confirmation of hypotheses by data strengthens their validity.
2. Repeated and widespread inconsistency of data with a hypothesis eventually leads to the rejection of that hypothesis.

Note that working scientists do not reject a promising hypothesis if only a few pieces of evidence do not seem to fit.[17] In fact, suspending judgment on hypotheses that are currently being investigated is a crucial tool in science, and it leads to an ever more refined understanding of the phenomena. However, if one is confident beyond reasonable doubt that other explanations can be ruled out, then the hypothesis under test is abandoned in favor of a more promising alternative.

In formulating hypotheses, scientists attempt to derive the simplest possible explanation that accounts for the data. This principle, known as Occam's razor (or parsimony), is equivalent to saying that science attempts to make the fewest possible assumptions in explaining how the world works.[18] If you do not need to invoke a particular assumption, if you can equally well explain the results in a more parsimonious way, there is no reason to make your theoretical edifice more cumbersome than is strictly necessary. This is not to say that the world could not be more complex than scientists can imagine. It is simply that there has to be a good reason to think that it is more complex; other-

[17] This is the contrast between naïve and sophisticated falsificationism. A naïve falsificationist would maintain that one piece of evidence contradicting a given hypothesis is enough to bring the hypothesis down. A more realistic view consists in realizing that testing scientific hypotheses is a complex affair because it depends on several ancillary assumptions that can also be questioned if the results do not match the main hypothesis. For example, it could be that the hypothesis in question is in fact correct and that the discrepancy with the data is due to faulty measurement, to the conditions not being appropriate for the test being conducted, or to inappropriate analyses of the resulting data.

[18] Principles of explanation typically come with existential commitments that must themselves be verified. One reason that intelligent design theory does not qualify as science is its violation of the principle of parsimony. ID invokes the supernatural, which is burdened at the outset with the need to establish the existence of the supernatural. There is no scientific methodology, and no satisfactory philosophical method, for doing this. ID theorists, therefore, are trying to introduce into the scientific enterprise an unverifiable explanatory principle, which amounts to adding a layer of explanation that is itself inherently (and probably permanently) problematic and yields neither explanatory adequacy nor pragmatic success in science.

wise there is no guide as to which extra hypotheses are warranted and which are not.

Let's consider an example: To understand how airplanes fly, we need to invoke the law of gravity and the principles of aerodynamics (among other things). We could also make special hypotheses concerning other as yet unknown principles, such as antigravity. But we have no evidence for, nor any need to invoke, antigravity. This does not mean that antigravity does not exist. However, if we call it into play for no apparent reason, then why not also invoke any number of other hypothetical phenomena, such as special properties of metals when lifted in the air, or a mystic role of Earth's magnetic field, or anything else? The point is that there is an infinite number of potential additional phenomena that could explain why airplanes fly. But we have no way of determining which of these alternatives to consider, so scientists choose not to consider any of them unless there is a compelling reason to do so. It really boils down to common sense.[19]

Any scientific theory or hypothesis that is repeatedly confirmed is provisionally accepted as valid, pending further information and tests. Any theory or hypothesis that has been overwhelmingly invalidated by empirical data is forever rejected. So, for example, the theory that the sun rotates around Earth is no longer a viable scientific theory, given the overwhelming evidence against it. Einstein's theory of relativity is the dominant paradigm in physics (together with quantum mechanics) because it has survived many empirical tests. This, however, may change in the future. In fact, a brilliant young physicist may set herself the goal of producing an even better theory than that of general relativity, and she may succeed. Her new theory will then be provisionally regarded as the best available, and so forth. Science, if you will, is always about "very likely maybes," never about absolute truth. This is a feature that both creationists and scientific fundamentalists apparently find disturbing and that explains quite a bit of the ideological posturing on both sides of the issue.

The *only* reason that all of the aforementioned aspects of scientific methods can be justified is that *they work.* There is no other external or internal justification of science available to us.[20] Now, the list given earlier in this section embodies most aspects of what practicing scientists recognize as the scientific method. However, the word *method* here is not used in the sense of a plan of action or order of events, a checklist that is prominently displayed in every laboratory and that each scientist keeps consciously consulting and marking off.

---

[19] In this book I have sometimes argued that science is not commonsensical, and occasionally the opposite. To clarify, I think that scientific *methods* are commonsensical (indeed, some of them are the result of the basic patterns of thought that naturally evolved in humans), but the *results* of science can be highly counterintuitive.

[20] Indeed, the failure of logical positivism highlighted by Weinberg and mentioned earlier consists in not realizing the impossibility of a program that attempts to justify the validity of science from first principles. This is also the shortcoming of a Wilson-like attempt at consilience of knowledge because science cannot justify the validity of its own practices by using the scientific method.

It is simply an approximate explanation in plain words of the principles governing the working approach of scientists as they proceed in their study of nature. At some point in every scientific undertaking, every one of these aspects becomes an operational principle—that is, operative in what scientists actually do.

A useful way to think about science is to notice the striking similarity between it and what a detective does. A scientist is really not different from a modern, technically savvy version of Sir Arthur Conan Doyle's Sherlock Holmes. Like Holmes, a scientist has to weigh the evidence, which sometimes completely excludes some suspects from consideration in the solution of the case, and in other instances strongly points toward the implication of a particular culprit. As with Sherlock Holmes, the scientist can never be completely certain about the resolution of a case. There is always the possibility that an important piece of evidence has not been found or has been improperly interpreted. This is true because of the strongly inductive nature of both science and detective work (although, as we have seen, deductive reason plays a role as well).

Unfortunately, nature does not make confessions, but even Holmes would not rely too heavily on a confession because it could be obtained by the subject's being pressured, or by other illegal or unreliable means. What brings Holmes to the solution of a case is usually the independent *convergence* of many kinds of evidence, all pointing toward the same solution. Similarly, the scientist relies on the very reasonable assumption that there is an outside world. If that is so, then our observations, experiments, and theories are but partial windows on the same underlying landscape. As more of these windows yield comparable and consistent glimpses into that reality, we become more confident (but never logically certain) in our knowledge of the landscape. This convergence of disparate facts as if they were indicating a single underlying reality is sometimes referred to as the property of "consilience" (the same word used in Wilson's book mentioned earlier (see note 3), but with a much more limited meaning).[21]

Consequently, the more pieces of independent evidence we can gather, or the more experimental manipulation we can carry out, the more comfortable scientists will feel about drawing conclusions about a particular subject of inquiry. Sherlock Holmes would certainly endorse such a cautious approach. The

[21] The word *consilience* was introduced by William Whewell in 1840 to explain the phenomenon in which often pieces of evidence from disparate sources "jump together" toward a common explanation, what he termed a consilience of induction. As he put it (in *Philos. Induct. Sc.* II. 230): "Accordingly the cases in which inductions from classes of facts altogether different have thus jumped together, belong only to the best established theories which the history of science contains. And, as I shall have occasion to refer to this particular feature in their evidence, I will take the liberty of describing it by a particular phrase; and will term it the Consilience of Inductions." Although Whewell's work on induction was overshadowed by the later contribution of John Stuart Mill, Whewell was the first philosopher to resurrect the importance of induction in science.

important point to consider is that the application of science rests on the possibility of testing hypotheses, regardless of which methods are employed to test these hypotheses in any given instance.

A related and equally crucial point to understand about science is that scientists' findings are continually, and indeterminately, subject to peer review. In other words, scientists are constantly checking on one another's findings—a scientific system of checks and balances. This is what is known as the peer review process, and it is an integral part of what makes science such a unique human endeavor: Peer review leads to self-correction. Science continually revises and updates its own (provisional) conclusions. According to the apt metaphor proposed by John Casti (in *Paradigms Lost*[22]), the mechanism of science is an "invisible boot" that kicks out ideas that do not work, even though individual scientists may be less objective or dispassionate than we would like them to be. The result is a fascinating search for better and more sophisticated explanations of reality, a search that so far has met with a degree of success that is simply unparalleled by any alternative method we have tried (including revealed religion, mysticism, and pseudoscience).

My earlier statement that the only reason we should use the scientific method is that it works may seem quite a weak link in the scientific edifice; therefore, it deserves further discussion. Many people have attempted to justify (or attack) the scientific method on external, usually philosophical, grounds. The idea is that if we want to invest such a large amount of labor and money on a particular kind of activity, we should know *why* it works. This sounds quite reasonable, except that there is no way to know why it works because there is no external, independent criterion of evaluation. Nobody has invented any independent and reliable way to understand reality, a way that is itself not a part of the reality we must try to understand (i.e., there is no such thing as what philosophers call a *God's eye view* available to humans). So although we can compare the effectiveness of science with that of other competing endeavors, there is no universal yardstick we can use to judge how any method of inquiry fares with the *real* truth. In fact, there is good reason to think that humankind will never gain such a universal reference point, because we would have to develop a complete knowledge of the world in order to construct the yardstick.

It should be clear, however, that the fact that the criteria of a practice emerge from the practice itself is quite common and legitimate, and such self-generated criteria are not in any sense fatal to the practice of science. On the contrary, the development of evaluative criteria in this way from within science is no different from the development of criteria from within any other practice, as an analogy will show. Consider a seemingly trivial, but actually quite relevant example. Shoemaking, being a competitive economic activity, requires evaluative criteria. How do shoemakers develop the criteria by which to judge the quali-

---

[22] J. Casti, *Paradigms Lost: Images of Man in the Mirror of Science* (New York: Morrow, 1989).

ty of their product relative to that of other shoemakers? They do so by paying attention to all of the factors they can discern that are, first, relevant to the *wearing* of shoes and to the basic need for them. Shoes are needed as protection for the feet from the numerous dangers they encounter while walking—sharp objects, hot surfaces, and atmospheric conditions such as cold. And, of course, wearing shoes affects one's mobility, so they must fit well enough that they themselves do not impede the mobility of the wearer. Thus the criteria for making shoes emerge from the process of shoemaking itself, on the basis of the experiences of the wearers of shoes.

Whatever ideas, methods, technologies, and so on produce the most comfortable, sufficiently protective footwear (which need not exclude aesthetic considerations) become the basis for evaluative criteria for judging the footwear. Similarly, when science takes on the tasks of explaining the natural world and solving practical problems, the needs of explanation and technology, as well as the successes and failures encountered during the process, themselves point the way to the criteria by which the success of science can be judged. There is nothing viciously circular about this. It is simply a rational way to go about solving any problem, and people in many different vocations and intellectual disciplines do it every day.

Why, then, are we so confident that science is by far the best tool developed thus far to uncover the mysteries of the world? Again, simply because it works far better *in practice* than any other method does. Science may be judged by its pragmatic success (e.g., its success at producing technological advancements) and by the adequacy of the explanations it yields (pretty much like successful shoemaking). To illustrate using another discipline, economists have produced one theory after another to explain why one type of economy (e.g., managed capitalism) should work better or worse than another (e.g., communism). However, there are many ways to construct such arguments, and one can cleverly defend one solution or another on theoretical grounds alone. Ultimately, all these discussions are made impotent by the fact that the most successful economic system will survive and be emulated in more countries. That Soviet-style communism was a bad theory might be difficult to demonstrate theoretically beyond any reasonable doubt. But the fact that the Soviet economy collapsed is in plain view for all to see (although it may have happened because of additional factors as well). Along the same lines, there is no a priori reason why revealed religion, pseudoscience, or philosophical discourse should not be equally successful or even more successful than science at understanding and controlling the natural world. But in practice they have not been, and therefore the burden of proof is on those who claim that these disciplines present a viable alternative to science.

Before moving on, let me clarify one more common source of misunderstanding about the nature of science: the differences among facts, hypotheses, theories, and laws. A *fact* in science is an observation, or the result of an experimental manipulation. Although facts can provide confirmation or refutation of a theory, a simple accumulation of facts does not automatically yield scientific knowledge. Facts need to be interpreted, and this interpretation is performed in the context of a framework of understanding, whether it is a generally accepted theory or a new theory in the making. As such, the interpretations remain provisional and subject to revision (which means that facts are not purely dispassionate and objective entities, a situation philosophers of science refer to as the *theory-ladenness of observation*). Facts are best thought of as continuously interacting with theories, where the current theory determines what kinds of observations to make and experiments to conduct (i.e., what kinds of facts to record). The theory in turn is retained or abandoned because of the collection of facts that it helped generate.

A *law* is a general statement reflecting the expectation that certain patterns of events will always occur if and whenever certain conditions are met, other things being equal. For example, the law of gravity says that two bodies will always attract each other in proportion to their masses and in inverse relation to the square of their distances. This does not tell us *why* bodies behave that way (for that, we need a theory of gravity, which is provided by Einstein's general theory of relativity, where gravity *is* a deformation of space–time). But we have never observed an exception to this rule, and it is therefore codified as a law. Whereas in biology there are few if any laws, in physics they are abundant. This difference may be due to the fact that biological systems are so complex (compared to atoms or even planets and stars) that it is impossible to have anything always happening under exactly the same conditions. That is why mathematical biology tends to be based much more on statistical approximations than on the kind of analytical, precise solutions typical of physics.[23]

A *hypothesis* is a human mental construct that is used to provide a reasonable, preliminary causal explanation of a set of facts. The usefulness of a hypothesis is measured by how well it can predict new facts that can be used to confirm or invalidate it.

A *theory* is a more mature, more complex, and more wide-ranging human mental construct than a hypothesis. It also is produced to provide a causal explanation of the world and to predict its future behavior, but theories span a much larger array of phenomena than simple hypotheses do (in fact, they include a variety of hypotheses as their building blocks). For example, in the early stages of the history of physics, scientists hypothesized the presence of ether as a medium through which light could travel because at the time it was inconceivable that anything could move in empty space. That hypothesis has now been abandoned because we have a more comprehensive theory of electromagnetism, which explains many properties of light, including the fact that it can travel in a vacuum.

---

[23] Notice, however, that subatomic physics also incorporates statistical principles, such as Heisenberg's indeterminacy principle. But this is true for entirely different reasons that are beautifully explained in Casti's book cited earlier (see note 22).

Science as a Bayesian Algorithm

There is another way to think about science and its power and limits that is gaining favor among scientists and philosophers alike, and that I think beautifully clears up many common misunderstandings that both scientists and the general public seem to have about the nature of science—the same misunderstandings that lead people to either scientism or antiscience. I am referring to an old and, until recently, rather obscure way to think about probability that was invented by Reverend Thomas Bayes back in 1763.

Bayes realized that when we consider the probability that a hypothesis is true, we base our judgment on our previous knowledge about the phenomenon under study (i.e., we use induction). We then assess new information in the light of this prior probability and modify our belief (meant as degree of confidence, not as blind faith) in the hypothesis based on the new information. This process can be repeated indefinitely, so that the degree of trust we have in any hypothesis is always due to the current (and ever changing) balance between what we knew before and the new knowledge that additional data contribute.

If the reader will indulge me, I would rather express Bayes's ideas in a simple mathematical formula. This is one equation that is easy to understand and that might change how you think not just about the creation–evolution issue, but about anything else you believe or don't believe. Here is the equation, known as Bayes's rule:

$$P(H \mid D) = \frac{P(D \mid H) \times P(H)}{[P(D \mid H) \times P(H)] + [P(D \mid \sim H) \times P(\sim H)]}$$

Where $P(H \mid D)$ (which reads, "the probability of H, given D") is the probability that the hypothesis is correct, given the available data; $P(D \mid H)$ is the probability that the data will be observed, given the hypothesis; $P(H)$ is the unconditional probability of the hypothesis (i.e., its probability before we knew of the new data); and the denominator is a product of the numerator plus an equivalent term that includes the probability to observe the data if the hypothesis is actually wrong (or, in the case of multiple hypotheses, the probabilities of observing the data if each additional hypothesis is correct; the ~ symbol stands for *not*). The denominator of the right-hand side of the equation is also known as the *likelihood* of all hypotheses being considered. The left-hand side of the equation is called the *posterior probability* of the hypothesis in question; the left-hand part of the numerator on the right side of the equation, $P(D \mid H)$, is known as the *conditional likelihood* of the hypothesis in question; and the right-hand part of the same numerator, $P(H)$, is called the *prior probability* of the hypothesis being considered. This sounds very complicated until we examine a particular example, so bear with me a few more minutes.[24]

---

[24] The example about to be presented is used in a nice explanation of Bayes's rule at http://library.thinkquest.org/10030/4pceapbr.htm?tqskip1=1&tqtime=0106.

A family has plans to go fishing on a Sunday afternoon, but their plans depend on the weather at noon on Sunday: If it is sunny, there is a 90 percent chance they will go fishing; if it is cloudy, the probability that they will go fishing drops to 50 percent; and if it is raining, the chance drops to 15 percent. The weather prediction at the point when we first consider the situation calls for a 10 percent chance of rain, a 25 percent chance of clouds, and a 65 percent chance of sunshine. The question is this: Given that we know the family eventually did go fishing, was the weather sunny, cloudy, or rainy? You will probably have your intuitions about this, and they may well be correct. But science goes beyond intuition to empirically based reasoning. Here is how Bayes would solve the problem:

1. Let's plug our preliminary assessment of the situation into Bayes's rule:
   - The probability of fishing (F) if it is sunny (S), $P(\mathrm{F} \mid \mathrm{S}) = 0.90$.
   - The probability of fishing if it is cloudy (C), $P(\mathrm{F} \mid \mathrm{C}) = 0.50$.
   - The probability of fishing if it is rainy (R), $P(\mathrm{F} \mid \mathrm{R}) = 0.15$.
2. The probability of each kind of weather, given the predictions of the weather report, can be summarized as follows:
   - The probability of sunny weather, $P(S) = 0.65$.
   - The probability of cloudy weather, $P(C) = 0.25$.
   - The probability of rainy weather, $P(R) = 0.10$.
3. Notice that the sum of the probabilities of each weather condition is 1 (i.e., 100 percent)—$P(S) + P(C) + P(R) = 0.65 + 0.25 + 0.10 = 1.00$—and that these hypotheses are mutually exclusive (in the sense that it was either sunny or cloudy or rainy, but not a combination of these).
4. The overall likelihood of going fishing (the denominator of the right side of Bayes's rule), $P(F)$ is 0.725: $[P(F \mid S) \times P(S)] + [P(F \mid C) \times P(C)] + [P(F \mid R) \times P(R)] = (0.90 \times 0.65) + (0.50 \times 0.25) + (0.15 \times 0.10)$.
5. We can now obtain the new conclusions about our hypotheses on the weather, given the prior and new information (the latter being that the family *did* go fishing). So given that the family went fishing, according to Bayes's rule,
   - The probability that the weather was sunny is as follows: $P(S \mid F) = [P(F \mid S) \times P(S)] / P(F) = (0.90 \times 0.65) / 0.725 = 0.807$
   - The probability that the weather was cloudy is as follows: $P(C \mid F) = [P(F \mid C) \times P(C)] / P(F) = (0.50 \times 0.25) / 0.725 = 0.172$
   - The probability that the weather was rainy is as follows: $P(R \mid F) = [P(F \mid R) \times P(R)] / P(F) = (0.15 \times 0.10) / 0.725 = 0.021$
6. Finally, note that $P(S \mid F) + P(C \mid F) + P(R \mid F) = 0.807 + 0.172 + 0.021 = 1.00$, because one of the hypotheses *must* be true (it was sunny or cloudy or rainy; no other possibilities are in the mix).

Bayes's rule, therefore, tells us that—given the prior knowledge of the situation we had and the new information that the family did go fishing—the likelihood that the weather was sunny was the highest. Well, you could have guessed that, no? Yes, in this simple case. But notice that Bayes's rule gives you additional information: First, it tells you what the *best* available estimate of the probabilities of all three hypotheses is. Consequently, it tells you how confident you can be that the weather was sunny (which is better than simply saying "it's more likely"). Also it is clear from the equations that the probability of the hypothesis that the weather was sunny *increased* with the new information (from 0.65 to 0.807). Finally, Bayes's theorem reminds us that our degree of confidence in any hypothesis is never either 0 or 100 percent, although it can come very close to those extremes.

Bayesian statistical analysis is a good metaphor (some philosophers of science would say a good *description*) of how science really works. More, it is a good description of how *any* logical inquiry into the world goes if it is based on a combination of hypotheses and data. The scientist (and in general the rationally thinking person) is always evaluating several hypotheses on the basis of her previous understanding and knowledge on the one hand and on new information gathered by observation or experiment on the other hand. Her judgment of the validity of a theory therefore changes constantly, although very rarely does it do so in a dramatic fashion. Bayesian analysis thus clearly shows why the extremes of antiscience and scientism are naïve positions: They correspond, respectively, to having certainty of the hypotheses that science never works (creationism) or always works (scientism). But attaching a probability of 0 or 1 to a given hypothesis, as the Bayesian framework makes clear, is the same as saying that our conclusions are true *no matter what the data say*; that is, we take them on faith.

Popularly Believed Things about Science That Are (Mostly) Not True

Now that we have a better understanding of the historical development, current practice, and philosophical underpinnings of science, it is useful to examine what sorts of misconceptions people harbor about science—misconceptions that are at the base of creationist critiques of evolutionary theory and of some scientistic responses to them.

Several myths about science are common among the general public and even among science students at the college level. Unfortunately, some scientists are in part responsible for the propagation of these myths, although philosophers and sociologists of science, as well as the media, certainly share the burden of responsibility.[25]

Perhaps one of the most pernicious myths about science is the belief that its practice leads to the Truth, a position maintained even by some scientists and the basis of what I have been referring to as scientism. As should be clear from my presentation of the scientific method, nothing could be, well, further

[25] An in-depth discussion of the most common misunderstandings of the nature of science can be found in a thorough article by W. F. McComas, "15 Myths of Science," *Skeptic* 5(2) (1997): 88–95.

from the truth. Since science is a method, not a body of results, the very statement that it leads to truth is misleading and misses the point of what a scientist's activity is all about.

Also, as I have already pointed out, by the nature of its methods science can reach only provisional, probabilistic conclusions, not absolute and immovable truths (see my discussion of Bayes's theorem in the preceding section). The parallel with the activities of Sherlock Holmes again comes in handy. A detective has finished her job (at least temporarily) when the evidence is sufficient to convict the suspect *beyond reasonable doubt.* That does *not* mean that anybody (detective, judge, or jury) is absolutely certain that the suspect is indeed the culprit. (The criterion for determining guilt is not, as sometimes is popularly thought, that a person must be determined guilty *beyond the shadow of a doubt.*) In that sense our expectations in matters of law are somehow much more realistic than our expectations in matters of science. Everybody understands quite well that we cannot keep deliberating until we are 100 percent certain of the guilt of somebody before convicting. That degree of certainty will never come, not even with a confession (since, as I pointed out earlier, there is always the possibility that the confession has been extorted by pressure, or it is made to cover somebody else dear to the accused). It is not clear why science is held to a much higher (in fact, impossible) standard.[26]

Also notice that in the same way that scientists *assume* there is a physical, objective reality out there, so Sherlock Holmes assumes that somebody did perpetrate the crime. Neither we nor Holmes *knows* that our assumptions are definitely true, but they make good sense, given our prior experience of moving around in the world and our familiarity with crime situations. The real question is not why science cannot get to the Truth, but whether any other known means of investigation can get further than science.

A second (related) myth about science is that it can prove things. Let me be clear about this: Neither science nor any other known human process can *prove* truths about the natural world (as opposed to some logical truths). In logic and in mathematics, the word *prove* has a special meaning that is confined to the kind of proof we get in deductive reasoning, in which we prove our conclusion by showing that it necessarily follows from our premises. In science, the best we can do in any logically conclusive sense is to *dis*prove something. Although Holmes can never irrefutably prove logically that Dr. Moriarty committed the crime, he can prove conclusively (logically)—given enough evidence—that somebody else did *not* do it.

Similarly, if a biologist hypothesizes that humans and dinosaurs never lived together, the hypothesis can be conclusively refuted by the discovery of di-

---

[26] It is highly ironic, I think, that creationist and lawyer Phillip Johnson has often remarked that the evidence for evolution is so bad that it would fail the legal test of being acceptable beyond reasonable doubt. On the contrary, I think any jury with an understanding of the evidence would have no doubt whatsoever in accepting evolution. The situation is quite different for creationism, including intelligent design: Those "theories" really do not come close to meeting the legal criterion.

nosaur and human prints in the same geological strata. But one cannot prove absolutely that these two kinds of animals did not live concurrently, because to do that one would have to have a complete fossil record of every animal at every instant and place in the history of Earth to make sure that dinosaurs and humans never showed up together anywhere at any time. Proving a negative statement is a much more difficult task than proving a positive. Even concerning positive statements, however, a definitive conclusion can never be reached (although science can accumulate evidence in their favor), simply because it is never possible to obtain *all* of the empirical evidence one would need.

For example, how do we *prove* Einstein's theory of relativity? We know that many experiments have been done that support the theory's predictions, and that is why the theory is currently accepted as valid. In principle, however, other theories might explain equally well the same evidence but simply not yet have occurred to any physicist's mind. This is not only possible; it has already happened: Einstein's theory explains all the data once explained by Newton's theory, and more. Perhaps two centuries down the line physics students will spend their introductory courses on Newton and Einstein, and then concentrate on the then current understanding of the world in which both have been superseded by the work of yet another physicist.

Notice that this does not mean that we can put no confidence in what science says. Science is a progressive, cumulative enterprise; that is, it builds on previous knowledge and theory to gain new knowledge and produce new, better theories. Better at what? At both explaining and predicting reality. Remember that the explaining by itself is not enough because many alternative hypotheses can explain the same facts, but only a small subset of these can actually predict what will happen under given circumstances. That is, only some theories actually work (and they do so within certain limits). As I will discuss at the end of this chapter, scientists' new ideas are not like new fashions or a new taste in art. As much as there is an element of subjectivity in science, the process can be likened—as Einstein did—to climbing a high mountain: The more we climb, the better our view of the entire landscape is. Every time a new view of the world (a new "paradigm," in the words of philosopher of science Thomas Kuhn[27]) takes hold of a scientific discipline, it is the result of continuing the climbing expeditions made by many other people. Other human endeavors such as art, by contrast, build an entirely new edifice—so to speak—on the dust generated by the demolition of the previous one.

A third pervasive misunderstanding about science is that it can be done only whenever direct experimental manipulation can be carried out. Earlier I pointed out that Galilei's contribution to the scientific method (the addition of experimental manipulation) was indeed fundamental, but it is not a necessary component of the method itself. One can think of experimentation as a very im-

[27] See Kuhn's classic book *The Structure of Scientific Revolutions* (Chicago: University of Chicago Press, 1962).

portant tool in the toolbox available to the scientist. But as Holmes could do his job even without a forensic lab or a computer, so scientists can investigate phenomena that are not amenable to manipulation (although this limitation obviously makes the task more difficult). Many examples will come to mind upon a moment's reflection. Geology and astronomy are mostly historical sciences; that is, we cannot really put stars and earthquakes into a laboratory and dissect them at will. Nevertheless, geologists are much more successful at predicting where to find water or oil under the surface than practitioners of dowsing are, and astronomers can predict eclipses while astrologers cannot. Again, success is the only valid measure of value here. Even in historical research we can gather quantitative data and test hypotheses, regardless of the obvious fact that we cannot conduct experiments.

Frank Sulloway's book *Born to Rebel*[28] provides a beautiful example of applying the scientific approach to historical questions, thereby making history amenable to more objective investigations.[29] Sulloway tackled the question of character differences between first- and last-born children within a family. The hypothesis had been floating around for a while that later-born children tend to be rebellious, while firstborns have a tendency to be conservative in their political and social attitudes. Instead of proceeding to argue in favor of or against the hypothesis by listing a series of handpicked anecdotes[30]—as is common practice in history and some social sciences—Sulloway did the hard work of collecting detailed historical information on 6,000 individuals and asking colleagues to rank their character attributes while being unaware of variables such as gender, social status, or birth order (i.e., he conducted a double-blind study).

Sulloway was then able to employ sophisticated statistical analyses to test the hypothesis of the influence of birth order on character and open-mindedness and found that, indeed, the major factor explaining variation in the tendency to be rebellious and to accept new ideas is birth order. The importance of birth order is, surprisingly, larger than that of gender and socioeconomic status (although the latter two are also statistically significant). One of the ideas whose acceptance was studied by Sulloway was the theory of evolution. Figure 4.2 shows that later-borns are much more likely to believe in the Darwinian theory than firstborns are, and that within each group younger people are more open-minded than older ones.[31]

---

[28] F. J. Sulloway, *Born to Rebel: Birth Order, Family Dynamics, and Creative Lives* (New York: Pantheon, 1996).

[29] See also M. Leier, "Can We Know What Really Happened in the Past?" *Skeptic* 4(4) (1996): 74–77; as well as Michael Shermer's *Why People Believe Weird Things: Pseudoscience, Superstition, and Other Confusions of Our Time* (New York: W. H. Freeman, 1997), particularly the chapters about Holocaust denial.

[30] Such an approach is really a very premature or incomplete form of induction.

[31] For the curious reader, Sulloway provides a test to estimate your own likelihood of being open-minded about new ideas. I took the test, and it turns out that—at age 34—I had 81 percent probability of accepting new ideas, which is fairly open-minded for a skeptic.

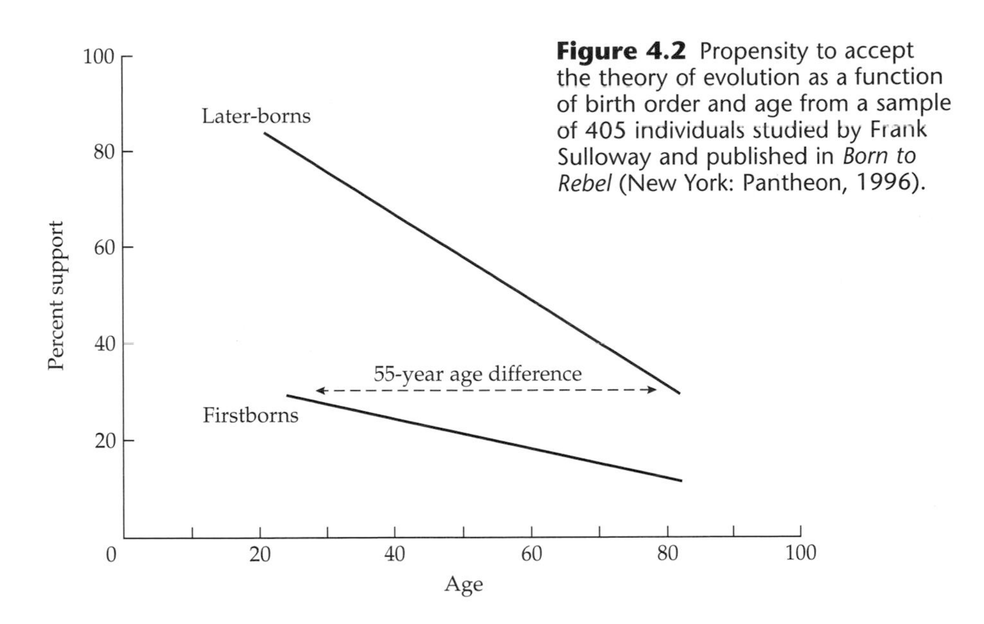

Figure 4.2 Propensity to accept the theory of evolution as a function of birth order and age from a sample of 405 individuals studied by Frank Sulloway and published in *Born to Rebel* (New York: Pantheon, 1996).

The essence of science is the testability of hypotheses, regardless of which specific methods can be applied to test them. As long as it is possible to gather data that will support or reject a stated hypothesis, so that our conclusions do not depend simply on anecdotes, we are doing science. In fact, anecdotal evidence should *never* be used for anything but to form preliminary hypotheses to be tested more rigorously. This is one elementary principle of investigation that seems to be hard for politicians to understand. Whenever a politician wants to promote a particular piece of legislation (regardless of partisanship), the best strategy is to collect powerfully emotional stories to convince the public and other politicians that the proposed legislation is the only hope to improve the situation. How many times have we heard of abuses of the welfare system (usually the political right complains here) or of environmentally irresponsible behavior on the part of big business (usually the political left's turn to complain)? Of course, it is almost always possible to find examples that fit one's view of the world or political agenda. The real question is this, How relevant are these examples compared to the situation at large? We should indeed reform welfare or regulate corporations more strictly *if* it turns out that the horror stories are representative of the majority of instances and are having a significant impact. Curiously, politicians and their staff hardly ever carry out this kind of research, which is so natural to a scientist.[32]

A fourth important point to clear up is the question of how scientists check on each other's results. The popular explanation (fueled by some scientists and

[32] One reason is that gathering actual data, as opposed to anecdotes, is time-consuming, costly, and exceedingly boring—which, by the way, summarizes pretty well the everyday activities of a scientist.

philosophers of science) is that experiments are replicated over and over by independent laboratories until new findings are confirmed or rejected. This is pretty far from the truth. For one thing, there simply are not enough scientists, laboratories, and especially funding for doing anything even remotely close to what the myth maintains. Second, and perhaps more importantly, there is no payoff for a scientist to simply repeat *exactly* what someone else has done. Should such an attempt succeed, the researcher who published the first results would be glorified (e.g., with a Nobel Prize), and everybody else would go down in history as a footnote under "independent confirmation." Despite what I just said, every scientific paper does include a detailed section called "Materials and Methods," in which as many details of the experimental procedure are reported as possible. This section is meant to disclose exactly how the results were obtained, so that *in principle* it would be possible to repeat them. Furthermore, by carefully examining the materials and methods section, critics can point to potential faults of the experimental result that may provide explanations alternative to the ones cherished by the authors of the report.

So who checks on whom? The checking is indirect and, indeed, much more fascinating than a simple repetition of already obtained results. The basic idea is that if a hypothesis or theory does shed new light on the underlying reality of a phenomenon, similar lights shown from different angles should illuminate the same reality. This should translate into a consistent pattern of results, even though the experiments were performed in different ways, tackling different aspects of the problem. If Sherlock Holmes is on the right track in identifying a suspect, Inspector Lestrade *might* be able to apprehend the same suspect on the basis of partially different clues (assuming he is as clever as Holmes—which of course he isn't, for narrative purposes). It is this consilience of evidence, as discussed earlier, that represents the most powerful reason to trust in science as a method to uncover reality.

Incidentally, this way of confirming results provides a powerful incentive to "repeat" what others have done. If a scientist confirms someone else's results on the basis of a different but congruent approach, the Nobel committee is likely to split the prize, and the scientific community is likely to reward the ingenuity of both researchers. Let us not forget that scientists are human beings, and they are motivated by a host of human impulses (including the pursuit of cash prizes or immortal glory), not just the pure search for truth.

A final, powerful myth about science is widespread among scientists themselves, and I am afraid it will be very difficult to eradicate. It is nevertheless, very insidious, because the creationists make one of their strongest and most novel points by exploiting it. The myth was clearly formulated by none other than Newton himself, when he said, "*Hypotheses non fingo*" ("I do not fake hypotheses"). He really meant *theories,* in the modern sense of the term. Sir Isaac was referring to the fact that his law of gravity is independent of what actually *causes* gravity. The law is valid because it is a generalization of empirically derived results. Newton did not have a theory to explain *why* gravity works that way. As I mentioned earlier, we had to wait until Einstein to have an acceptable theory of gravity. However, the implication of Newton's infamous sentence is not that *he* was unable to provide a theory of gravity, but that doing so is not a scientist's business and is more akin to philosophical speculation. This is entirely wrong, and in fact Newton himself did not follow his own precepts, producing an entirely speculative theory of light (which, incidentally, proved extremely successful at explaining the phenomenon of light).

There is a better way to reformulate Newton's disdain for speculation. This is what I referred to earlier as Occam's razor. What should really be avoided at all costs is an unnecessary multiplication of hypotheses and assumptions. If we can explain things with a simpler, less speculative set of axioms and assumptions, that is the way to go, until and if a more complex theoretical construct is made necessary by new data.

Let us take the example of evolution. As I have been arguing throughout this book, evolutionary theory is in direct contradiction with a literal interpretation of the Bible. This, however, does not mean that evolution denies any and all forms of the existence of God. Many scientists are indeed believers in some form of divinity, and there is nothing in biology that *directly* contradicts them (e.g., God could have started the evolutionary process by originating life, and then let it operate by mutation and natural selection). But such a position, while personally deserving of utmost respect, is indeed nonscientific. The scientifically justified conclusion would be that, until we need to incorporate God into the scenario because the evidence points that way, we *provisionally* exclude a deity from consideration.

This conclusion is neatly summarized by an apocryphal quote attributed to Pierre-Simon de Laplace, the astronomer who first proposed a modern theory of the origin of the solar system. The story goes that Laplace explained his theory to his former pupil, the future emperor Napoleon (something that did historically occur). Napoleon then supposedly asked about the role of God. Laplace's legendary answer was, "I do not need that hypothesis anymore." That is, since we can explain the origin of the solar system without God, the intervention of a divinity is ruled out because it is unnecessary; it serves no useful explanatory purpose.[33] This, of course, is not simply suspending judgment; rather, it is making a provisional *negative* judgment as far as the particular problem at hand (the origin of the solar system) is concerned—namely, that supernatural phenomena played no scientifically discernible role in the natural phenomena that science is investigating. This judgment is perfectly consistent with the scientific method.[34]

---

[33] Furthermore, the invocation of a supernatural entity introduces the need for verification of that entity's existence, which is a real metaphysical problem.

[34] Notice the subtle but important point that Laplace would not claim to have demonstrated the nonexistence of God, but simply to have made a good case for the lack of *necessity* of a divine intervention in the formation of the solar system.

It is rather unsettling that many scientists perceive this judgment as "forcing" science on a turf not its own.[35] No scientist would ever accept from a student or from a colleague a divine explanation to account for the results of an experiment. This is not because it couldn't be that way, but because, as Carl Sagan put it while paraphrasing philosopher David Hume, extraordinary claims warrant extraordinary evidence. If the evidence is not there, a *provisional* negative answer is the only *scientifically* valid option.

The Limits of Science

No discussion of science and its methods would be fair and complete without a consideration of two more points: the limits of science and the alternatives available to the human quest for knowledge other than science. A discussion of these two points will necessarily lead us to a brief consideration of the concept of reality itself.

First and foremost, **science is not about final (ultimate) answers.** As we have seen, the scientific method can *support* certain conclusions and provide evidence against others. But all such efforts are always seen as provisional, always subject to revision. This, however, does not mean that anything goes, contrary to the typical reaction from creationists whenever I use this disclaimer in public lectures ("Ah, but if you cannot provide final answers, then . . ."). Even in religion, there are no truly ultimate answers. One can legitimately ask about the origin of God, for example, and so on ad infinitum. The only truly ultimate answer would be one regarding which, once obtained, there would be not even the logical possibility of asking another question, so one stops asking questions when it makes sense to do so. And for claims that are based on the assumption of something's existence, it makes sense to stop asking questions when, given all the cognitive and technological means available to us, there is not even the theoretical possibility of obtaining the data that would answer the question. Therefore, the fact that science provides *provisional* answers simply means that we can attach probabilities, likelihoods (in a Bayesian sense), to theories and conclusions. This is a very powerful outcome of scientific analyses. Even though you cannot say for certain that the sun will not stop shining tomorrow, astronomical research suggests that it will keep doing so for a few billion more years. And the likelihood of this actually happening is, well, astronomically high (in fact, you could make your own calculation of such likelihood using Bayes's theorem).

Many people (including some scientistically oriented scientists) find this lack of absolutes in science frightening and unsettling, and understandably so. It seems that humans need to feel in control over their world, and to gather an-

---

[35] See, for example, E. Scott, "Two Kinds of Materialism," *Free Inquiry* 18 (1998): 20; but compare with R. Dawkins, "When Religion Steps on Science's Turf" *Free Inquiry* 18 (1998): 18–19, for a different opinion.

swers about it is one way to increase one's sense of control.[36] As fundamentalist Danny Phillips once put it, "Science changes every day, and for us starting believing in something that changes every day, over something that's stood fast for over 3000 years, that's hideous, and we don't need to do that."[37] Phillips does indeed have a very good point, to which I shall return in the next section.

Science cannot draw conclusions about things it cannot measure or manipulate experimentally. That is why science is a poor tool to decide beauty contests or to make aesthetic judgments on artistic matters. Nevertheless, these other things are indubitably real. As I will discuss shortly, there is a limited sense in which science can provide insights into morals, art, and emotions, but it certainly is not to the degree and in the same sense in which scientists can tackle the fundamentals of matter or the molecular structure of living organisms.

Science and scientists are committed to unemotional judgments by their choice to understand things rationally. This "limit" needs further clarification. Yes, science as an activity is *supposed* to be detached and devoid of emotional content. This, however, is hardly a limitation when the objective is to understand, not to empathize. Furthermore, scientists are certainly not unemotional. Not only can they become personally attached to their theories or objects of study (which can hinder the progress of science), but obviously they also share the same basic emotions of any other human being. We fall in love, get married, love our children, cry at the movies, become silly in good company, and deeply mourn the loss not only of friends and relatives, but also of dogs and sometimes even cats.

Another sense in which this limitation is held against science is to suggest that knowledge somehow "spoils" the "true" appreciation of the natural world itself. The poet accuses the scientist of being unable to feel the beauty of the moon high in the sky because of the "cold" knowledge that it is a planet made of rocks and scarred by craters. As Isaac Asimov splendidly put it in Martin Gardner's collection *Great Essays in Science,* this is simply nonsense.[38] If anything, a scientist has arguably a *heightened* aesthetic sense, being perfectly capable of fully feeling and enjoying a romantic moment, and also (not necessarily simultaneously) of thinking about the nature of the reality underlying that moment. Understanding does not dampen, but can deepen, aesthetic appreciation. Understanding the physical processes that produce the beautiful shapes of snowflakes makes them no less delightful.[39]

---

[36] It is interesting to note that split-brain experiments show that the human mind's striving for answers is so powerful that it will *make them up* if it cannot gather them otherwise. See M. S. Gazzaniga, "The Split-Brain Revisited," *Scientific American* 279 (July 1998): 50–55, and Chapter 8 in this book.

[37] On KMRT-TV, Rocky Mountains, in a program entitled *The Jesus Train.*

[38] M. Gardiner (ed.), *Great Essays in Science* (Buffalo, NY: Prometheus, 1994). See also Richard Dawkins's *Unweaving the Rainbow* (see note 7).

[39] The same point can also be made by consideration of the converse—namely, that understanding the cause and prevention of smallpox has not lessened our fear and loathing of the disease itself. Emotional responses still accompany scientific understanding.

Science is a limited, bounded enterprise. This claim is rarely made, but J. Horgan has skillfully elaborated on it.[40] The basic premise is that science is about uncovering the fundamental mechanisms underlying the functioning of the world, not about accumulating detailed knowledge on every single occurrence in the world. Horgan argues that although occurrences may be infinite, there are only a limited number of fundamental principles to be understood. Indeed, the very essence of the scientific enterprise is that we can generalize to larger and larger degrees, and collapse specific principles into ever more encompassing ones. The Holy Grail of modern physics, for example, is the "theory of everything," in which one fundamental law, one basic equation, can be used to derive all the others as special cases.[41] Einstein worked the last years of his life toward achieving such a goal, and so far there is no reason to believe that it is not achievable in principle.

Horgan, I think, has a very good point. However, he seems to imply that this decline of science has already started, and that we will soon witness the disappearance of science (and consequently of funding opportunities) from the center of human attention, as soon as scientists have answered all the basic questions about the universe. This, as John Casti pointed out,[42] may be akin to Mark Twain's famous warning that rumors of his death were a bit premature. Twain eventually died, of course, but not quite as soon as some of his detractors may have wished. The same is very likely true for science. The rapid progress made in the last couple of centuries notwithstanding, there are plenty of "basic" questions regarding which we have barely scratched the surface. Just think of the origin of life, the formation and ultimate fate of the universe, and the basis of human consciousness to realize how deep a divide still stands between present science as an exciting enterprise and future science as a boring rehearsal of what everybody already knows.

The Alternatives to Science

If science has limits, either provisional or absolute, what alternatives can we turn to in order to better understand reality? An exhaustive discussion of this topic would probably take a whole book, but a few points need to be made so that we can better follow the evolution–creation debate.

First, it should be obvious (but apparently it isn't to many creationists) that if we do not have an explanation for a phenomenon or a collection of facts, this does not automatically imply that the intervention of a supernatural force provides the correct account. It simply means that we currently do not know or do not understand what is going on. We may never understand it, but this still does

[40] See, for example, J. Horgan, "The End of Science," *Skeptic* 4(3) (1996): 26–33.

[41] On the quest for a unified theory of physics, see S. Weinberg, *Dreams of a Final Theory* (New York: Pantheon Books, 1992); and B. Greene, *The Elegant Universe: Superstrings, Hidden Dimensions, and the Quest for the Ultimate Theory* (New York: Norton, 1999).

[42] In J. Casti, "Dreams of the End of Science," *Skeptic* 4(3) (1996): 34–35.

not mean that anything more than natural laws are at work. More generally, the absence of a solution within a given explanatory framework (e.g., science) is no point in favor of any particular alternative framework. If an alternative *method* of explaining the universe is valid, it has to stand on its own merits, for it cannot be adopted simply by default when the first option does not work.

One of the most ancient and intellectually respected alternative forms of inquiry is represented by *philosophy.* In fact, science originated as a branch of philosophy, known as *natural philosophy,* and as we have seen, philosophy still has much to say about science itself and the nature of the scientific method. One of the main tools of philosophers is the use of logic (although they most certainly do not ignore the importance of observation and other modes of experience). They construct and explore the consequences of various logically consistent views of the world; internal logical consistency is a primary philosophical criterion. However, there is a problem with any philosophical approach that emphasizes logical consistency at the expense of empirical observation: Although it is likely that there is only one commonly experienced reality, there are infinite logically consistent ways of explaining it. If logic is our primary guide, then it will be difficult to identify which sets of explanations best approximate the real world.

If the reader has ever taken a course in philosophy, perhaps she will have experienced my own mix of excitement and frustration at studying the major philosophers (which usually means the ones characterizing the history of Western philosophical thought). One spends a lot of time understanding Plato's arguments, only to find out that Epicurus's are equally convincing, yet markedly different. And the same goes for Aristotle, Descartes, Locke, Hume, Kant, Marx, Schopenhauer, and all the rest. Certainly, some of the systems of the world proposed by these thinkers are better than others, in the sense that they are more complete or more internally consistent. But which one is *real?* To answer that question, we would need to confront a given philosophical system with data from the outside world. But when we put together logic and evidence, we come close to reinventing the scientific method and moving progressively from philosophy to science.[43]

A more radical alternative, of course, is the total abandonment of logic. This approach has been proposed by various philosophers (from St. Augustine to Paul Feyerabend), and especially by *mystics* of all times and places (Jesus, Muhammad, and to some extent Confucius and Buddha). And why not? If one tool (science or philosophy) does not work, it is perfectly legitimate to try out

[43] The dividing line between philosophy and science might not be as clearly marked as is often thought. The American pragmatic naturalists, for example, based their arguments—indeed, their entire worldview—on the findings of modern science, but they were not doing science. They constructed arguments about other things—for example, art or education—by using logic, certainly, but also by taking into account evidence provided by scientific disciplines such as psychology. John Dewey was notable for this. He wrote a book, in fact, called *Experience and Nature,* that exemplifies this approach. Dewey argued that the processes of reasoning that are most salient in science are really those that people use in many other areas of inquiry because they are common to all rational attempts to solve problems, both intellectual and practical.

another one. However, there are two problems with the abandonment of logic in favor of mysticism. First, doing so seems contradictory to the way the universe actually is when viewed in terms of common practical experience. St. Augustine may have believed that divine illumination and faith were the preludes to understanding God's world, but he certainly was forced, like everyone else, to take the stairs to the ground floor instead of walking out the window. The point is that even though his religious views were strongly mystical, he had to conduct himself consistently with the laws of physics and the realities of corporeal existence like everyone else. Admittedly, there is no a priori reason why the world should behave logically and predictably, but apparently it does.

The rules of logic reflect the relative stability of human experience, which in turn reflects the relative stability of nature itself. The laws of physics do not change at every turn of the stone. Our lives are characterized by continuity, and even random events do not violate Newtonian mechanics or Einstein's relativity. It *could* be otherwise; there is no logically necessary reason to expect absolute and eternal stability in our experience of nature. Our dreams, for example, lack continuity and a logical structure (and it is no coincidence that many Eastern mystic traditions liken life to a dreamlike tapestry). But common experience shows us otherwise in very compelling ways.

The motion picture *Dark City* (1997, directed by Alex Proyas) presents another alternative. In the movie an entire city and its inhabitants radically change their appearance, behavior, and memories, with no continuity from one night to the other, and no apparent logic. It turns out that the city is being manipulated by a group of aliens intent on experimenting with human behavior in order to understand it. As far as we can tell, however, nothing of the sort is going on with *our* reality (unless the aliens are really astute). Therefore, to abandon logic in order to understand a universe that seems very logical is, well, an illogical thing to do.

The second problem with the mystical approach is that it has not worked. Again, we come back to the only ultimate criterion that can more or less objectively determine if a method of inquiry is worthy of our support or not. Mystics and religionists have not discovered a single new thing about the world in thousands of years of attempts. The reason is that mystical explanations are rather vague, and they are *post hoc*—that is, after the fact. Although mystical principles can indeed provide an aesthetically pleasing interpretation of why things are or have been, and they can even provide psychological comfort, they fail to predict how things *will* be, and prediction is the only test we have that enables us to discriminate between adequate and inadequate models of the world. That does not mean that one is not entitled to keep trying, but it does imply that—at the moment—science has a deep edge over mysticism, and things are not likely to change dramatically in the near future.

One of the problems that will face the reader throughout this book is the attempt to distinguish between science and *pseudoscience,* a difficulty that philosopher Karl Popper called the "demarcation problem."[44] This is a general problem affecting us in everyday life. How do I know that the "scientific" claims concerning the latest pill against obesity are actually backed by solid evidence? Does acupuncture really work? Should I consult my horoscope before leaving home? In the case of the creation–evolution debate, an abundance of claims (not necessarily all coming from one side) are unsubstantiated, or at least questionable. Readers should therefore sharpen their sense of critical thinking, turn up their baloney detectors so to speak, and proceed with caution.

John Casti (in *Paradigms Lost;* see note 22) provided an excellent summary of the characteristics of pseudoscience, which I will briefly discuss here as a user-friendly guide to the evolution–creation claims and counterclaims:

- *Anachronistic thinking.* If an argument is based on the wisdom of the ancients (who, remember, knew much less about the world than any junior high school graduate today should), or on the use of outmoded scientific terminology, there is good reason to be suspicious.
- *Search for mysteries.* Whereas the objective of science is to solve mysteries, pseudoscience tends to emphasize the existence and supposed *unsolvability* of mysteries. This is a rather sterile position because if a mystery is by definition insoluble, then why waste one's time thinking about it?
- *Appeal to myths.* This is the idea that ancient myths *must* be based on some kind of real events, which became distorted in the course of oral transmission from generation to generation. Although this can certainly happen, just because some cultures share (usually superficially) similar myths, such commonalities do not imply that the underlying events are the same, or even that they ever happened. An alternative explanation is that human minds tend to work in a similar fashion and therefore provide similar explanations for things they do not understand.
- *Dismissive approach to evidence.* Evidence is the cornerstone that sets aside science from any other human intellectual endeavor, including (to a large extent) philosophy. Given its pivotal role, admissible evidence has to be solid and reliable. If we cite a "fact," we have to be reasonably sure that it indeed corresponds to a verifiable piece of evidence. Hearsay is not admissible. Moreover, solidly researched evidence must, in any reasonable approach, be respected, not distorted, minimized, or ignored.

---

[44] Indeed, Popper proposed his criterion of falsification as a way to distinguish between science and pseudoscience. The idea is that science makes falsifiable predictions, while pseudoscience does not because one can always go back and modify the prediction in an ad hoc fashion so that it fits the facts. Unfortunately for Popper, the demarcation problem is not so simple to resolve, mostly because science itself does not follow what I have termed "naïve" falsificationism. A better way to think about this problem is as a continuum from "hard" sciences such as physics and chemistry (where experimental manipulation is possible), to "soft" ones like biology and geology (where the element of historicity becomes more significant), to proto-scientific disciplines (most of the social sciences, for which overarching theories are often lacking or difficult to support empirically), to clear pseudosciences such as astrology and parapsychology (where not only is the theory unsound when compared to anything else we know about the functioning of the universe, but the empirical evidence clearly rejects the claims of the discipline's practitioners).

- *Appeal to irrefutable hypotheses.* Scientific progress can be made only if a hypothesis is at least potentially open to being disproved. If your hypothesis is not refutable (i.e., falsifiable) *no matter what the evidence,* then it is useless (of course, it may still be true, but there is no way to verify it).
- *Spurious similarities.* A very insidious trap in human thinking is drawing parallels between concepts or phenomena that seem reasonable and that require an in-depth analysis to be verified or discarded. For example, one can draw mystical significance from the fact that one's car plate number is the same as one's home address. But a moment of reflection would easily lead one to conclude that this is simply a coincidence. In general, similarities can yield genuine insights into the matter under consideration, but they require a higher standard of verification than the one provided by a first intuition.
- *Explanation by scenario.* It is pretty easy, if one has just a little bit of imagination, to explain something by telling a story, that is, by imagining a reasonable scenario. Scientists are sometimes guilty of this practice (which is widespread, for example, among evolutionary psychologists who try to explain current human behavior by appealing to speculative behavioral scenarios involving our prehuman or early human ancestors). In fact, scenarios can be useful because they may point the inquiry in the right direction. However, when scenarios remain just-so stories, not backed by data, they are not useful tools because many scenarios can be proposed to explain the same data, but presumably only one is correct.
- *Literary interpretation instead of empirically grounded research.* This occurs when the proponent of a pseudoscientific position claims that statements by scientists are open to alternative, *equally valid* interpretations. This approach treats scientific literature as one might consider a novel or a painting: No one interpretation (not even the one espoused by the author!) is necessarily better than any other. In science, this is a far cry from the reality of things. The more precise and unambiguous scientific statements are, the more useful they are. Ideally, a scientific hypothesis or theory should have one and only one possible interpretation, and this is either correct or not.
- *Refusal to revise.* One of the hallmarks of pseudoscience is the refusal to revise one's own positions in the face of new evidence. No matter how many studies are conducted on the ineffectiveness of astrology, astrologers continue to repeat the same arguments in support of their trade. Science is a process of a completely different nature, where the primary element is continuous revision and correction to accommodate new evidence.
- *Shifting the burden of proof to the other side.* The reader should be wary of statements such as, "But it has not been disproved." First, there are simply not enough scientists and funding to verify or disprove every claim that has ever been made. A lack of disproof, however, is not positive evidence of a given claim, but simply of our ignorance (or disinterest) on the matter. Second, when one proposes an alternative to a very well established theory, the burden of proof is logically and squarely on the side of the newcomer. When

Copernicus suggested that Earth rotates around the sun, and not vice versa, people did not believe him just because nobody had proven him wrong (on the contrary, most people did not even consider his arguments!). Other astronomers demanded evidence, and it took more than a century for the theory to be accepted.

- *Acceptance of a theory as legitimate simply because it's new, alternative, or daring.* This is the *Galilei* effect. Proponents of new theories are fond of recalling the many examples of scientists who had been derided, ignored, or worse, persecuted, because of their radical theories, which then proved to be correct. What this line of reasoning ignores, of course, is the fact that for every Galilei who eventually succeeded, there were thousands of crackpots who did not. For every example of a daring, new scientific theory that ends up being accepted, there are many, many examples of wrong theories, forever rejected and confined to the limbo of pseudoscientific history. Novelty per se is not evidence.

One last question needs to be discussed while we are considering alternatives to science. Why should someone who believes in God engage in scientific discourse that might threaten his religious beliefs? The answer is that if he feels uncomfortable about it, he probably shouldn't. However, as soon as a religious person engages in a debate based on evidence and logic, he automatically accepts the rules of the scientific game and cannot honestly call for any exception. If you admit that you believe God created the heavens and earth as stated in the Bible *regardless* of any evidence to the contrary, you have created for yourself an unassailable position. Because you are making a statement of faith not grounded in logic and evidence, nobody can seriously challenge it on the basis of logic and evidence (other than perhaps to point out that it is illogical to believe something without evidence). But if your position is that nature reveals the details of God's creation and that science actually confirms what the Bible says about physics and biology, you are obligated to discuss and accept the evidence—in whatever direction it may point, even if it undermines your own position.

This is why "creation science" is so vulnerable to scientific objections. We will see in Chapter 5 that creationists invoke known laws of nature, such as the second principle of thermodynamics, as evidence against the theory of evolution. That is fair as far as it goes. But then one has to accept the outcome of such challenges, whatever it is, and if physicists do not find anything in their laws that contradicts evolution, one has to honestly admit it and move on. Furthermore, creationists (or anybody else engaged in honest intellectual debate) cannot claim exception from the very same laws they invoke against their opponents. For example, one can argue that the second principle of thermodynamics contradicts the existence of an arbitrary Creator of the universe as described in the Bible, an objection that will not be accepted by fundamentalists who claim that God is outside of the laws of nature. But it is simply unfair to use an argument against someone else's position and then to prohibit one's opponent from doing the same to oneself.

Of course, the debate between theists and nontheists pre-dates the creation–evolution controversy and even the very birth of science. Thomas Aquinas (1225–1274) was one of the great church *apologists*. An apologist is someone who attempts to make a logical argument for the existence of God (and, in the case of Aquinas, in particular for the Christian God). Apologetics has always been controversial among exponents of the church precisely because it opens the way to inquiry and doubt. It is all well and good if the apologist can argue persuasively for the characteristics of God endorsed by a particular religion and the word of a specific sacred text, but what if somebody else argues even more persuasively on behalf of the other camp? What if the evidence that is available can be interpreted in a different way or, worse yet, clearly points to a different conclusion?

This book, obviously, is for people who like to think for themselves, weigh arguments, and consider the evidence. It does not make the pretense of being neutral. I am an evolutionary biologist, and I think I have very good reasons for being one. But the arguments laid out here (and the counterarguments I discuss on the creationist side) are there for anybody to see and weigh. The reader can check assertions against independent sources (often cited in the text), and can accept or reject parts or the whole of my arguments. The process should be a stimulating intellectual exercise with an open-ended outcome. That is what learning and critical thinking are all about. The only person who will definitely not benefit from this book is the one who refuses to engage in critical examination. If your position is, "I believe in X, no matter what" (where X is the Bible, the absolute power of science, or whatever), you may as well shut the book now and rush back to the store; you may still qualify for a refund.

On Reality

When it comes to the limits, real or perceived, of science, many people—including some scientists—seem to perceive things in one of two ways. On the one hand, there is the *physical* world, clearly the realm of science. We can conduct investigations of physical phenomena, and the marvelous fruits of this possibility are enjoyed every day by humans the world over in the form of technology. On the other hand, there is the *nonphysical* world, sometimes referred to as *spiritual*, or characterized as an *alternative reality*. This is, by definition in the minds of many, *beyond* the realm of science, and is rather the province of religion, mysticism, or other alternative ways of knowing. This position is adopted by some agnostic scientists, such as Stephen J. Gould.[45] The position is in fact at least as old as Augustine's (354–430 C.E.) distinction between a "city of man" and a "city of God," though Augustine's preference is clearly for the latter, while Gould more Solomonically divides merits between the two.

---

[45] For example, see S. J. Gould, "Nonoverlapping Magisteria," *Natural History* March 1997: 16–22.

I would like to submit here that there are in fact *three* realms of experience, only one of which is the domain of most science as currently practiced, though—I think—science is by no means diminished by this tripartition. The three domains are as follows:

1. *The physical world.* This is (for some philosophers) a hypothetical but (given compelling practical considerations—like the need to avoid running into walls) very likely existing realm, made of whatever is "out there," meaning what has existence and properties independent of the human mind. The objective of science is to uncover as much as possible about this world. Although other human enterprises have attempted to do this as well (philosophy, literally interpreted revealed religion, pseudoscience) none has succeeded to an extent even remotely approaching what science has to show to its credit.
2. *The emotional, or psychological, world.* This, as far as we know, exists only where humans are concerned, although it may exist for other organisms equipped with complex nervous systems (possibly some other primates, maybe extraterrestrial intelligences). Parts of this realm are not only emotions in the strict sense (love, anger, hate, compassion, and so on), but also all of the consequences of these emotions that we take for granted. These include morals, and therefore laws, but also art in the various forms of literature, visual arts, and music. The role of science in this realm is ambiguous. There is no real reason, in principle, why art or emotions could not be "explained" by science. For example, biology is certainly potentially capable of providing insights into why emotions occur, both from a mechanistic viewpoint (by identifying which hormones cause certain responses) and from a historical and comparative one (by explaining human emotions in the context of the emotions and reactions of other animals). On the other hand, science is not equipped (and it does not pretend) to make value judgments on these matters. So although mathematicians are able to study the internal structure of Bach's cantatas, they will never arrogate for themselves the right to tell somebody that those cantatas are better or worse than Mozart's symphonies. Analogously, comparative anthropologists have much to say about the ways in which different human societies choose their moral rules, but they are not in a position to say that a given set of rules is absolutely good (though some work better than others under certain circumstances and given specific individual and societal goals).
3. *The supernatural realm.* Here, science does not have anything to contribute. Because by definition the supernatural is unpredictable on the basis of natural laws, science cannot yield an iota of insight into what a God is going to do or why.[46]

---

[46] For special exceptions, see M. Pigliucci, "Science and the Falsifiability Question in Theology," *Skeptic* 6(2) (1998): 66–73. See also T. Edis, "Taking Creationism Seriously," *Skeptic* 6(2) (1998): 56–65.

This classification should make several things clear. First, as I have been arguing throughout this chapter, science is by far the best method to inquire into the realm of physical reality. Any claim to the contrary has to be substantiated by strong evidence because it would be completely irrational to accept it at face value. To do so would be like throwing away a perfectly valid tool from your toolbox in exchange for something completely new, about which the only information you have is the salesperson's assurance that it works miraculously. And remember that the salesperson is the *only* one who is going to surely make a profit from the sale. This absolutely and unquestionably does *not* imply that science can solve every problem related to the physical world (scientism), or that humans will eventually know all there is to know. It simply means that of all the tools we have devised so far, this is the one that works best at providing knowledge of the physical world.

Second, there is indeed a realm of human experience, the emotional or psychological, where science has a very circumscribed role to play (though certainly not nil). This domain is as "real" as the physical world; in fact, it is in some sense part of the physical world itself. After all, not only did Picasso paint on very physical canvas and with very real brushes and colors, but the painting was directed by his mind, certainly a very real characteristic of his equally physical brain.[47] Regardless of the ability of science to illuminate us about morals and the beauty of art, the latter are both very tangibly a part of the human condition (and therefore of "reality"). Every human being in the world can compare Mozart and Bach anytime she wants. Everybody can invent moral precepts or choose to live under someone else's version of morality. All humans, to the best of our knowledge, can at least potentially experience love and a wealth of other emotions, and can certainly compare their experiences (otherwise, talk shows would go out of business—not an entirely distasteful proposition).

Third, and most important to our discussion of the nature and limits of science, there is indeed possibly a realm in which science has very little business, the supernatural. However, notice that this is the only domain whose existence is questionable, to say the least. Clearly, not every human being claims to experience supernatural phenomena, and even for the ones who claim they do, such phenomena certainly represent uncommon and unpredictable occurrences.

Every reader's belief is his own prerogative, and it cannot be judged by any objective means, much like the beauty (or lack thereof) of a Dali painting. But if we engage in rational discourse, the third realm cannot fairly be compared to either of the other two. It is not the fault of science that we cannot investigate the supernatural. The matter is better framed as the inherent inability

[47] Of course, one can argue that the mind is not simply an outgrowth of the brain and is not entirely physical. Although this is certainly possible, it flies in the face of all available evidence, and such a claim is not based on any evidence to the contrary. Again, it is simply a matter of the burden of proof, not of final pronouncements. Philosophers of mind refer to the minimalist physical theory of mind as the "no ectoplasm" clause; that is, the mind is physical at least in the sense that if you take the brain away, the mind goes with it.

(unwillingness?) of the supernatural to fairly and squarely manifest itself impartially and detectibly to any human consciousness.[48]

But, the reader might object, regardless of where the fault lies, this still does not mean that there *is* no such thing as the supernatural. Of course not, but how do we know what the supernatural is? If by definition it is beyond any rational means of inquiry, then every conceivable version of the supernatural is equally valid. Why, then, believe in astrology but not in telepathy, in the Judeo-Christian religion but not the Muslim one? The point is, if we have no guide, it is reasonable (albeit perhaps psychologically and emotionally unsatisfactory) to suspend judgment and refuse to endorse any particular creed over any other. That is a very scientific attitude indeed. Ludwig Wittgenstein, one of the twentieth century's most cunning logicians and philosophers, put it best in his *Tractatus Logico-Philosophicus:* "What we cannot speak about we must pass over in silence." I think that this is, ultimately, not just what every scientist, but any rational and conscientious human being, should stick to. For to speak of things without evidence is not only intellectually dishonest; it is harmful to the search for truth, and it likely ends up directly or indirectly hurting other people, either psychologically or—in some cases—physically.

Unfortunately, very few humans seem compelled to follow Wittgenstein's suggestion. In this book I make a careful attempt to speak only of what we know and to frankly admit what we do not. I hope the reader, whatever her opinions on evolution and creationism are, will consider this an honest and fruitful approach to the controversy.

Additional Reading

Casti, J. L. 1989. *Paradigms Lost.* New York: Avon Books. A wonderfully delightful tour de force on science, its methods, and some of the big unanswered questions concerning life and the structure of the universe.

Lakatos, I. 1977. *The Methodology of Scientific Research Programmes.* Cambridge: Cambridge University Press. A more technical treatment of the scientific method.

Lindberg, D. C. 1992. *The Beginnings of Western Science.* Chicago: University of Chicago Press. A detailed and fascinating history of Western science and how it developed into the modern conception of what science is.

Popper, K. R. 1968. *Conjectures and Refutations: The Growth of Scientific Knowledge.* New York: Harper & Row. Another more in-depth view of the philosophy of science, from a different perspective than Lakatos's.

Schick, T.-J. 2000. *Readings in the Philosophy of Science: From Positivism to Postmodernism.* Mountain View, CA: Mayfield. A comprehensive overview of philosophy of science based on primary (but accessible) sources.

---

[48] This, incidentally, is a serious theological problem, analogous to the perennial question in ufology: Why don't the aliens just quit playing with us and land in the middle of a large city with plenty of eyewitnesses? Curiously, ufologists use the same argument adopted by some religionists: The aliens (or God) are so much smarter than we are that we can't possibly comprehend their plans and why their behavior seems so irrational. How convenient and utterly unconvincing.

– 5 –

Creationist Fallacies

Fallacy: n., misleading argument . . . flaw that vitiates syllogism . . . delusion,
error; unsoundness, delusiveness (of arguments, or beliefs).
From the Latin fallere, to deceive.

– *Oxford English Dictionary* –

From a drop of water a logician could infer the possibility of an Atlantic
or a Niagara without having seen or heard of one or the other.
So all life is a great chain, the nature of which is known
whenever we are shown a link of it.

– *Sherlock Holmes, in Arthur Conan Doyle's "A Study in Scarlet"* –

Are creationists simply people with defective abilities to reason? Obviously not, as is amply demonstrated by the fact that they conduct normal lives in a complex world of computers, cellular phones, and frozen yogurts. Indeed, to assume that all there is to the controversy is nothing more than misunderstandings that can be cleared up by a few lectures on how science works is the greatest fallacy that scientists entering this debate often commit. I refer to it as the *rationalistic fallacy,* and I will discuss it at some length in Chapter 7. Nevertheless, it is instructive to examine some of the most common errors of reasoning committed by creationists when they discuss evolution or science.

The following pages are not meant either as an exhaustive list or as a way to poke fun at the other side of the debate. My intent is rather to analyze the major problems with the way creationists think and to ask why they think that way. The latter question is obviously closely related to the problem of anti-intellectualism in its many facets, which I discussed in Chapter 3. I have touched on some of these fallacies in other parts of the book, so some of the following sections will be rather brief and simply summarize the problem and why it is relevant to our discussion. In other cases this chapter will present the main treatment of the topic in question.

Fallacy 1: Science Must Be Ethical, or It Is Not True

It should be obvious even to the casual observer that creationists do not have a problem with scientific results per se, but only with those scientific results that threaten their particular worldview, or are perceived as doing so. A quintessential icon of this view is the so-called tree of unbelief (Figure 5.1), which has appeared in several creationist publications in a variety of versions. The picture is emotionally powerful, capable of generating strong reactions from both supporters and opponents of evolution.

The tree typically represents the fruits of what is often characterized as "philosophical" evolution (obviously in an attempt to distinguish the practice from "true" science). The root of the tree is constituted by unbelief (in God) in the broadest sense. A quote from Matthew (7:18) eloquently states, "A good tree cannot bring forth evil; neither can a corrupt tree bring forth good fruit." The botanical metaphor leaves a lot to be desired, but the message is clear. The figure then juxtaposes the general public against Christians and shows the "scientific" creationism axe cutting the tree down.

The most instructive thing we get out of this figure comes from looking at some of the alleged bad fruits of philosophical evolution. We find everything from hard rock and inflation to sex education and "dirty" books, from relativism and terrorism to homosexuality and genetic engineering. How, of course, all of this is related to evolutionary theory is not explained. Indeed, it would take a very fervid imagination to weave credible just-so stories to connect all those dots.

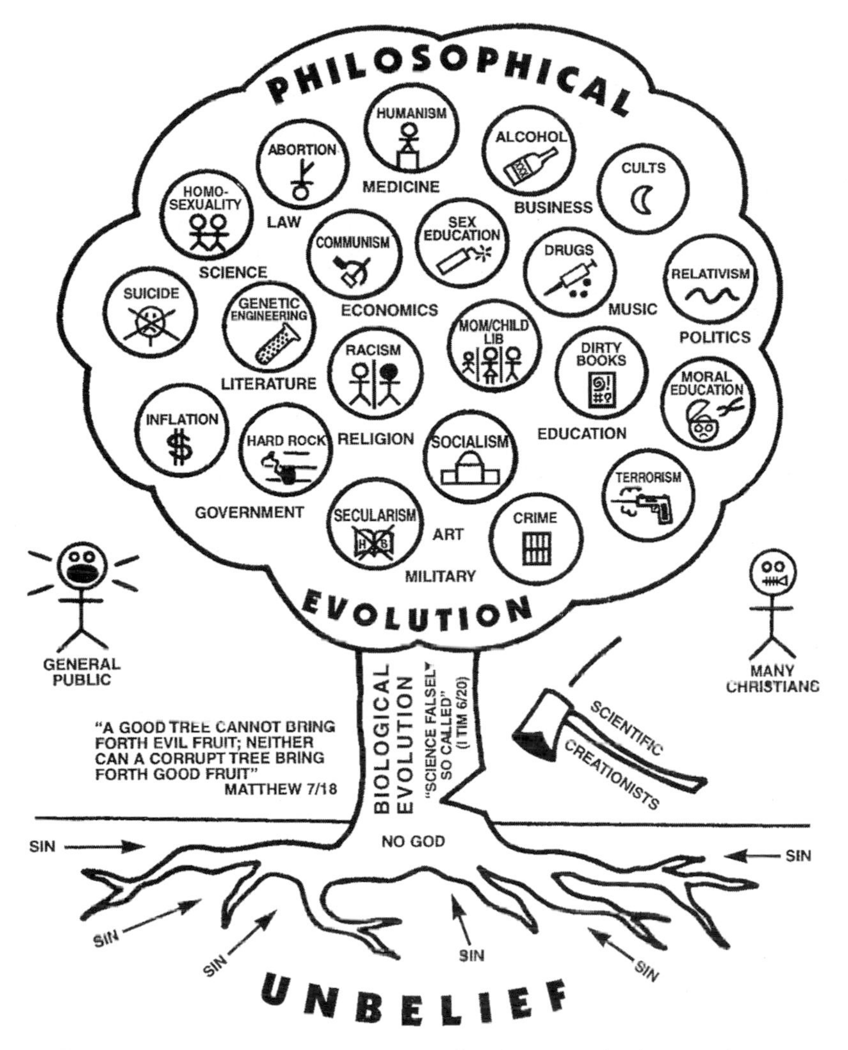

Figure 5.1 The "tree of unbelief," representing how creationists see the connections among atheism, evolution, and a series of real or alleged maladies of modern society. (Courtesy of the Skeptic Society)

But let us grant here for the sake of discussion that there is a connection of some sort between evolutionary biology and *some* questionable practices that have marked the recent history of human society. One such example could be eugenics,[1] the idea that we can and should improve the human stock in a manner similar to what we do with livestock and crop plants.

[1] For a lively and accessible discussion of the relationship between Darwinism and eugenics, see Chapter 6 of E. Caudill, *Darwinian Myths* (Knoxville: University of Tennessee Press, 1997).

Eugenics was founded by Darwin's cousin Francis Galton (Figure 5.2), who linked it directly to his relative's theory of common descent with modification. Galton's ideas were very influential for several decades and even led to the establishment of a Eugenics Record Office in the United States in 1911, under the direction of Charles B. Davenport. The office was housed at the prestigious Cold Spring Harbor Laboratory and at one point had accumulated 750,000 records of families and pedigrees. It published a regular newsletter and organized conferences at the American Museum of Natural History. To make the connection with breeding more clear, the Eugenics Record Office held events at local state fairs, where people were gathering to observe and buy the results of the artificial selection of animals.

Eugenicists maintained many scientifically absurd notions—for example, that town life and alcohol could ruin the genetic stock of a family, or that blacks and whites have recognizably distinct chromosomes. But what made the movement particularly despicable was the direct effect it had on many peoples' lives. Eugenicists convinced the U.S. immigration authorities to administer intelligence tests to potential immigrants, regardless of their native language (no matter what these tests actually measure—a topic on which controversy has raged since their inception—a subject can surely not perform well if the questions are asked in a language that he does not understand, and the results of such a test obviously tell nothing about his intelligence). Even more astonishing was the eugenicists' plan to sterilize criminals and mentally defective people—and the fact that 30 states[2] partially implemented such plans in the period leading up to the Second World War.[3] Table 5.1 shows a summary timeline of the eugenics movement in the United States.

[2] Interestingly, none of these states was found in the South, the cradle of American creationism. They were instead concentrated on the West Coast, in the north-central region, and in the Northeast.

[3] It is interesting to note that this was happening in the United States, the epicenter of Western democracy and the future enemy of the Nazi regime—which was later condemned by the United States precisely for planning and executing eugenics programs on a large scale.

Figure 5.2 Francis Galton, Darwin's cousin and founder of the eugenics movement.

Table 5.1 *Timeline of the practical consequences of the eugenics movement in the United States*

1907	Indiana is the first state allowing involuntary sterilization of sex offenders, mentally retarded, epileptics and habitual criminals.
1913	29 of the 48 states passed laws forbidding interracial marriages. Penalties included up to ten years in prison and $2,000 fines.
1924	The Johnson Immigration Act limits the yearly inflow of people to 2% of the existing population. Country quotas are established, restricting immigration from southern European countries.
1927	The Supreme Court upholds the forced sterilization of retarded individuals in the Buck vs. Bell case. Famously, the majority opinion prepared by Justice Oliver Wendell Holmes declares: "Three generations of imbeciles is enough."
1930	A conservative figure of 15,000 Americans have been sterilized because of eugenic laws.

The Cold Spring Harbor office was finally closed in 1945 (*after* the Nazi-led Holocaust of Jews in Europe), and eugenics has since become mostly relegated to history books.[4] Yet the question remains: What is the connection of eugenics to evolutionary theory, and—more to the point—can we blame the latter for the former?

Contrary to a rather politically correct fashion among biologists, I do not think that eugenics was entirely an aberration based on pseudoscience. We know that many of the claims made by eugenicists were based on sloppy research or were simply unfounded. Nonetheless, it is undeniable that eugenics is a logical consequence—applied to humans—of our understanding of genetics and natural selection. Humans do breed, like most other organisms, and our genes behave in the same manner as those of cattle or crop plants. There is no relevant difference there. As for responding to selection—natural or artificial—it is obvious that humans have been shaped in part by past natural selection, and it is equally clear that we would respond to a breeding program in pretty much the same way any large mammal would (i.e., slowly, given the long generation time). In this sense, therefore, eugenics *can* be practiced and it would (more or less) work.

The real question, of course, is this: *Should* eugenics be practiced? The answer, given the moral standards of many human societies, is a resounding no. But this is an ethical, not a scientific, conclusion. The choice is not based on the science of eugenics—or of genetics and evolutionary biology—but rather on what we as human beings wish to do. The same disconnection between science and ethics is found in many other fields—for example, atomic physics. The fundamental research that enabled us to understand how atoms are constructed

[4] I say "mostly" because Cold Spring Harbor actually tried to reopen the office in 1980 and because vestiges of the movement can be found on the World Wide Web—for example, at http://www.eugenics.net/.

was sound science. That same knowledge has subsequently given us power to split the atom and liberate an amount of energy sufficient to kill hundreds of thousands of our fellow humans at once, and probably to blow up the whole planet, if we wished to do so. No matter how strongly one might disapprove of the atomic bomb or of eugenics, it is simply a non sequitur to infer that the science that led to those practices or applications is bad science.

I am in no way suggesting that scientists are devoid of ethical responsibilities. Some of these responsibilities are rather direct: A scientist who works for the government and is involved in a secret weapon development project knows that every single idea he will produce could be used at any time as an instrument of death. Scientists who are disturbed by that fact simply should not engage in such research. However, scientists also have indirect responsibilities. They may not be working on the Human Genome Project with the intent of eventually helping health insurance companies to deny coverage to high-risk patients, but there is a possibility that their research will lead to just that sort of objectionable application. In these cases, however, the responsibility of the scientist is rather remote, given that it is very difficult to predict the practical consequences of any piece of scientific research.

A scientist, then, has more or less the same duty as any other citizen to participate in the public debate on the use of his research, perhaps heightened by the fact that he knows better than most other people what the research was about and how it was conducted. But the primary responsibility for the good or bad use of genetic information or atomic energy goes to our elected officials (in a democracy), which ultimately means to the people themselves. The fact that creationists might not like what they perceive to be the fruits of evolution has no connection to the scientific validity of evolutionary theory. To change the world, people need to become involved politically, not deny that the planet Earth rotates around the sun.

Fallacy 2: Scientific Discussions Are a Sign of Internal Crisis

"If evolutionists disagree among themselves, how can they pretend to tell the truth?" This line is typically thrown at me during debates or in the midst of correspondence. It is perhaps one of the most frustrating arguments because it betrays a fundamental misunderstanding of the process of science itself (see Chapter 4).

From a religious fundamentalist standpoint, there cannot be waffling or uncertainty in the world: Everything is clear, down to the most minute details. Indeed, when disputes over religious matters arise, the behavior of mainstream and fundamentalist denominations is, by and large, significantly different. Whereas most Christian denominations, for example, recognize that many theological questions are difficult and depend on subjective interpretations of Scripture, fundamentalist groups splinter over very minor points of doctrine, precisely because they see any form of disagreement as a sign of fatal corruption.

Enter science as a process that, as we have seen, is always characterized by debates and by the succession of new hypotheses and general theories to (hopefully) improve on the previous ones. From a fundamentalist perspective, it must look as though scientists literally do not know what they are talking about. From a scientific viewpoint, it is this deep-seated misunderstanding by the fundamentalists of how science works that is at the root of so much trouble.

To consider the problem in more depth, let me turn to one specific example, perhaps the one that is most commonly brought up by creationists: the debate between Darwinian gradualism and the so-called theory of punctuated equilibria advanced in 1972 by Niles Eldredge and Stephen Gould.[5] A bit of background and history are necessary to appreciate the point.

One of Darwin's contentions had been that evolution proceeds at a slow pace, in a fashion similar to how mountains are built and erased over geological time. Darwin was markedly influenced by the British geologist Charles Lyell, who had based his fundamentally important *Principles of Geology*[6] on precisely such an assumption of continuity between historical and current phenomena (an idea known as the principle of uniformitarianism in geology as well as in other sciences). According to Darwin and Lyell, science does not need to invoke special mechanisms to explain past events. All that is necessary is to extrapolate backward to the the effects of the same phenomena we see at work today. It was a reasonable and powerful assumption, but not one that was necessary for the internal logic of the theory of evolution. In fact, Thomas Huxley ("Darwin's bulldog") warned that this was an unnecessary restriction on the workings of nature.

Huxley was vindicated well before Eldredge and Gould proposed their idea of punctuated equilibria. Paleontologists discovered that occasionally very unusual events do occur and shape the course of evolution on Earth in dramatic new directions. Although it is not clear if the dinosaurs would have eventually become extinct anyway (at the time of their extinction they had passed the peak of their diversity on Earth), surely the collision of a large meteor with Earth 65 million years ago did not help them or countless other life forms that were wiped out. Meteoritic impacts are not gradual phenomena that occur every day. Of course, the discovery of evidence revealing the occasional exception certainly does not negate the rule: Most evolution still occurs during the intervals between mass extinctions or celestial interventions. Evolutionary biology is a pluralistic scientific theory, and it has to accommodate many causal mechanisms that affect a variety of biological phenomena.[7]

---

[5] The original paper is N. Eldredge and S. J. Gould, "Punctuated Equilibria: An Alternative to Phyletic Gradualism," in *Models in Paleobiology,* T. J. M. Schopf (ed.), pp. 82–115 (San Francisco, CA: Freeman, Cooper, and Co., 1972). For a more accessible assessment of the controversy, see M. Shermer, "Gould's Dangerous Idea: Contingency, Necessity, and the Nature of History," *Skeptic* 4(1) (1996): 91–95.

[6] C. Lyell, *Principles of Geology* (London: J. Murray, 1830–1833).

[7] For an in-depth discussion of the fundamental role of pluralism in science, see J. Dupré, *The Disorder of Things: Metaphysical Foundations of the Disunity of Science* (Cambridge, MA: Harvard University Press, 1993).

What Eldredge and Gould proposed in the early 1970s was yet another step toward pluralism (though that was not clear from the often ambiguous language they initially used). Ironically, they were addressing (not directly) one of the typical creationist objections: that the fossil record is rather spotty, making it difficult to study the derivation of new taxa from older ones (see Chapter 6). The classical explanation, maintained by most paleontologists and going back to Darwin himself, is that because fossilization is a rare event, one would expect to find only sparse fossils here and there. Eldredge and Gould turned the problem around and suggested that perhaps what we see is what we get: Maybe the fossil record is much more complete than it looks, and the gaps are due to a real biological phenomenon, not to the absence of data. It was just possible that—to put it as Gould once did—evolution is a bit like baseball (or cricket, if you are on the other shore of the Atlantic): Nothing happens most of the time, but occasionally the action is rapid and decisive. The long periods of inaction were called periods of "stasis," and they were "punctuated" by bursts of change that led to the next state of equilibrium. Hence the name of the theory.

This was not as extraordinary a suggestion as it might seem. First, other paleontologists had already advanced the possibility that evolution can run in different modes at different speeds. One of them was George Gaylord Simpson, who contributed most to the so-called neo-Darwinian synthesis of the 1930s and 1940s that is still the accepted paradigm in the field.[8] Second, Eldredge and Gould themselves had derived their hypothesis from one of the prevailing views on the process of formation of new species: the allopatric speciation mechanism proposed by Ernst Mayr. According to Mayr, new species originate as detached fragments of existing populations, away from the strong competition of their parents (*allopatric* means "in a different place"). The process involves small, geographically isolated, peripheral populations, most of which simply go extinct. But the few that succeed evolve for some time in isolation from the parental stock and therefore become distinct in appearance and genetic constitution. Occasionally the daughter populations (now constituting new species) may rapidly expand and replace the old guard, probably because of altered environmental conditions to which they are better suited. If one looked at this process over geological time, Eldredge and Gould argued, the fossil record would present the sudden disappearance of one form and its substitution by another one that would seem to have materialized from nowhere (because small populations are even less likely than large ones to leave fossils).

What happened next was a remarkable episode in the history of science, which provides a wonderful window on how real, messy, but ultimately successful the working of science is. Eldredge and Gould's ideas were initially harshly criticized, partly because of an interesting case of guilt by association. Gould has always had an interest in the history of science, and he wrote the

[8] On the neo-Darwinian synthesis, see E. Mayr and W. B. Provine, *The Evolutionary Synthesis: Perspectives on the Unification of Biology* (Cambridge, MA: Harvard University Press, 1980).

foreword for the new printing of a classic 1940 book by German geneticist Richard Goldschmidt: *The Material Basis of Evolution*.[9] Toward the end of the book (Chapter 7), Goldschmidt proposed his now discredited theory of "hopeful monsters": He suggested that major rearrangements of the genetic material ("genetic revolutions") may occasionally bring about the sudden appearance of novel forms of animals and plants, thereby making gradual evolution and intermediate links completely unnecessary.[10] Gould (correctly) suggested in his foreword that many of Goldschmidt's ideas were still current and should be reconsidered, but many people—partly because of Gould's own florid prose—deduced that he was arguing that genomic revolutions and hopeful monsters were the biological bases of punctuated equilibria.

Despite the hopeful monster misunderstanding, Eldredge and Gould's proposal was exceedingly successful as a scientific hypothesis. It did exactly what one expects good ideas to do in science: It spurred a new research program, produced hundreds of papers, engaged a whole community of scientists in intense cross-disciplinary talk, and eventually resulted in a broadened view of the fundamental patterns and processes of evolutionary change. The currently accepted version of punctuated equilibria is less radical than the original proposal, and the mechanisms that explain stasis and punctuation episodes are an elaboration on the original ones but do not include hopeful (or, rather, hopeless) monsters. Indeed, several papers have demonstrated that much of the controversy was due to a confusion of terminology: When paleontologists (like Eldredge and Gould) use the word *sudden,* they do not mean quite the same thing that a population biologist might. Something is sudden in geology if it happens over a few tens of thousands of years, but that timescale includes plenty of classic Darwinian, gradual change. The lasting contribution of punctuated equilibria is the demonstration that indeed evolutionary change can proceed at very different paces, and that the fossil record is—if not complete—at least much more useful as a tool to understand the evolutionary process than it was previously thought.

What is interesting for our purposes here is the use that creationists have made of punctuated equilibria. To his immense chagrin, Gould is paraded as an antievolutionist, as the destroyer of Darwinism—despite his own repeated statements that he is no such thing. Even more ironically, creationists like Duane Gish can then turn around and attack Gould himself for having proposed the ludicrous idea of the hopeful monster (which he didn't). One of Gish's famous jokes (which he has repeated at every debate at which I have

[9] R. Goldschmidt, *The Material Basis of Evolution* (New Haven, CT: Yale University Press, 1940).

[10] Goldschmidt's ideas have been unjustly discounted because his critics focused too much on the hopeful monsters, which are discussed actually in only a few pages right at the end of the book. Many of his other suggestions represented astounding insights into aspects of developmental and evolutionary biology that we are still grappling with today. For a discussion of Goldschmidt's contributions, see Chapter 2 of C. D. Schlichting and M. Pigliucci, *Phenotypic Evolution: A Reaction Norm Perspective* (Sunderland, MA: Sinauer, 1998).

had the pleasure to encounter him) is that mama reptile surely must have been surprised when she looked into her nest and found a bird among her offspring (a very crude caricature of the original idea of hopeful monsters as proposed by Goldschmidt). Furthermore, many creationists make fun of the idea of genomic revolutions (and justly so), yet they will also quote cosmologist Fred Hoyle as another "scientific" detractor of evolution. Interestingly, Hoyle proposes periodic genomic revolutions as an alternative (and amply discredited by modern molecular biology) source of evolutionary change (e.g., in his *Mathematics of Evolution*[11]). It truly seems that creationists want to have their cake and eat it too!

Perhaps the most surprising thing about this particular fallacy is how blind creationists are to the trouble in their own house. Whereas they consider small scientific controversies to be fatal blows to the theory of evolution, they do not seem bothered by such "minor" internal disagreements as the age of Earth or the role of microevolution. Some creationists, as we saw in Chapter 2, believe Earth to be about 6,000 years old; others accept the geologists' figure of 4.5 *billion* years. That is a difference of 750,000 times, enough to send any scientist into a frenzy if it applied to the predictions of any scientific theory with which they were working. In addition, some creationists (mostly intelligent design proponents) have no problem allowing for even large amounts of evolution by natural selection. Michael Behe, for example (see Chapter 2), acknowledges that the eye evolved gradually by natural selection, a statement that would horrify most rank-and-file members of the creationism movement.

Yet it is amazing how sharing a common ideology makes creationists gloss over their huge differences and focus with a sneer of triumph on truly minor disagreements among evolutionists. This is perhaps one of the most damning pieces of evidence that creationism is not a science. What counts in creationism is only the general idea that a personal God exists in the universe. How creationists arrive at that idea is something they confine to their internal theological squabbles, not to be discussed before the public. One never knows, people might start noticing the inconsistencies . . .

Fallacy 3: "It's Just a Theory"

I cannot recount how many times I have heard this phrase used to qualify evolution. The idea seems to be that since evolution is a theory, it is not a fact. And science, as we all know, is about facts, not "mere" theories.

Once again we are faced with a fundamental misunderstanding of science, one for which professional scientists and educators are surely at least in part to blame. The *Oxford Dictionary* gives several definitions of *theory*. The first one comes close to what a scientist would recognize the theory of evolution to be: "System of ideas explaining something, especially one based on general prin-

---

[11] F. Hoyle, *Mathematics of Evolution* (Memphis, TN: Acorn, 1999).

ciples independent of the facts, phenomena, etc. to be explained." Examples are given: the atomic theory, the theory of gravitation, and the theory of evolution itself. The second definition is the one that most creationists seem to refer to: "Speculative (especially fanciful) view" as in "one of my pet theories."

In Chapter 4 we discussed the scientific method, and we will examine in Chapter 7 some of the fallacies that scientists commit when explaining to the general public what they do. One of these misunderstandings is evident even in the *Oxford* definition just given: Scientific theories are *not* independent of the facts. Furthermore, the "facts" are not independent of the theory either. This is what philosophers of science refer to as the *theory-ladenness of observation.*

The idea is that what counts as a fact depends in part on what theory one is using to make sense of the world. This is not just typical of science, but of the way most animals approach the world in which they live. If we are interested in finding food and we eat only fruits, the world is automatically divided into *fruit* and *nonfruit.* In a sense, we (or any frugivorous animal) are using a "theory" of the world and we are filtering our observations accordingly. We might miss some facts that would be relevant—for example, if we do not realize that something that does not look like a typical fruit in fact belongs to that category.

Analogously, scientists do not take in the whole world with billions of "facts" and then wait until theories emerge through a sort of intellectual seepage. They look for a subset of the world that appears to be relevant to the theory at hand. But since this choice is influenced by the theory itself, facts and theories are interwoven and cannot be Solomonically separated as the current mythology of science wants them to be.

Of course, this sort of discourse is music to the creationist's ear: If one admits that the distinction between theory and facts is muddy, then science is not objective and therefore is no better than any other just-so story about the world. Indeed, as we saw in Chapter 3, this is exactly the conclusion reached by postmodernists and other types of anti-intellectuals. But it in no way follows from the premise.

The dream of a purely objective science was pursued by an early school of philosophers of science known as logical positivists (though I am not aware of any alternative group labeling itself "illogical" positivist). Their naïve (but understandable at the time) conception of science was soon faced with insurmountable problems such as the theory ladenness of observations mentioned here and the problem of induction discussed in Chapter 4. This, however, does not mean that there is *no* degree of objectivity in science, or that science does not make progress. It simply means that—like all real-world human activities—things are a bit more complicated than we would like them to be.

Although scientists do focus on facts and interpret observations on the basis of their current theories, this choice is not arbitrary, and it is always made with an eye to the possibility that the current theory is in fact wrong. Even if one refuses to think of scientists as genuinely interested in the pursuit of knowledge and truth (and most of them are), one must consider that they have a strong incentive to demonstrate that somebody else's theory is wrong. That is the surest and fastest way to fame and glory (if not money). Furthermore, the bigger the target one can shoot down, the higher the stakes and incentives, and few targets are as big as Darwin's theory of evolution.

But as far as this particular creationist fallacy is concerned, subtle questions of philosophy of science are not at the core of the controversy. Rather the problem seems to be that most people do not understand that a theory (in science) is not just a hunch somebody has about something, a half-finished thought that comes to someone upon briefly considering a given subject matter. Scientific theories are complex and elaborate statements—often, but not always, in mathematical form—that provide a general explanatory framework for large portions of the world. That, of course, does not mean that they cannot be wrong (they often are, at least in some detail), but it does mean that one cannot discount them on the basis that they are "just" theories. Moreover, one cannot pretend to consider a statement such as "an intelligent designer did it" as a "theory" in the scientific (or any other) sense. The latter is no theory at all. In fact, it isn't even a hunch because it does not provide even potentially an *explanation* of the phenomena at hand. It is simply another way to say, "I don't know."[12]

This discussion of the difference between theories and fact is particularly relevant to one of the major political contentions of creationists: the idea that evolutionary teaching in public schools should be accompanied by "disclaimers" warning students that they are reading about a theory and not about scientific facts (see the history of court challenges in Chapter 1). Such a warning, as it should now be clear, is based on a fundamental misunderstanding of what science is. If taken seriously, one would have to provide disclaimers for any other "theory" as well, including the germ theory, cell theory, the theory of relativity, and even quantum mechanics. A much better strategy would be to teach our students how science actually works and how to think critically about science, pseudoscience, and the philosophical assumptions that underpin them.

Fallacy 4: Natural Phenomena Mean Randomness

For some reason many people, not just creationists, seem to think that if something is natural then it must also be random (in the sense of being undirected and therefore, in the minds of those with a misunderstanding or ignorance of natural selection, clearly not designed). This is the basis for one of the most persistent fallacies of creationism: that evolution cannot be true because it purports to explain complexity in the biological world by means of random accidents.

In Chapter 2 I tackled the more sophisticated version of this problem posed by intelligent design proponent William Dembski, and I have argued that there

[12] Most people do not understand the difference between a theory and a hypothesis, which is relevant to our discussion. A *hypothesis* is a preliminary, tentative, untested explanation; a *theory* is a hypothesis transformed into a more general set of statements both by consistent confirmation and by unsuccessful efforts to disprove it.

are many possible types of design in nature, only some of which require any intelligent agency at all. In fact, the simplest way to think about this problem is to go back to the basic definition of adaptive evolution (i.e., evolution that leads to organisms apparently designed to cope with certain environments and functions):[13]

Adaptation = Mutation + Natural Selection

That is, adaptive evolution is the result of (at least) *two* forces. One of them, mutation, is truly random as far as we can tell. This randomness derives from the fact that mutations occur with a frequency that is not related to how useful they would be for the organism.[14] Selection, on the other hand, is anything *but* random because by definition it excludes certain variants (ultimately originated by mutation) and favors others. Thus one can think of adaptive evolution as the product of the engine of natural selection fueled by the gas of mutation. The process as a whole is not random, although one of its components is.

Yet one can become exceedingly frustrated with creationists who—on purpose or not—completely ignore one's explanations and proceed down their path without the least perturbation. As I mentioned already, I have had the pleasure of debating creationist Duane Gish of the Institute for Creation Research five times (at present count). During one of our last encounters I noticed that he kept repeating that evolutionists are a bunch of simpletons because they would want you to believe that complexity arises out of randomness. Given widespread nodding in the audience, I took some time to explain the equation above, even projecting it on a large screen with accompanying labels of *random* (for mutation) and *not random* (for selection). I thought that would clear the air of at least that particular problem. I was, of course, naïve. Gish repeated *exactly* the same point about randomness 15 minutes later, during the rebuttal period. It was as if I had not spoken at all (though some members of the audience did sense the awkwardness of the situation). I was left with only two possible conclusions: Either Gish—holder of a Ph.D. in biochemistry—does not understand an extremely simple concept that is taught in junior high school, or he in-

---

[13] Evolution in the broadest sense, as changes in the frequencies of genes within a population (regardless of whether these changes are adaptive, maladaptive, or neutral from the viewpoint of the organism), does not, of course, require natural selection. In fact, classical theory in population genetics recognizes five fundamental mechanisms that can change gene frequencies: selection, migration (from or to other populations), mutation, recombination (through some sort of sexual reproduction), and what is called genetic drift—the tendency of gene frequencies to fluctuate wildly (and randomly) in very small populations (a sort of statistical sampling "error" by Mother Nature). More details can be found in D. Hartl, *A Primer of Population Genetics*, 3rd ed. (Sunderland, MA: Sinauer, 2000).

[14] This does *not* mean that all mutations occur with the same frequency. That frequency depends on a variety of factors, including the position of the gene in the chromosome and within the cell nucleus, where the DNA is housed. In fact, some mutations are found in natural populations much more often than others, and this higher frequency can bias the course of evolution in particular directions. The important point, however, is that such directions are unrelated to the degree of adaptation of the organism to its changing environmental conditions.

tentionally ignored it because it didn't square with his theological agenda. I leave the reader to ponder the answer.

Fallacy 5: The World Can Be Understood by Common Sense

Thomas Huxley, the nineteenth-century biologist who did more than anybody else to spread understanding of Darwin's theory, was a very popular and engaging speaker. He traveled all over England talking about science to general audiences, and he brought in large crowds. Science has always attracted the curiosity of nonscientists, probably because wonder about the natural world is an innate (evolved?) component of human nature. Huxley, however, unwittingly shared with creationists a fallacious assumption: that the world is relatively simple and intuitive, and that science is just common sense writ large.

This is the basis for popular disbelief of some apparently outrageous statements that evolutionary biologists make and for belief in some seemingly obvious, but entirely incorrect, statements that creationists make. An example of the first kind is the difficulty that most people have with swallowing the idea that—ultimately—organic life comes from inorganic matter. Again, I owe it to Duane Gish for coming up with a good joke about it. He usually addresses his audience by saying that evolutionists would like you to believe that hydrogen is an odorless, inert gas that—given enough time—becomes people. Well, this is a simplification of course, but all we know from astronomy, cosmology, physics, chemistry, biochemistry, and biophysics points to the fact that we are the products of a universe in which there was once nothing except hydrogen and helium. Carbon, for example, necessary for organic compounds, was produced in the stars formed from these first elements, indicating that life does come from inorganic matter. Indeed, we were once, in a manner of speaking, hydrogen atoms.

The only reason this latter statement seems outrageous and is rejected out of hand by people who have not read a single word of physics or chemistry is that it has obvious theological implications. Some kinds of gods are excluded from the realm of possibility if we accept that living beings come from inorganic compounds, that "life comes from nonlife." Yet science is full of *other* such counterintuitive statements that not only go unchallenged by the general public, but are in fact popularly accepted as profound insights into the nature of the world. For example, physics also tells us that the apparently solid objects that make up our world, including the chair you may be sitting on and this very book you are reading, are in fact made of mostly empty space. (One might note, however, that some books seem to be made of more empty space than others.) This is a direct consequence of what we know about atoms and subatomic particles, and its oddness does not seem to bother too many people. And it shouldn't. The point is that the idea that living organisms ultimately originate from inorganic compounds is the same kind of statement; it belongs in the same category as other equally counterintuitive ideas that people have no trouble accepting.

Science has done a great job of eradicating (or at least greatly diminishing) all sorts of "intuitive" but erroneous beliefs we used to have about the world. Nowadays, even a junior high school student knows that if he rotates an object attached to a string around his body and then lets it go, it will fly in a straight line instead of maintaining a rotational motion. But Aristotle and many other smart people throughout the ages thought the latter to be the correct answer.

It is equally "commonsensical" to believe that there was a worldwide flood 4,000 years ago and that the Grand Canyon was carved by the floodwaters. On the surface, this idea does seem to make sense. We know that the canyon carries the signs of large amounts of water having run through it. We know that dramatic changes in the landscape can be caused by sudden geological turmoil—like the famous eruption of the Mount St. Helens volcano[15] that occurred at 8:32 A.M. on May 18, 1980 (Figure 5.3). If one adds the description of the flood

[15] An informative point-by-point explanation of creationists' fallacies concerning the St. Helens eruption can be found at http://www.talk-origins.org/faqs/mtsthelens.html. To learn something about the geology and history of the area, go to http://www.fs.fed.us/gpnf/mshnvm/.

(a)

(b)

Figure 5.3 Mt. St. Helen's volcano, before (a) and after (b) the 1980 eruption. Notice how the clear mountain lake was displaced by a massive landslide and filled with debris and ash. (Photos by J. Hughes [a] and J. Franklin [b], USDA Forest Service)

in the Bible and the fact that various cultures have similar myths, isn't it just a matter of common sense to conclude that Noah's flood really happened?

Such a conclusion may seem obvious to common sense, but it is wrong, just as the commonsense observation of the sun's rotating around Earth, though intuitively "obvious," is wrong. The geological evidence is very clear: The Grand Canyon formed slowly over a period of hundreds of millions of years, and the flood myths are indeed common to many cultures throughout the world, but they are limited to people who lived near floodplains. Their stories therefore probably do refer to real events, but events that were local to those respective geographical areas. Cultures incorporate into their mythologies what is important to them, and the floods of the Nile, for example, were of vital importance to Egyptian civilization.

The message in this case is that common sense can be deceptive and that we simply need to do a bit more work to understand complex matters. There is a good reason for this and—ironically—it is an evolutionary one. As we shall see in Chapter 8, the human brain is a marvelous instrument, but it, too, evolved as a tool for survival. Some of our cognitive abilities were probably built by natural selection to aid in the everyday struggle to find food and mates and to avoid becoming food for other beasts. It is very likely that a rudimentary understanding of cause and effect represented an advantage and that is why we are capable of deductive reasoning. In fact, research on infants has shown that we have innate expectations about certain cause-and-effect relationships. Very young babies who observe an object moving on a table are surprised if the object goes beyond the table and does not fall down. This sort of expectation is evident at a stage before any cultural influence or scientific notion could possibly have entered their minds.[16] That is, the human brain has a minimal set of proto-scientific "theories" about the world already built in, and these are theories that served us well throughout our history.

However, surely natural selection did not equip us to understand magic tricks, geological phenomena, or quantum mechanics. Such understanding was simply not relevant to our everyday survival (and still isn't), so we should not be surprised that our brains have a hard time wrapping themselves around the unusual concepts that come out of science. That is why science—unfortunately but marvelously—is *not* just common sense writ large, as Huxley thought. On the contrary, the best science is counterintuitive and surprising, and the people who discover things about how the universe really works deserve our credit and respect, not our scorn because their insights don't square well with the content of a particular book written when humans understood very little of the universe and their place in it.

---

[16] On experiments concerning human infants' understanding of the physical world, see Chapter 25 in E. Margolis and S. Laurence, eds., *Concepts: Core Readings* (Cambridge, MA: MIT Press, 1999).

Fallacy 6: We Win by Default

This is related to the previous fallacy, in that it originates from the same simplified view of the world. One gets the clear impression in talking to creationists that it is either *us* or *them*. People go either to heaven or to hell; purgatory is only for misguided Catholics (remember that modern Christian fundamentalism is a Protestant phenomenon).

A nice black-and-white choice may work for theology, but it certainly does not work for the world of nature and science. Any scientific theory can be wrong, and in fact to some extent they may all be (in the sense that even the best theories only approximate reality through the limited brainpower and cognitive abilities of human beings). But the fact that a particular theory is wrong does not lend victory to the alternative by default. The reason is that there is never only *one* alternative. Indeed, even if we had several competing theories to explain a certain range of phenomena, they could *all* be wrong and we would have to start from scratch.

Furthermore, as Francis Bacon clearly put it back in 1620,[17] it is not enough to show that a particular explanation is wrong (what he called the *pars destruens*). One must also be able to advance a *better* alternative (the *pars construens*). This has become accepted wisdom in the philosophy of science. Calling any scientific theory "true" betrays a misunderstanding of what a scientific theory does and, consequently, of how it must be evaluated. Although a single proposition—for example, the assertion of a particular fact—may be either true or false, it makes no sense to label a theory itself as true or false because a theory serves to *explain* a collection of facts or body of data; it does not itself convey data. Just as a hypothesis serves as a *preliminary, tentative* explanation of a body of data, so the continuing acquisition of compatible data, paired with unsuccessful efforts to find data that *disprove* the hypothesis, eventually transforms the hypothesis into a stable, well-grounded explanation. This well-grounded explanation is what science calls a theory.

Given that science is an open-ended process in which new data are continuously accumulated, no theory is ever accepted in science as a *final* explanation in any ultimate sense. It continues to be accepted as long as the data are consistent with it and no one offers a better explanation of the data. Consequently, scientific theories cannot be shown to be "true" (though they can be conclusively shown to be mistaken or inadequate). At any moment in time, we need to choose among the available alternatives and bet on the one that performs better at explaining the world around us (see the explanation of Bayesian approaches in Chapter 4). This means that even if evolutionary theory as currently accepted is wrong in some fundamental way (and it is hard to see how this could be), the victory does not go either to traditional creationism or to intelligent design creationism, because both clearly fail to provide a better explanation of nature.

---

[17] See F. Bacon, *Novum Organum* (Cambridge: Cambridge University Press, [1620] 2000).

Should Darwin fall, like Newton did before him, we would owe that development to the next Einstein, not to the creationist movement. If science had stopped every time it could not elucidate something, and had fallen back instead on divine explanations, we would still think that Zeus's temper tantrums are the cause of thunder and lightning.

Fallacy 7: Living Beings Are Obviously Designed

This fallacy, which I dealt with extensively in Chapter 2, is listed here largely for the sake of completeness. The basic argument is really another application of the "common sense" approach to science that characterizes the fifth fallacy. As the classic example goes, if one sees a watch, one immediately thinks of a watchmaker. Analogously, the obvious fit between organisms and their environments must have been the result of a purposeful design. This is precisely why Richard Dawkins's book devoted entirely to debunking this notion is entitled *The Blind Watchmaker*.[18]

Darwin himself recognized this potential problem for the acceptance of his theory, even though his idea was precisely that design *can* occur in nature without intelligence, by the undirected force of natural selection. Cunningly, Darwin addressed this objection by focusing on the defects, rather than the alleged perfection, of living organisms—humans included.

In fact, it is amazing what *poorly* designed organisms we all are. Although we wonder at the complexity of the human eye, we cannot understand why it includes a blind spot that could have easily been avoided (and in fact is not present in the squid's and octopus's eyes, which are otherwise very similar to the vertebrate version). Analyses of biochemical pathways show that living organisms do not make the most efficient use of the complex networks of enzymes that ensure their everyday metabolism. Perhaps the most convincing instances of bad design are found in human beings themselves, as discussed in Chapter 2 (see Figure 2.2).

It is important to understand that this sort of anti-intelligent design argument has nothing to do with making "arrogant" statements on the perfectibility of God's creation. It is rather a relatively simple matter of engineering analysis. Olshansky, Carnes, and Butler,[19] for example, have performed just such an analysis, asking the question of how we could improve the basic design of a human being to cope with the aging process. The result is something that not only an intelligent designer with unlimited means, but even a modest genetic engineer of the future, could readily accomplish (see Figure 2.2). The reason we humans don't happen to be well designed for our later years is that the post-

[18] R. Dawkins, *The Blind Watchmaker: Why the Evidence of Evolution Reveals a Universe without Design* (New York: Norton, 1996).

[19] In S. J. Olshansky, B. A. Carnes, and R. N. Butler, "If Humans Were Built to Last," *Scientific American* 284(3) (2001): 50–55.

reproductive period is irrelevant from the point of view of natural selection. By then humans have fulfilled (or not) the basic purpose of every living being: to reproduce. Our personal comfort afterward simply cannot enter into the equation because it doesn't affect the survival of our genes.[20] Indeed, if it did, this would be one piece of evidence that would be very difficult to explain for the theory of evolution.

Although one cannot *prove* that living beings were *not* intelligently designed, the evidence is much more coherently explained by the naturalistic hypothesis. And a coherent explanation in agreement with the data is all that we can hope for in these matters.

Fallacy 8: It's a Debate about Origins

This is also a recurring fallacy in debates on creation–evolution, and one that I already dealt with in Chapter 2 (see also Chapters 6 on the origin of life and 7 on the origin of the universe). Briefly, the problem is that creationists do not make a distinction between different *origins* debates. For them the origin of the universe, the origin of life, and the origin of species are all one and the same.

Of course, they are not. Evolutionary biology deals only with the origin of species, and even that is only a relatively minor part of what interests evolutionary biologists. Darwinian theories have absolutely nothing to say about either the origin of life or the origin of the universe—the first one being a problem for biochemistry and biophysics, the second a problem for physics and cosmology. Again, therefore, this fallacy reflects a deep misunderstanding of the nature of science, one that scientists themselves need to correct on every possible occasion.

Fallacy 9: Scientific Findings Are Independent of Each Other

Creationist attacks have certainly not focused only on evolutionary biology. Young-Earth creationists have aimed their fire at the theory of radioactive decay, the idea that the speed of light in a vacuum is a constant,[21] the reconstruction of the geological column, the idea of slow accumulation of sediments, and essentially anything else that conflicts with the assumption of a recent episode of special creation.

On the other hand, creationists claim to be highly respectful of "good" science and are allegedly not trying to question the whole edifice of scientific

---

[20] Except perhaps to the extent that we are still aiding our kin (which would be more distantly viewed by natural selection).

[21] In fact, it may turn out that the speed of light is not a constant after all, though not in the sense that creationists maintain, and not in a way that will help reduce the age of the universe to 6,000 years. See J. Magueijo, "Brave New Cosmos: Plan B for the Cosmos," *Scientific American* 284(1) (2001): 58–59.

method and knowledge (with the declared exception of the Wedge movement, which we considered in Chapter 2).

But this is an impossible situation, yet another case in which creationists want to have their cake and eat it too. Science is not a disconnected assemblage of facts derived independently of one another. Scientific disciplines are connected in intricate ways, so a major advance or discovery in one area of science often has a ripple effect into other areas. Questioning the age of Earth denies not only the theory of evolution, but also geology and subatomic physics, because all these branches of knowledge are interrelated. The world may not be a fundamentally coherent unity, as philosopher John Dupré has maintained in his *The Disorder of Things,*[22] but the various aspects of nature studied by science are related in important and intelligible ways that science has quite successfully illuminated.

A good way to think about the situation is that science is a complex fabric with threads that, while not connected to every other thread, are attached to many others, sometimes to very distant ones. This fabric represents our understanding of reality. Occasionally a major rip is made in the fabric when we realize that a fundamental explanation was wrong. However, the challenge for a new theory is not only to replace the missing piece, but to reconnect the new thread (explanation) to all the other threads left hanging. It is this interwoven property of scientific theories that makes it impossible to question only radioactive decay and not the entire atomic theory. But creationists wouldn't dare question the latter; it is simply too well established by observations and experiments (and too complicated to understand, given its heavy mathematical formulation).

In a twist of irony that is not a rare phenomenon in this dispute, creationists do make use of the interconnectedness of science—when it appears to be convenient for their purposes. In Chapter 6 we shall discuss the never-ending controversy surrounding the second principle of thermodynamics. I maintain that creationists simply do not—or refuse to—understand the second principle, and I will try to explain why. The point here is that they claim that all sciences must agree on one fundamental reality, and if physics contradicts evolutionary biology, one of the two has to go. Because the second principle is one of the most solidly established laws of physics and of science in general . . . well, the reader will get my drift.

This sort of selective picking and choosing how to consider science is not legitimate in a serious intellectual debate, which is why the creation–evolution controversy doesn't qualify as such. Creationists have to make up their minds: They either attack science as an enterprise, or they have to stop being selective *with no other guide than their religious ideology* on what's supposed to be right and wrong about the findings of science. As I have already mentioned, the most intellectually advanced component of creationism—intelligent design

[22] See note 7.

"theory"—has indeed chosen to attack science as an enterprise, which is why this debate has implications for far more than just academia or even religion. It is about the control of a nation's resources and education.

Fallacy 10: Education Must Be Democratic

We have encountered this fallacy when discussing some forms of anti-intellectualism (see Chapter 2), and we have seen that it led directly to the infamous Scopes trial of 1925, the first legal confrontation in this long dispute (see Chapter 1).

The idea is simple: Because the taxpayers support public education, they should decide what is taught and what is not. This idea is simultaneously an example of tyranny of the majority, infringement of academic freedom, and bad educational policy.

Suppose that the Flat Earth Society (based in California) gains enough support to sweep the nation with its followers. They become an important force in local and national elections, and eventually demand a revision of all science curricula in astronomy: Schools should stop teaching that nonsense about a round Earth and warn students that if they travel far enough, they will fall off the edge of the planet. This scenario seems laughable; indeed, that is why people in virtually every other industrialized country are laughing at this state of affairs in the United States: The scientific status of creationism is in no way superior to flat Earthism.

Education is not about allowing equal time for every wacky theory that is proposed. It is about teaching the best of what we know about the world. More importantly, it is about teaching students how to think critically so that the next generation will be able to supersede current scientific notions—which they will be enabled to do not by means of ideological commitments, but through the solid science that they themselves will have produced. Alas, as we shall see in Chapter 8, we are still doing an abysmal job of teaching critical thinking. That is no excuse, however, for doing even worse and allowing creationists to have their way.

Fallacy 11: Science Is a Religion

This is perhaps the most astounding piece of reasoning of them all because it can be espoused simultaneously by people who support something they call *"scientific" creationism.* It is yet another example of creationist doublethink: If science is a religion, they maintain, then we should not teach either evolution or creationism in public schools. But creationism presents itself as a science precisely for the purpose of bringing religion into science classes throughout the country. Which is it? More importantly, what is the actual relationship between science and religion?

It is important to explore this topic in some detail here because it is so fundamental to the creation–evolution debate. As we will see, there are plenty of possible positions to take, and I will provide the reader with the essential background to start thinking about the problem in a less simplistic way than most of us are led by the popular press to adopt. I think there are three distinct components of the science and religion (S&R) discussion that need to be considered carefully:

1. S&R discussions, especially in the United States, carry practical consequences that may negatively affect science disproportionately over religion.
2. Discussing S&R has repercussions for the cherished value of freedom of speech for scientists (be they theistic or not), nontheists, and religionists alike.
3. The relationship between science and religion is a legitimate area of philosophical inquiry that must be informed by both religion (theology) and science.

The first point regarding the practical repercussions of the S&R discussion is a particularly murky one. It boils down to the fact that attacks on religion are considered politically incorrect, and that scientists are especially aware that their funding depends almost entirely on public financing through various federal agencies, such as the National Science Foundation and the National Institutes of Health. Because federal funding is controlled by politicians, who in turn have a very undemocratic tendency to respond to every nuance of opinion among their constituents as gauged by the latest poll, it follows that no matter what one's opinion as a scientist on matters of the spirit is, it is wiser to stick to one's job and avoid upsetting one's prince and benefactor. This is all the more so because of two other things that we all know about scientists: The overwhelming majority of them (about 60 percent of general scientists and 93 percent of top scientists[23]) do not believe in a personal God, and the reason they become scientists is to pursue questions for which science is a particularly good tool. Most of these questions are rather more mundane than the existence of God.

The result of this awkward mix of circumstances is that many prominent scientists and educators cease to believe in a personal God because of their understanding of science and of its implications, but come out in public with conciliatory statements to the effect that there is no possible contradiction between the two.[24] The best that one can say therefore is that there is a philosophical (or logical), if not scientific, contradiction between science and some aspects (or some types) of religion (see below), but it is not in scientists' interest to start an unholy war that they would lose hands down, given the religious and political climate of the United States. Therefore, if asked, one can answer with the universally convenient, "No comment," and live at peace with one's conscience and funding sources.

---

[23] See E. J. Larson and L. Witham, "Scientists Are Still Keeping the Faith," *Nature* 386 (1997): 435–436; and E. J. Larson and L. Witham, "Leading Scientists Still Reject God," *Nature* 394 (1998): 313.

[24] See W. Provine, "Scientists, Face It! Science and Religion Are Incompatible," *Scientist* 9(5) (1988): 10.

The second important component of the S&R discussion is rarely raised directly within the S&R debate, but it clearly lurks behind some of the responses one gets when talking or writing about it. Let me make it as clear as possible: No self-respecting scientist or educator—religious or not—would want to limit the freedom of speech or expression of any party, including creationists. There is a fundamental, if not often fully appreciated, distinction between openly criticizing a position, which is part of the very idea of free speech, and attempting to somehow coerce people into believing what one thinks is correct or to limit their ability to believe and practice what one thinks is not true. Whereas religious fundamentalists often do not respect this distinction, most religious progressives, agnostics, and nontheists follow it. The point should therefore be clear that discussions about science and religion, or evolution and creationism, deal with free inquiry and education, and in no sense are they meant to limit anybody's free speech. To say this another way, asking schools to limit what is taught in a science classroom to what is pertinent to the subject matter is sound educational policy, not censorship. Moreover, criticizing religion from a philosophical standpoint is an exercise in free speech that is valuable to our society, and it is not meant—and should not be interpreted—as a call to shut down churches and synagogues.

In order to continue our discussion of the legitimate philosophical, scientific, and religious aspects of the S&R discussion (the third component), we need a frame of reference as a guide. What I present here is an elaboration of a classification scheme proposed by Michael Shermer.[25] Shermer suggests that there are three contrasting worldviews, or "models," that people can adopt when thinking about science and religion. According to the *same worlds* model, there is only one reality, and science and religion are two different ways of looking at it. Eventually, both will converge on a set of logically consistent answers, within the limited capabilities of human beings to pursue such fundamental questions. A second possibility is represented by the *separate worlds* model, in which science and religion not only are different kinds of human activities, but pursue entirely separate goals. In this case, asking about the similarities and differences between science and religion is the philosophical equivalent of comparing apples and oranges, so it shouldn't be done. Finally, the *conflicting worlds* model asserts that there is only one reality (as the same worlds scenario also acknowledges) but that science and religion collide head-on when it comes to determining the shape that reality takes: Either one or the other is correct, but not both (or possibly neither).

Using Shermer's model as a starting point for thinking about S&R, I realized that something was missing. One cannot reasonably talk about the conflict between science and religion unless one also specifies what is meant by religion or God (usually there is less controversy over what is meant by science,

[25] In M. Shermer, *How We Believe: The Search for God in an Age of Science* (New York: Freeman, 1999).

though some philosophers and social scientists would surely disagree). So what makes Shermer's picture incomplete is the very important fact that different people have different gods or conceptions of God. I am not referring to the relatively minor variations of the idea of God within the major monotheistic religions, but to the fact that God can be one of many radically different things and that unless we specify which God we are talking about, we will not make much further progress. For example, whereas traditional Christianity conceives God to be a person, physicist Stephen Hawking thinks of God as nothing more than the mathematical laws governing the universe. These differences must be made clear, and the interest of clarity is not served by the use of the same word, *God,* for both (and indeed many other) conceptions.

My tentative solution to the problem is therefore presented in Figure 5.4. Here the panoply of positions concerning the S&R debate is arranged along two axes: On the abscissa we have the level of contrast between science and religion, which goes from none (same worlds model) to moderate (separate worlds) to high (conflicting worlds). On the ordinate is the "fuzziness" of the concept of God, which ranges from a very well defined, personal God who intervenes in everyday human affairs, to the more imprecise concept of a naturalistic God who acts through or is equivalent to the laws of physics, to the

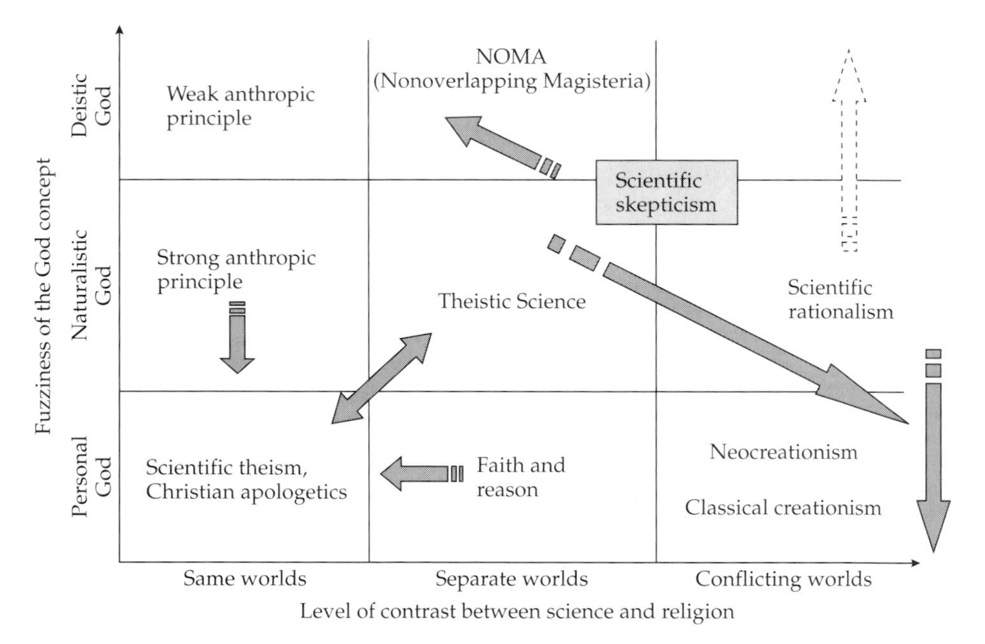

Figure 5.4 A classification scheme of attitudes toward science and religion based on two criteria (represented by the horizontal and vertical axes): the level of perceived contrast between science and religion, and the type of concept of God one subscribes to.

most esoteric position of deism characterized by an undefined God who created the universe but has not interfered with it since its creation.

Of course, all of these conceptions of God can take many specific forms. However, the common denominator of the belief in a personal God is the idea that He (or She) intervenes in individual lives, performs miracles, or otherwise shows direct concern for us mortals. A naturalistic God, on the other hand, is a bit more detached: If He intervenes at all, it is through the tortuous ways of the natural laws that He himself designed for this universe. Finally deism, broadly speaking, is the idea that God created the universe and then for all effective purposes retired, perhaps to enjoy the fruits of such labor. This kind of God does not interfere, even indirectly, in human affairs, but simply answers the fundamental question of why there is something instead of nothing.

Let us start at the lower left of the diagram, with the position I labeled *scientific theism,* which is adopted by people who think that science and religion deal with the same world and believe in a personal God. One of these people is Sir John Templeton, a British citizen and native of Tennessee (U.S.), who has invested $800 million of his personal fortune into furthering a better understanding of religion through the scientific method. The Templeton Foundation has sponsored a panoply of activities resulting in articles, books, and conferences whose goal is to "discover spiritual information."[26] According to Sir John, science has made incredible progress in discovering truths about the natural world. Therefore, its powerful methods should be useful to religion in order to augment our knowledge of God and all things spiritual. Templeton is putting his money where his mouth is by funding several scientific projects (at the rate of hundreds of thousands of dollars each), as well as by giving out the Templeton Prize, which is currently (monetarily) heftier than the Nobel.

Examples of the science-to-religion connection that Templeton envisions are illuminating. His foundation has given hard cash to Pietro Pietrini of the National Institute of Neurological Disorders and Stroke to study "imaging brain activity in forgiving people" ($125,000); Lee Dugatkin of the University of Louisville has been awarded $62,757 for research on "evolutionary and Judaic approaches to forgiving behavior"; Herbert Benson of Harvard has been aided in searching for an answer to the question, "Does intercessory prayer help sick people?"; and Frans de Waal of Emory University is investigating "forgiveness" among primates. All of these projects are being conducted on grants from Templeton.

Scientific theism is actually a very old and revered position within the S&R universe, having its roots in the classical Christian apologetics of St. Thomas Aquinas and continuing today through the efforts of individual theologians like Alvin Plantinga and William Craig. If, however, one believes in a more remote kind of God but wishes to retain the concept of science and religion un-

---

[26] On the Templeton Foundation, see C. Holden, "Subjecting Belief to the Scientific Method," *Science* 284 (1999): 1257–1259.

covering the same truth, the choice is not limited to scientific theism. Two other positions are possible, depending on whether one subscribes to a naturalistic or to a deistic God. For lack of better terms, I labeled them the *strong anthropic principle* and the *weak anthropic principle,* respectively; the latter is also known as the *God of the Big Bang.*

The weak anthropic principle says basically that there can be very little variation in the known constants and laws of physics if the universe is a place friendly to life as we know it.[27] As is, this is a rather trivial observation,[28] but if one wants to read philosophical implications into it, then it is a small leap of faith to claim that the *purpose* of the universe is to allow life to exist. This brings us to the strong anthropic principle, which infers the activity of an intelligent designer behind the whole shebang.[29] Several physicists and cosmologists have played with different versions of the anthropic principle, including Frank Tipler, one of the original proponents of it.

The anthropic principle is difficult to counter on purely philosophical grounds. At its most trivial, it simply begs the question and makes rather strange statements about causation (akin to, "We are here because we are here"). This makes it not particularly useful as a scientific hypothesis because it is difficult to see how one could derive falsifiable predictions from the principle. The assumptions of the anthropic principle (even in its "weak" version) have, nonetheless, been effectively attacked on positive scientific grounds by demonstration of the fact that many more possible universes could support some sort of life—an attack that has weakened the "improbability" argument on which the principle is based.[30] A more fatal blow may come in the near future from superstring theory, the current working hypothesis for the reconciliation of the theories of relativity and quantum mechanics.[31]

Although all these positions are compatible with Shermer's same worlds scenario, it is clear that a scientist feels more and more comfortable toward the upper end of the ordinate in my diagram—that is, the more fuzzy and distant the concept of God becomes (notice that one can adopt a strong anthropic prin-

---

[27] For a very clear discussion of the anthropic principle, see J. L. Casti, *Paradigms Lost* (New York: Avon, 1989).

[28] However, it is often overlooked that clearly some of the "amazingly perfect" conditions that allow life to exist are not "fine-tuned or perfect." For example, why would plants increase their productivity under higher CO_2 concentrations than the ones actually encountered in Earth's atmosphere, implying that either the atmosphere or the plants could be better suited to the task of producing life? Furthermore, some "perfect" conditions did not even exist before the initial origin of life (oxygen levels, for example), and were produced by the activity of living organisms themselves.

[29] For a presentation of the strong anthropic principle, see F. J. Tipler, *The Physics of Immortality: Modern Cosmology, God and the Resurrection of the Dead* (New York: Doubleday, 1995).

[30] Against the anthropic principle, see V. J. Stenger, "Cosmythology: Was the Universe Designed to Produce Us?" *Skeptic* 4(2) (1996): 36–41; and B. J. Leikind, "Do Recent Discoveries in Science Offer Evidence for the Existence of God?" *Skeptic* 5(2) (1997): 66–69.

[31] See Brian Greene's book *The Elegant Universe: Superstrings, Hidden Dimensions, and the Quest for the Ultimate Theory* (New York: Norton, 1999).

ciple scenario and slip toward a personal God at the same time, as indicated by the arrow in the figure). This observation by itself, I think, points toward a fundamental degree of discomfort between science and religion.

When we examine the portion of the graph in Figure 5.4 that falls in the area identified by Shermer as the domain of the separate worlds model, we deal with a range of characters that go from agnostic evolutionary biologist Stephen J. Gould (Harvard) to the pope himself (John Paul II), passing through the ambiguous position of the charismatic Huston Smith (the acclaimed author of *The World's Religions*[32]) (see Chapter 4). Let's see how this variation is again accounted for by the different concepts of God that these positions reflect.

Several scientists, philosophers, and nontheists loosely fall into the position labeled by Gould as "NOMA," or "nonoverlapping magisteria."[33] The idea is basically that science deals with facts and religion deals with morality; the first focuses on what is, the latter on what ought to be. Citing what in philosophy is known as the *naturalistic fallacy*,[34] the principle that one cannot derive what ought to be from what is, Gould concludes that science and religion are forever separate. Another way to look at NOMA is articulated by Eugenie Scott of the National Center for Science Education[35] when she makes the distinction between methodological and philosophical naturalism.[36] According to Scott, science adopts naturalism only in a methodological sense as a convenient tool for conducting research. To deny the existence of God, however, one has to be a naturalist in the philosophical sense of the term; that is, one has to conclude that the physical world is all there is. Therefore, science cannot inform us as

[32] H. Smith, *The World's Religions: Our Greatest Wisdom Traditions* (San Francisco, CA: HarperSanFrancisco, 1991).

[33] On the idea of nonoverlapping magisteria, see S. J. Gould, "Nonoverlapping Magisteria," *Natural History* 106(2) (1997): 16–22; and S. J. Gould, *Rocks of Ages* (New York: Ballantine, 1999); as well as my critique in M. Pigliucci, "Gould's Separate 'Magisteria': Two Views. A review of *Rocks of Ages: Science and Religion in the Fullness of Life* by S. J. Gould, Ballantine, 1999," *Skeptical Inquirer* 23(6) (1999): 53–56.

[34] The term *naturalistic fallacy* is usually attributed to philosopher G. E. Moore in *Principia Ethica* (Cambridge: Cambridge University Press, 1903). However, in the sense discussed here, which is generally accepted, the concept actually goes back to skeptic philosopher David Hume (see A. Flew, *A Dictionary of Philosophy* [New York: Gramercy, 1999]). The fallacy consists of deducing conclusions about what ought to be from premises that state only what is the case. As Hume put it in his *Treatise of Human Nature,* (III(i)1): "In every system of morality, which I have hitherto met with, I have always remark'd, that the author proceeds for some time in the ordinary way of reasoning . . . when of a sudden I am surpris'd to find, that instead of the usual copulations of propositions, *is,* and *is not,* I meet with no proposition that is not connected with an *ought* or an *ought not.* This change is imperceptible, but is, however, of the last consequence. For as this *ought* or *ought not,* expresses some new relation or affirmation, 'tis necessary that it shou'd be observ'd and explain'd; and at the same time that a reason should be given, for what seems altogether inconceivable, how this new relation can be a deduction from others, which are entirely different from it."

[35] See E. C. Scott, "The 'Science and Religion Movement'," *Skeptical Inquirer* 23(4) (1999): 29–31.

[36] For an in-depth discussion of the difference between methodological and philosophical naturalism, see B. Forrest, "Methodological Naturalism and Philosophical Naturalism: Clarifying the Connection," *Philo* 3 (200): 7–29.

to the existence of God because naturalism is not a scientific conclusion, but only an assumption of the scientific method. If science does not have anything to say about God (and obviously, also according to Scott, religion is incapable of informing science about the natural world), then NOMA or a similar position logically follows.

Moving down the God axis in Figure 5.4, we come to what I have termed *theistic science* (as opposed to scientific theism, which we have already encountered). It is not exactly clear how well theologian Huston Smith fits into this category, but his position is the closest I could find to represent the land between NOMA and the pope (the latter is associated with the *faith and reason* quadrant, to which I will turn next). Notice the diagonal arrow bridging theistic science and scientific theism, which could to some extent represent two sides of the same coin. Smith argues against scientism, an idea that can be defined in different ways and that was discussed in Chapter 4. I argued there that scientism is the concept that science can and will resolve every question or problem in any realm if given enough time and resources. I do not think that most professional researchers readily subscribe to it, but I know of individuals who seem to. Smith, however, thinks of scientism[37] as the idea that the scientific method is the best way to investigate reality. According to Smith, there are other equally valid or superior ways, which include intuition and religious revelation. The important point is, however, that these latter two alternatives are not available within science, so certain aspects of "reality" are excluded from scientific investigation.

Although the area occupied by theistic science is borderline and intermixed with different degrees of scientific theism and NOMA, the general idea is that according to theistic science it is perfectly sensible to say that there is a God as well as a physical universe. The distinctive point of theistic science is that the God behind the universe works in very subtle ways and entirely through natural laws, so it is impossible, or at least very difficult, to infer His presence (unlike the case of the anthropic principle already discussed, where an intelligent designer is the only possible conclusion). As the reader can see, then, the center of the diagram in Figure 5.4 is a rather gray area from which one can easily move to most other positions in the diagram (especially toward the left) by introducing one or more caveats. If applied to evolution in particular, theistic science translates into theistic evolution. This is the idea that evolutionary theory is by and large correct (therefore science is on solid ground), but it includes the added twist that evolution is the (rather inefficient and clumsy) way that God works. This is what Reverend Barry Lynn (Americans United for Separation of Church and State) may have meant when he concluded the 1997 PBS *Firing Line* debate for the evolution side by suggesting that the Word (God) in the beginning may simply have been "Evolve!"

[37] For a good example of Smith's position on science and religion, see H. Smith, "The Place of Science," in *Forgotten Truth: The Common Vision of the World's Religions*, pp. 96–117 (San Francisco, CA: HarperSanFrancisco, 1993).

The Pope's position is labeled here *faith and reason* (from the title of the encyclical *Fides et Ratio*)[38] and assumes the personal God of Catholics but includes an element of fuzziness as well. It is accompanied by an arrow pointing left in Figure 5.4 because one could think of it as a variation of the same worlds model that does not go quite as far as scientific theism à la Templeton. In a letter written to the Pontifical Academy of Sciences,[39] the pope declared that Christians should accept the findings of modern science, and evolutionary theory in particular. The reason is that, in his words, "Truth cannot contradict Truth" (which is why this position could be construed as leaning toward the left side of the diagram). However, he drew a line at the origin of the human soul, which of course had to be injected into the body directly by God. This creates a rather abrupt discontinuity because it introduces an arbitrary dualism within the process of human evolution—a stratagem that does not sit very well with science—as Richard Dawkins pointed out.[40] John Paul II then published *Fides et Ratio,* referred to earlier, which is a scholarly contribution arguing that science and faith can be used to uncover parallel realities for which each is best equipped—pretty much what Gould states are the foundations of NOMA. It is because of this position and the implied dualism that I would situate the Pope toward the center of our diagram.

Within the separate (or almost separate) worlds, therefore, one can go from essentially no conflict between science and religion if a deistic God is considered, to a position that is logically possible but increasingly inconsistent with the principle of parsimony known as Occam's razor (that simpler hypotheses should be preferred whenever the evidence does not warrant more complex explanations). Depending on how much importance one accords to the philosophical foundations of science, this area of the S&R space can be more or less comfortably inhabited by both moderate scientists and moderate religionists.

The lower right-hand corner of Figure 5.4 characterizes two positions that we have discussed throughout the book and obviously do not need any further introduction here: (1) *classical creationism* as embodied, for example, by Duane Gish and his fellows at the Institute for Creation Research; and (2) the *neocreationism* movement represented, among others, by Michael Behe, William Dembski, Phillip Johnson, and other associates of the Discovery Institute.

No matter what kind of creationism is espoused, the creationist is very likely to believe in a personal God and in a fundamental conflict between science and religion (otherwise there wouldn't be a creation–evolution controversy to begin with). The main difference between Gish's group and Johnson's ensemble is that the latter are more sophisticated philosophically and make a slicker use of scientific terminology and pseudoscientific concepts. They are also much

[38] The full text of the Pope's encyclical is available at http://www.cin.org/jp2/fides.html.

[39] See "The Pope's Message on Evolution and Four Commentaries," *Quarterly Review of Biology* 72 (December 1997): 381–406.

[40] See R. Dawkins, "You Can't Have It Both Ways: Irreconcilable Differences?" *Skeptical Inquirer* 23(4) (1999): 62–64.

more politically savvy, though they do not enjoy the grassroots support of some traditional creationists because they ironically tend to be seen by some of the latter as "too intellectual."

Finally, let me consider the two main versions of modern skepticism. I am referring to what in Figure 5.4 are labeled *scientific skepticism* and *scientific rationalism,* positions associated with people such as Carl Sagan and Richard Dawkins. First, notice that both skeptical positions are rather unusual in that they span more than one quadrant, diagonally in the case of scientific skepticism, vertically for scientific rationalism. Scientific skepticism is the position that skepticism is possible only in regard to questions and claims that can be investigated empirically (i.e., scientifically). However, scientific skepticism potentially embarks on a slippery slope because testable religious claims, such as those of creationists, faith healers, and miracle men are amenable to scientific skepticism, so religion is not entirely out of the scope of skeptical inquiry. Furthermore, it is difficult to distinguish between religious and pseudoscientific claims of a variety of sorts: Beliefs in leprechauns, alien abductions, ESP, reincarnation, or the existence of God—all equally lack objective supporting evidence. From this perspective, separating out the latter two beliefs and labeling them as *religion*—thereby exempting them from critical analysis—is problematic at best. Scientific skepticism therefore converges toward scientific rationalism when one considers personal gods that intervene in everyday life, but moves toward a NOMA-like position if God is defined in a distant and less comprehensible fashion.

One of the most convincing arguments adduced by scientific skeptics to keep religion out of skeptical inquiry is that a believer can always come up with unfalsifiable, ad hoc explanations of any inconsistency in a religious belief. Although this is certainly true, this is an equally valid critique of, say, skeptical inquiry into paranormal phenomena, which is routinely carried out nonetheless. After all, true believers in the paranormal often say that the reason a medium failed a controlled test was the negative vibrations produced by the skeptic, or some other clearly ad hoc explanation. In his excellent book *Leaps of Faith,*[41] Nicholas Humphrey even reports that paranormalists have come up with a negative theory of ESP "predicting" that the frequency of genuine paranormal phenomena is inversely proportional to attempts at empirically investigating them! This is not very different from the statements of at least some religionists bent on "proving" the existence of their particular God.

As much as one might raise more or less subtle philosophical points concerning scientific skepticism, there is another, more practical side to this position, which also makes for a convergence toward NOMA. As in the case of scientists, even most skeptics simply do not want to be entangled in religious disputes and prefer instead to concentrate their energy on debunking UFOs

---

[41] N. Humphrey, *Leaps of Faith: Science, Miracles, and the Search for Supernatural Consolation* (New York: BasicBooks, 1996).

and paranormal phenomena, a position dictated at least by the scarcity of available resources and by the realization that the God question is a bit too sensitive for too many people.

The last position depicted in Figure 5.4 is that of scientific rationalism. Within this framework, one concludes the nonexistence of any God, not because of certain knowledge, but because of a sliding scale of methods. At one extreme we can confidently rebut the personal God of creationists on firm empirical grounds: Science is sufficient to conclude beyond reasonable doubt that there never was a worldwide flood and that the evolutionary sequence of the cosmos does not follow either of the two versions of Genesis. The more we move toward a deistic and fuzzily defined God, however, the more scientific rationalism reaches into its toolbox and shifts from empirical science to logical philosophy informed by science. Ultimately the most convincing arguments against a deistic God are summarized by what is sometimes referred to as *Hume's dictum* (extraordinary claims require extraordinary evidence, as Carl Sagan paraphrased it) and the already encountered Occam's razor. These are philosophical arguments, although they also constitute the bedrock of all of science, as we saw in Chapter 4. The reason we trust in these two principles is that their application in the empirical sciences has led to such spectacular successes throughout the last three centuries. Where, if anywhere, they will lead us when applied to religion, remains to be seen.

The scientific rationalist is surely on less firm ground the more she moves vertically upward in Figure 5.4, but that is not necessarily a fatal flaw, because no reasonable person asserts her positions as definitive truths. The main difference between scientific rationalism and scientific skepticism (which is more akin to the philosophical position known as empiricism and espoused by many English philosophers between the seventeenth and nineteenth centuries) comes down to a matter of which trade-offs one is more willing to accept. Scientific skepticism trades off the breadth of its inquiry (which is limited) for the power of its methods (which, being based on empirical science, are the most powerful devised by humankind so far). Scientific rationalism, on the other hand, retains as much of the power of science as possible, but uses other instruments—such as philosophy and logic—to expand the scope of its inquiry.

It should be obvious from this survey that there are many ways to slice the S&R question, undeniably more than I have discussed or can even think of. As I mentioned at the outset, the point is not to censure any particular position, but rather to explore their differences from a logical as well as a psychological perspective. In fact, it is almost as interesting to debate the question at hand as it is to wonder why some people subscribe to one point of view or another.[42]

[42] Michael Shermer's findings on this point in *How We Believe: The Search for God in an Age of Science* (New York: Freeman, 1999) are particularly illuminating. Apparently most people think that they believe in God for rational reasons (which are usually variants of the intelligent design argument), and they think other people believe because of emotional needs. Obviously, something is amiss in this picture . . .

Whereas the most nontheistic position is represented by scientific rationalism, scientific skepticism, NOMA, and even some very weak forms of the anthropic principle are certainly difficult to definitely exclude, and enough intelligent people adopt them to provide some cause for reflection. The more we move from the upper right to the lower left or lower right quadrants of Figure 5.4, however, the more difficult one's position becomes to defend empirically or rationally.

The two axes of the figure characterize the degree of personality of the God one believes in and the conflict one feels between that concept of God and the world as science explains it. As such, this diagram defines a series of fuzzy, slowly intergrading areas of thinking that can help us understand both the relationship between science and religion and the human protagonists of this debate. Where one sees oneself and others in Figure 5.4 is bound to shape one's life trajectory in this world and one's interactions with other people.

Additional Reading

Dawkins, R. 1996. *The Blind Watchmaker: Why the Evidence of Evolution Reveals a Universe without Design* (New York: Norton). Another clear and lucid explanation of the process of natural selection by Richard Dawkins. Also a good answer to intelligent design arguments.

Gould, S. J. 1996. *The Mismeasure of Man* (New York: Norton). A classic critique of the sometimes facile science that goes into the study of human nature. An excellent introduction to the fact that the nature of science is to be open to continuous discussion and revision, contrary to the creationists' assumption.

Humphrey, N. 1996. *Leaps of Faith: Science, Miracles, and the Search for Supernatural Consolation* (New York: BasicBooks). An excellent book on pseudoscientific thinking and its fallacies.

Longino, H. E. 1990. *Science as Social Knowledge: Values and Objectivity in Scientific Inquiry* (Princeton, NJ: Princeton University Press). A somewhat technical but very clear book on the philosophical foundations of the scientific enterprise.

Shermer, M. 1999. *How We Believe: The Search for God in an Age of Science* (New York: Freeman). A delightful book on why people believe in God and why they think *others* do.

– 6 –

Three Major Controversies

One's ideas must be as broad as Nature if they are to interpret Nature.

– *Sherlock Holmes, in Arthur Conan Doyle's "A Study in Scarlet"* –

There is one thing even more vital to science than intelligent methods; and that is, the sincere desire to find out the truth, whatever it may be.

– *Charles Sanders Pierce* –

The goal of this book is not to provide a laundry list of creationist complaints and scientific answers, partly because that would be an exercise of limited use, but also because it has been done so much more thoroughly by other people.[1] However, some of the specific issues under debate must be addressed simply because they help the reader form an accurate picture of what the debate is about as well as why it has developed. In Chapter 5 I examined several typical fallacies in creationist reasoning (and in Chapter 7 I will do the same for the approach taken by scientists and educators). In this chapter I wish to focus on three major components of the controversy that recur in practically every outlet where creationism and evolution are discussed: the second principle of thermodynamics, the origin of life, and the Cambrian explosion (usually coupled with the question of the incompleteness of the fossil record).

I will examine each of these in considerable detail because they are paradigmatic of what is so difficult for scientists to explain to the public and of what is so easy for creationists to misunderstand and bend in their favor. Each section of this chapter begins with a short statement of the creationist and evolutionist claims and ends with a conceptual summary of both positions to help the reader make side-by-side comparisons. Of course, books have been written on each of these topics, and I refer the reader to the list of references at the end of the chapter as well as to the footnotes for a more in-depth treatment of the scientific and often philosophical issues prompted by these areas of research and controversy.

The Second Principle of Thermodynamics

Creationist Claim: The science of thermodynamics, in particular its second principle, makes evolution impossible.

Evolutionist Claim: Evolutionary change not only does not contradict any known law of physics, but in fact is made possible by the principles of thermodynamics.

Classical thermodynamics is a branch of physics that deals with the behavior of any system capable of exchanging energy with other systems. This means that thermodynamics really covers everything in the universe because everything is made of energy/matter (the two equivalent states of all things that exist, according to Einstein's theory of general relativity). It is called *classical* thermodynamics because it was developed before the modern era of quantum mechanics, and its mathematical formulation does not require any of the sophisticated and (to the layperson) somewhat abstract tools of modern physics.

Thermodynamics is a *statistical* theory. That is, it describes not exactly what every single atom or molecule in a given system actually does, but only the av-

---

[1] The best and ever growing collection of point-by-point rebuttals of creationist claims is the Talk.Origins Archive at http://www.talkorigins.org/.

erage, or general, behavior of all the particles that make up that system. So, for example, to apply thermodynamics to the understanding of what happens inside a car engine, one does not need to keep track of every single molecule of gasoline. This is fortunate because otherwise one would have to measure each molecule's position and velocity before being able to tell how and with what efficiency the engine was going to work. Instead, the principles of thermodynamics require a physicist to know only how much gasoline is in the engine, what the operating temperature is, and what the pressure of the system is, to derive a satisfactory description of the working of the engine itself.

Such simplicity, of course, as with everything in science, comes at a cost. Thermodynamic theory makes a series of simplifying assumptions about the real world, and its results are only as good as such assumptions are reasonable. Two major assumptions of thermodynamics must be understood. First, each component of the system (in our example, each molecule of gasoline) is assumed to be equal to each of the others. This assumption makes it possible to ignore the potentially idiosyncratic behavior of each molecule and focus on the (average) properties of the *ensemble* of molecules—that is, on the whole amount of gas present in the engine. Second, the systems studied by classical thermodynamics are assumed to be at *or near* equilibrium. This means that thermodynamics is not very good at telling us what happens under unusual circumstances, or when the system's characteristics are changing dramatically over short periods of time. But the theory is excellent for explaining what happens during the "normal" functioning of the system, provided that the changes inside the system itself are slow and gradual. In other words, thermodynamics can tell us pretty well what is happening in the engine once it is running and the driver is keeping it at more or less the same velocity (or if she is altering velocity slowly). But the theory is much less satisfactory in extreme situations, such as when one turns the engine on or off, or keeps accelerating and decelerating wildly. *The more sudden the change in the system, the less accurate thermodynamics is in telling us what is going on.* In technical terms, the accuracy of the theory diminishes when the system is far from equilibrium.

Modern thermodynamics has four laws, although usually only the second one (sometimes also the first) is invoked in creation–evolution debates. The so-called *zeroth law* (because historically it was added after the first three, yet it provides the conceptual foundations for them) simply defines what temperature is. Given that any number of systems (objects) that come into contact with one another exchange heat until eventually they reach an equilibrium, the quantity that stabilizes when that equilibrium is reached is temperature.[2]

[2] The reader may recall other definitions of temperature from introductory physics. For example, temperature is proportional to the average velocity of the particles in a system. In mathematical terms, in a system in equilibrium, $pV = nRT$, where p = pressure, V = volume, n = moles of gas (the number of molecules), R is the universal gas constant, and T = temperature. So not only temperature stabilizes when equilibrium is reached, but so do also pressure and volume.

The *first law* (or principle) states that energy can neither be created nor destroyed, but only transformed (this principle is also known as the principle of *conservation of energy*). This means that any energy can be used to do work, but the total amount of work can never be larger than the amount of energy with which one starts (and, in fact, it is lower according to the next law).

The *second law* of thermodynamics says that the degree of disorder (*entropy*) of a closed system cannot decrease. This statement is a little esoteric and requires a bit more explanation. To put it another way, every closed system is bound to become increasingly disordered over time. For example, if you close down your room and isolate it from the rest of the universe (i.e., you don't clean it regularly), it will become dusty, and objects will eventually start falling apart. The only way you can keep it clean and orderly is by *working on it*—that is, by spending energy once in a while to dust it and make sure that the books don't become moldy or the bed doesn't rust. Does that mean you are violating the second principle of thermodynamics by keeping your room in order? No, you have simply violated one of its fundamental assumptions: that the room is a *closed system*. Once you start doing work in the room, the system becomes open, and it is now made not only of you and the room, but also of everything that contributes to giving you the energy that you are using to set things straight in the room.

As another example, consider the refrigerator that keeps your food edible. The electric generator to which the refrigerator is connected, the food industry that provides you with what you put into the refrigerator, and even the photosynthetic apparatus by which plants fix solar energy and transform it into the food that you can eat are all parts of the larger system. The second principle is not violated as long as the entropy (disorder) of the *entire* system is increased, although local pockets of the system (such as your refrigerator) may temporarily experience a decrease in entropy.[3]

The *third principle* (which is actually the fourth, if we count the zeroth) simply says that there is a minimum temperature in the universe, called *absolute zero* (–273.2°C, or –459.7°F), below which no physical system can go. What would happen if something could reach that temperature? Well, all movements of atoms and subatomic particles would cease, and (because of the consequences of the other principles) the entire universe would be forever frozen at a standstill. This idea is called the *thermal death* scenario of the end of the uni-

---

[3] Incidentally, the second principle also predicts that there can never be a *perpetual motion machine*—that is, a machine that, once started, can keep going forever. The reason is that any transformation of energy into work is never 100 percent efficient because the system as a whole has to increase its entropy. Therefore, even in machines with infinitesimal friction and very efficient engines, heat is lost while the work is being done. Since heat is the lowest possible form of energy (i.e., it has the lowest capacity to do work), it acts as a sort of last station on the energy chain. Eventually, all the energy in the universe will be transformed into heat, and no work will ever be possible again. A sad but inescapable consequence of this principle is that you cannot count on ever buying a car that, once started, will keep going forever without need for refueling.

verse, and according to one possible outcome of the predictions of modern physics, it is what will eventually happen several billion years from now.

So why—according to creationists—does evolution contradict the principles of thermodynamics, especially the second one? The basic reasoning can be summarized like this:

1. The second law of thermodynamics predicts that the amount of disorder is bound to increase and can never decrease.
2. Living organisms, on the other hand, are a clear example of highly ordered systems. Therefore, their very existence contradicts the second principle, unless supernatural intervention is called upon to maintain their existence.
3. Furthermore, evolution is a process whose main characteristic is to increase life's complexity over time.[4] Therefore it flies in the face of the second principle, which predicts that eventually the universe will reach the state of maximum disorder known as thermal death.
4. Hence, either the second law or evolution is wrong. Since the second law is accepted by every physicist, it has been repeatedly confirmed experimentally, and we do not know of any exception to the principle, evolution must be discarded.

There are several more facets to the creationist argument. One of them is that the Bible allegedly predicts the incompatibility of evolution and thermodynamics. Although scientists do not accept the Bible as a scientific text, if indeed that book contains assertions that can be reasonably shown to be pertinent to the debate, they must be considered. This is especially so if such assertions predict something that contradicts a scientific theory and is then demonstrated to be correct. The key passage in the Bible that is deemed relevant by some creationists is found in Ecclesiastes 3:14. "I know that, whatsoever God doeth, it shall be forever: nothing can be put to it, nor anything taken from it." This is interpreted as an alternative statement of the first principle of thermodynamics: No energy can be created or destroyed because that would add to or detract from God's work. The sentence in Ecclesiastes is relevant to the evolution debate because the first law would imply that the universe could not possibly have come out of nothing (i.e., created itself), since energy cannot appear out of no energy. Therefore, there must have been a Creator. If there was a Creator, then any contradiction between the second law of thermodynamics and the existence of biological organisms can be explained by supernatural intervention (i.e., God makes life's complexity possible by suspending the second principle by His will).

The creationist line of reasoning on evolution and thermodynamics encompasses two more, equally important, points. First, they (e.g., Duane Gish of the Institute for Creation Research) maintain that the second law does not apply only to closed systems, but can be generalized to open systems as well. There-

[4] "Evolution is change outward and upward," according to creationist H. M. Morris, in "Evolution, Thermodynamics, and Entropy," *Impact* no. 3 (1973), an Institute for Creation Research pamphlet.

fore, *all* systems in the universe, no matter how complex or extended, are bound to increase their own entropy (level of disorder). This point is crucial because if it is true, these systems would include living organisms and the process of evolution (which, as we shall see shortly, are open rather than closed systems).

Second, the creationists argue, even if we allow for the possibility of local reductions of disorder in open systems, there are many more conditions that must be met before evolution can occur. In particular, Morris maintains[5] that the simple availability of energy from outside a given system (e.g., solar energy entering the biosphere) is not enough to allow the evolutionary process to take place. He asserts that two more conditions are needed. On the one hand, there must be a "program," so that the process can be directed toward a particular outcome (e.g., the appearance of humans). On the other hand, there has to be a "conversion mechanism" that transforms raw energy (e.g., from the sun) into high-quality energy that can be used by the system to do work. For example, according to Morris, building a house does not violate the second principle of thermodynamics because not only are there materials and energy coming from outside the house (i.e., it is an open system), but there is also a directing program (in the form of the blueprints for the house) and a conversion mechanism (represented by the workers who actually use energy and materials to build the house according to the program). If one has just bricks and energy, the house will never be built. Although the biosphere has materials (provided by Earth) and energy (provided mostly by the sun), Morris maintains that it lacks both a program and a conversion system. Therefore, evolution is impossible.

For a long time there has indeed been some misunderstanding even among evolutionary biologists about the relationship between evolution and thermodynamics. This confusion is more than likely due to the fact that biologists are biologists, not physicists (misunderstandings of biology on the part of physicists are also known to occur). In fact, a few biologists still contribute to the overall confusion, and it is no wonder that the general public cannot see through the jargon and the rhetoric on both sides. However, the matter was very clearly put to rest at least as early as 1971 by Nobel Prize–winning biochemist Jacques Monod. In his *Chance and Necessity*,[6] Monod eloquently explained not only that there is no contradiction between any principle of physics and biology, but also, in fact, that evolution is possible *because* of the laws of thermodynamics.

The evolutionist argument is based on several points. First, the fact that living organisms are ordered systems does not necessarily mean that they violate the second principle. If that were the case, not only could evolution not have happened, but also a fertilized egg could not develop into a fish, or any other animal. Furthermore, we could not build roads and houses, or for that matter

[5] H. M. Morris, "Entropy and Open Systems," *Impact* no. 40 (1976), an Institute for Creation Research pamphlet.

[6] J. Monod, *Chance and Necessity; An Essay on the Natural Philosophy of Modern Biology*, translated from the French by A. Wainhouse (New York: Knopf, 1971).

even keep our bedrooms clean. All of these are examples of (natural and human) orderly systems, characterized by a much reduced level of entropy compared to their surroundings. None of these, however, violates the second principle because the resulting order is obtained at the cost of increasing the disorder of the universe at large.

Let us discuss in more detail the example of the refrigerator mentioned earlier and illustrated in Figure 6.1. One of the things happening inside the refrigerator is that the water you put into the freezer compartment becomes ice. Ice is a crystalline substance and therefore a much more orderly form of water than liquid water. To put it into thermodynamic terms, your ice has formed because the water has *lost* entropy! But if the second principle says that a decrease in entropy is impossible, how could that have happened? The reason is that your refrigerator is actually using energy from an outlet. That energy is coming from somewhere outside the refrigerator, and it is produced, and carried to your house, by an increase in the entropy of the surroundings (e.g., at a power plant

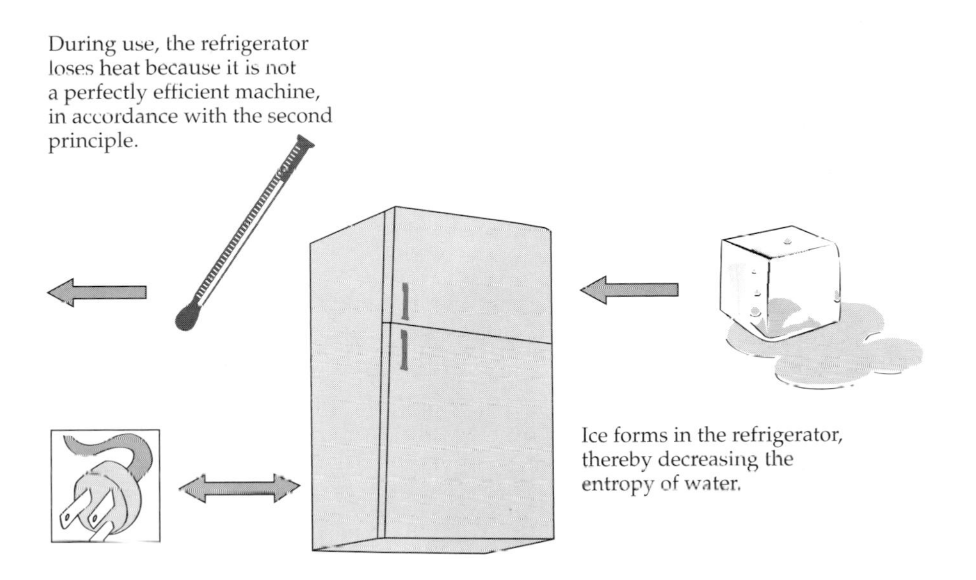

Figure 6.1 A schematic explanation of why a refrigerator, a system that dramatically decreases entropy, does not violate the second principle of thermodynamics. Note that the fact that humans designed the refrigerator does not invalidate its use as an example in favor of evolution. If we were able to construct something that *did* violate the second principle, we would have a powerful argument against evolution by natural means.

where the entropy of coal or atoms is being increased by their being burned or split, respectively). Furthermore, the refrigerator (also in accordance with the second principle) is not a perpetual machine, and it loses energy in the form of unusable heat (if you check, the back of your refrigerator is warm, and that is why it needs a ventilation system). This dissipated energy also increases the entropy of the environment. Such an increase in entropy, combined with the one occurring during the production and transfer of energy to the refrigerator, is *higher* than the decrease in entropy caused by the crystallization of water.

What does this have to do with evolution? In order for evolution not to be at odds with thermodynamics, whatever order is created during the evolutionary process simply must be more than offset by the disorder created by the same process. This is exactly what happens because the entire biosphere receives huge quantities of energy from outside sources, mostly the sun.[7] The production and transfer of energy from the sun to the biosphere increases entropy by a large amount throughout the solar system—exactly as the transfer of energy to your refrigerator does. Moreover, while energy flows through different organisms (from plants to the animals that eat them, for example), it generates large quantities of heat, which is not usable for further work and thus contributes to the general increase in entropy. This is why, incidentally, only a small number of carnivores can be sustained by a large number of herbivores, which in turn need a huge quantity of plants: because at each step in the food chain, (a lot of) energy is irretrievably lost.[8]

The situation is analogous to changing money if you have to pay a commission. You can go to a bank, or to an exchange agency, and get a certain number of British pounds for your American dollars. However, you have to pay a small percentage of the total to the bank as a commission. Now, let us suppose that as soon as you walk out of the bank you change your mind: You decide you don't want to go on vacation to England after all, and you would rather spend your money in New York City. So you go back and change your pounds into dollars. But you pay a *second* commission, thereby reducing your total holdings. If you keep being uncertain about where to have your dream vacation, and you keep going in and out of the exchange office, you will pretty soon have no vacation at all—because there will be no money left.

Thermodynamics says simply that any exchange of energy is always accompanied by a loss that is never recoverable. The lost heat is nature's equivalent to your small—but always to be paid—commission to the exchange office. Now this absolute law does not imply that you cannot increase your pot of money again. For example, you can have some money wired to you by your parents (who are outside the closed system represented by you and the bank) and restart the process of making up your mind about the vacation. What you *cannot* do is avoid

[7] Some microorganisms use alternative sources of energy, such as special chemical reactions, or the thermal energy emanating from very hot spots on Earth's surface or from its oceans' floors.

[8] This also means that if we really want to populate the planet with billions of people—a questionable strategy—it would be much more efficient if all of us were vegetarians.

paying the commission every time you go to the exchange office. Analogously, evolution can occur *and* does forge new order in nature, but it has to pay the thermodynamic commission every single step of the way. The hard cash has to come from somewhere else (like your parents' money), specifically from the sun.

A second important point to consider is that, contrary to the statement by Morris quoted in note 4, evolution is not defined as "change outward and upward." First of all, "out" and "up" with respect to what? Even though most biologists would agree that evolutionary change has been accompanied by an increase, in one form or another, in the complexity of living organisms, there are many exceptions to the pattern. For example, parasites are usually *less* complex than their close evolutionary relatives, and cave animals tend to *lose* their eyes, which are among the most complex structures in the animal kingdom. Since the appearance of parasites and of forms of life adapted to caves is the result of evolutionary processes, it follows that evolution is not *necessarily* a process that increases complexity.

Nevertheless, evolution does increase complexity in many cases. No one doubts that multicellular organisms are more complex than unicellular ones, and multicellularity certainly evolved from unicellularity (several times, in fact). In addition, animals with complex nervous systems, such as all vertebrates, evolved from animals with much simpler nervous systems, such as the class of invertebrates known as echinoderms (which includes sea urchins, sea cucumbers, and sea stars). Therefore, why evolution increases complexity *most* of the time is a legitimate question. Stephen J. Gould has articulated the best explanation for this observation:[9] It is the only way to proceed from simple beginnings, as one would expect in the case of a natural phenomenon. After all, in order to have *multi*cellular organisms, evolution had to invent a *single* cell first. To allow for the marvelous complexity of the human brain, a mindless, purposeless process such as natural selection had to go through many other simpler forms first and build upon each attempt.[10]

Gould explains the process by analogy with the situation of a drunken man who starts wandering at random when he leaves the bar. We can predict that he will eventually end up on the curb and away from the wall of the bar from which he came, even though he is not making any conscious attempt to go in that direction. The reason is that the wall is a starting point from which one can only get away. It is equally likely that the first steps will bring the man away from or back to the wall, but he will never be able to go farther back than the wall itself. Once a few steps have been taken from the wall, the random walk (a technical term in statistics to describe precisely this kind of behavior) will inevitably lead farther and farther away from the starting point, and eventually all the way down to the curb.

[9] S. J. Gould, *Full House: The Spread of Excellence from Plato to Darwin* (New York: Harmony, 1996).

[10] This is not to say that the human brain was the "goal" of natural selection, of course. You could use the same argument to understand the evolution of feathers in birds from less sophisticated structures present in reptiles.

It must be stressed that this bumpy progression from less to more complex forms is exactly what one would predict if evolution were a natural, not a directed, phenomenon. The drunken man does not *want* to go to the curb; he simply ends up there. Had he (intelligently) designed to go there, he would have (as best he could) walked straight and without hesitation. This is what Richard Dawkins refers to as "climbing Mount Improbable."[11] It would be a truly unexplainable phenomenon if very complex life forms popped out of nothing without any previous history connecting them to simpler forms. Then the creationist explanation of the world would have an advantage over naturalistic evolution. In fact, the quotation from Ecclesiastes referred to earlier could be interpreted as being in contradiction with the second principle of thermodynamics. Ecclesiastes 3:14 says that "nothing can be put to it [God's creation], nor anything taken from it." But creationists acknowledge that the second principle implies that order is being reduced in the universe as a whole. Therefore, the continuous deterioration of energy and matter *is* taking something from God's creation.

One important point raised by creationists and summarized earlier is their dismissal of the "closed system excuse." They maintain that the claim of biologists that the second principle is valid only for closed (isolated) systems is a ludicrous and convenient excuse. Their reasoning goes along these lines: True, physicists have arrived at the second principle (in fact, at *all* of thermodynamics theory) by considering the useful simplification of closed systems (near equilibrium). But all real systems are in fact open (and in a state far from equilibrium). Therefore, the principle *must* apply by extension to all open systems (i.e., to all existing objects); otherwise it would be useless.

There are several important conceptual errors in this way of thinking. First, theoretical physics, and even most of theoretical biology, *always* deals with highly simplified, "idealized" systems. This approach is taken for the very good reason that these models of the world are easier to understand and describe mathematically than the much more complex and messy real world. However, it is not true that these models are useless if they do not describe the full extent of reality. They are first approximations from which more complex and satisfactory models can be derived. For example, current research in thermodynamics is attempting to move beyond the near-equilibrium situations described by classical thermodynamics, but it can do so only because the simpler circumstances are already well understood. On the other hand, simplified models do not necessarily capture all of the important features of real situations, so they cannot simply be extended by fiat. In fact, it is curious (and a little presumptuous) that creationists who are not physicists are offhandedly extending the principles of thermodynamics to situations where the physicists themselves don't venture.

In the case of the controversy over open versus closed systems, thermodynamic theory applies strictly only to closed systems. It can, however, be extended to open systems, as long as these are then considered in *isolation* from

[11] R. Dawkins, *Climbing Mount Improbable* (New York: Norton, 1996).

the rest of the universe; that is, they are themselves turned into closed systems. This process can be reiterated several times, until one considers the whole universe as a closed system. The important point to understand is that evolution does not violate the second principle, as long as one does not consider the biosphere a closed system but allows for flow of energy and materials from outside (to deny this would be to deny overwhelming empirical evidence that it is warmer in the sunlight than in the shade).

In addition, as mentioned earlier, there is yet another argument advanced by creationists concerning thermodynamics—one that is much more sophisticated and not adequately considered even by many biologists (who, let us remember, are usually not physicists). The argument is that even allowing for energy and materials to come into the biosphere from outside, thereby rendering it an open system, does not necessarily imply the existence of a process such as evolution that creates new order. There is no doubt that creationists are completely right in this sense. However, this is by no means a difficulty for evolution. Let me explain where the problem actually lies.

The creationist analogy that we considered earlier is to the construction of a house—a process that reduces entropy because it creates order. Creationists say that energy (e.g., the electricity and fuel used by the construction company) and materials (bricks or wood) do not make a house. Two more elements are needed: a program to guide the construction process (the blueprints) and a conversion mechanism that translates this program into the final product (the workers who can read the blueprints and assemble the materials in the proper way, while using the correct amount of energy). Therefore, just saying that Earth and the sun provide materials and energy to the biosphere does not explain the existence of an evolutionary process that creates order.

That is entirely correct, and the analogy with the house also makes clear where the problem lies. The materials and energy that go into building the house are what scientists call *necessary but not sufficient* conditions. That is, you do need both to build a house, but you also need other elements—namely, the blueprints and the workers. Analogously, evolution needs a flow of energy (from the sun) and materials (from Earth) to occur, but these are not the whole story. A program and a conversion mechanism are indeed necessary. I will discuss in a moment the evolutionary analogs of blueprints and workers, but let me first make clear that—no doubt unbeknownst to the reader—we have shifted the focus here (a typical creationist tactic[12]). The original creationist charge was that evolution was *incompatible* with the second principle, not that it needed more than thermodynamics to occur. The situation is analogous to maintaining that you *cannot* build a house if you don't have energy and materials. This is fair enough. However, when you point out that both energy and materials are in fact there, the charge shifts to, "But you need more than that; you

[12] As pointed out in Chapter 4 of Philip Kitcher's *Abusing Science: The Case against Creationism* (Cambridge, MA: MIT Press, 1982).

have to have a plan and someone to execute it." Well, nobody would argue with that either. But this second condition is an entirely different question and in no way related to the original one. Confusing the two makes for sloppy reasoning and obfuscates the whole issue.

Nevertheless, science does have an answer. In fact, one of the fascinating insights of modern biology is exactly that we know how the roles of blueprint and energy conversion mechanism have been played throughout the history of life on our planet. The blueprint is provided by the genetic instructions encoded in the DNA (or other nucleic acids, such as RNA) of any living organism. The conversion mechanism is (mostly, but not uniquely) photosynthesis. Living beings that belong to a particular category, including the green plants, are capable of harvesting energy from the sun and transforming it into other forms (mostly sugars) that they and other organisms can use to grow and reproduce and—ultimately—to evolve.

Now the creationist can legitimately come back and ask, "Where did DNA and photosynthesis come from?" That is yet another very good question on which scientists are working very hard, and to which we have only partial answers. But notice that this new question again shifts the subject, making creationist claims a moving target. Now we have satisfactory answers to the relationship between thermodynamics and evolution, and—additionally—we have even shown how evolution has all necessary *and* sufficient elements to proceed without violating any law of physics. The origination of those elements is a separate question, albeit a valid one. We will take a look at this thorny issue (the origin of life) shortly.

One final point, to which I alluded at the beginning of this chapter, must be considered about the thermodynamics controversy. A fairly universal rule of discourse (and debate) is that one has to be consistent in the use of logical arguments. However, many ideas are logically consistent but nevertheless untrue (e.g., it is logically consistent to think of unicorns as real animals, but they don't exist). This is why we need a combination of logic and evidence—the scientific method—to make progress. At a minimum, however, logically *in*consistent arguments should simply be unacceptable to any thinking person. Therefore, using logical reasoning to attack a position and then refusing to explain one's own position according to the same logic amounts to a double standard, to having one's cake and eating it too.

That is exactly what creationists do when they invoke the topic of thermodynamics.[13] By attempting to make a case against evolution based on the laws of physics, they are implicitly accepting those laws as a relevant part of the debate. To then introduce God as a principle of explanation at any point in the argument, or to reject the scientific argument because it does not square with Scripture, is to violate the rules of logic and evidence that constitute the entire

[13] See the Web site "Frequently Encountered Criticisms in Evolution vs. Creation," at http://www2.uic.edu/~vuletic/cefec.html.

foundation of reasoned debate. Having begun by appealing to physics, the creationists' turning to the existence of a supernatural God definitely violates not only all four principles of thermodynamics, but *all* laws that physics has established so far! It is highly unfair to pick and choose which aspects of the problem have to adhere to standards of reason and evidence, and then to suddenly change the rules of the game in the middle if play. This point is especially relevant to the next segment of my discussion.

Conceptual Summary

Creationism	Evolutionism
The second law of thermodynamics predicts a decrease in order in the universe. Since evolution causes an increase in order, evolution must be rejected.	Evolution is compatible with the second law because the increase in order is confined to a small portion of the universe, while the universe as a whole increases its disorder.
Evolution is a process that continually increases life's complexity.	Evolution can increase complexity, but it does not have to. Parasites, for example, often evolve by simplifying their morphology and life cycle.
The Bible predicts the first principle of thermodynamics (conservation of energy) (in Ecclesiastes 3:14).	The Bible contradicts (at least) the second principle of thermodynamics.
The second law applies equally to closed and open systems.	The second law applies only to closed systems. It can be approximately extended to open systems by successive iterations.
The flow of energy and materials is not sufficient to cause evolution. A program and an energy conversion mechanism are also needed.	The program is encoded in the DNA (or other nucleic acids) of living organisms, and the conversion mechanism is (mostly) provided by photosynthesis, the process by which plants and other organisms transform solar energy into usable biological energy (mostly sugars).
One can legitimately invoke different principles of explanation (either natural or supernatural), depending on which aspect of the argument is being considered.	Any valid discussion has to be framed within consistent rules of logic and evidence. If evolution is charged with violating the principles of physics, creationism has to defend itself from the same accusation.

Where Do We Come From? We Still Have Few Clues to the Origin of Life

Creationist Claim: Since evolutionists cannot explain the origin of life, evolutionary theory must be wrong, and there must have been an act of direct creation by a supernatural being.

Evolutionist Claim: The origin of life is a complex problem with few clues available. It may or may not be solved by science, but that implies nothing either about the validity of evolutionary theory or the existence of supernatural beings.

The following is a rather skeptical view of the origin-of-life field, which on the surface should make a creationist very happy. On the contrary, my take on the subject is meant to show that the problem is really quite difficult, but also that we can nonetheless attack it from a rational and naturalistic perspective. The fact that we have not succeeded so far implies nothing about the future and surely does not give the other side victory by default (see Chapter 5 on the latter fallacy). But if the reader really wants to understand what is going on instead of simply adopting a priori ideological positions, I suggest that a healthy immersion in and appreciation of the limits of scientific research are essential parts of the treatment.

The proper purview of science is to answer our questions about the natural world in a rational manner. One of the fundamentals of the scientific method is the repeatability of the phenomena under investigation. Here lies perhaps the most difficult aspect of the endless quest for the origin of life on Earth. Contrary to what is maintained by creationists, this issue clearly is a question about the natural world, in fact perhaps one of the ultimate questions (together with the origin of the universe itself). Yet the events we are attempting to investigate are by definition unique. Life may well have originated multiple times in the universe, including perhaps in our galactic neighborhood, but so far we have only one example to go by. Our Earth is the only place that we know for certain harbors life as we conceive it.

Before getting to the heart of the problem, let us answer an even more fundamental question: Why do we care? I can propose three reasons. First, definitely ascertaining that life originated by natural means would certainly have profound implications for any religious belief, which is why Christian and other fundamentalists are so opposed to the very idea of a nonsupernatural explanation of the mystery. Second, arriving at the conclusion that life originated elsewhere in the universe and was then somehow "imported" to Earth (one of the possibilities that I will consider in this discussion) would automatically imply the existence of life as a phenomenon occurring elsewhere in the universe, and therefore the fact that living beings are not unique to our planet. It is hard to conceive of a more compelling blow to anthropocentrism since Copernicus and Galilei swept Earth away from the center of the universe a few centuries ago. Third, and perhaps more relevant, humankind would finally have a decent answer to the question, "Where did we come from?" This question, like it or not, has been vexing our philosophy, art, and science since the beginning of recorded history (and probably much earlier than that). If that doesn't sound like enough of a reason to ponder the controversy over the origin of life, the neurons in charge of one's sense of curiosity are definitely in need of some repair.

Let me start by clearing the field of one important misconception. There is no such thing as a modern-day "primitive" organism that we can examine to tell what our earliest ancestors looked like. True, there are many "simple" organisms around today, from viruses to bacteria to slime molds. But slime molds are *eukaryotes* (albeit of a taxonomically very uncertain position); that is, their cellular structure and metabolism are basically no different from those of an animal or a plant (they actually look like fungi, though they are not even closely related to them). They are much too complex to help us understand life's beginnings. Bacteria are *prokaryotes;* that is, their cells are indeed simpler than those of most other living beings. Yet bacteria have been around for more than 3 billion years, and they have become very sophisticated reproductive machines, characterized by an incredibly efficient metabolism and the ability to withstand environmental changes. It is not by chance that they have proliferated for so long. So we cannot use bacteria as a model of primitive life either.

Finally, viruses are indeed among the simplest living creatures in existence—so simple, in fact, that some biologists even doubt that they really qualify as "living." Evolutionarily speaking, however, viruses are late arrivals on the Darwinian stage. Viruses are short pieces of nucleic acids wrapped in a protein. They live only inside host cells, they originate from preexisting host cells, and they depend entirely on the host's metabolism to reproduce. Quite obviously, since our problem is to understand how the first living organisms came about, we cannot utilize as a model something that cannot survive outside an already existing cell. Yes, we are looking for something simple, but it must also be self-sufficient, and *truly* primitive.

There is, of course, another possibility. No serious scientific discussion of any topic should include supernatural explanations, since, as we have seen, the necessary basic assumption of science is that the world can be explained entirely in physical terms, without recourse to godlike entities. However, we are considering here creationism and its claims, so the God question must enter this debate as well. After all, as that fictional archetype of rationalism known as Sherlock Holmes said, "When you have eliminated the impossible, whatever remains, however improbable, must be the truth," and the truth just might imply a supernatural origin of life.[14]

Indeed, even some scientists of decent reputation, such as the British astrophysicist Fred Hoyle, have gone on record precisely suggesting a supernatural beginning to all life on Earth. Hoyle, together with his colleague Chandra Wickramasinghe, suggested that a sort of silicon chip creator goes around the

[14] Incidentally, it is ironic that Holmes's creator, Sir Arthur Conan Doyle, believed in spirits and poltergeists—but that's another story. See M. Polidoro, "Houdini and Conan Doyle: The Story of a Strange Friendship," *Skeptical Inquirer* 22(2) (1998): 40–46.

universe sprinkling the seeds of life here and there, though for what purpose is not at all clear. An entirely different yet congruent argument is the one advanced by creationists such as Duane Gish. In Gish's case, of course, what we have is the classic God of the Bible, who created the universe and humankind with a personal touch, and did so in the span of only six days.

There is one crucial problem with both Hoyle's and Gish's positions, of course: There isn't a single shred of evidence supporting either of them. Furthermore, at least Gish's claims are falsifiable and are demonstrably false: One of the cardinal points of his theory is that Earth is only a few thousand years old, but geology long ago demonstrated that the real time frame of Earth is measured in billions of years. Hoyle's proposition suffers from the hallmark of all nonscientific statements: It is not disprovable. Not a single experiment could falsify the British physicist's hypothesis, which means that it lies by definition outside the realm of science (which of course does not make it an impossibility).

Should we then reject outright any possibility of special creation of life? Well, no. As much as it is implausible, it is still theoretically possible. There are two points to bear in mind, however, before we opt for a Hoyle-like supernatural explanation of the origin of life. First, we must really have no clue about how life originated *on Earth* by natural means, and even in principle there must be no natural means of ever finding out. As we will see, though the situation is messy, it is not that desperate. Second, as I have already pointed out, the mere fact that we cannot currently (or even ever) explain something does not constitute *positive* evidence for a supernatural explanation. To assume so is to commit the *ad ignorantiam* fallacy—that is, to use the ignorance of one's opponent as evidence of the correctness of one's own position. After all, for a long time we did not know that natural phenomena could cause lightning, but eventually the theory based on Zeus's anger was shown by the acquisition of (natural) knowledge to be wrong. Consequently, even if we had no better answer, it would still be up to the "supernaturalists" to provide at least a shred of positive evidence for their own position. Without that, the next best option would simply be a provisional, and perhaps even therapeutic, "I don't know."

The next class of explanations about the ultimate provenance of life is that—as any good old-fashioned science fiction movie or magazine of the 1950s would have proclaimed—it is not of this world. Interestingly, Hoyle and Wickramasinghe have made their contribution in this realm too, by suggesting that life was brought to this planet courtesy of an interstellar cloud of gas and dust, or perhaps by a comet.[15] Yet another British scientist (and also an ex-physicist— but I'm sure this is a coincidence), the Nobel laureate Francis Crick, joined the ranks of the extraterrestrialists. Crick suggested a scenario that envisions extraterrestrials "seeding" the galaxy,[16] much in the fashion of Hoyle's silicon chip creator.

---

[15] See F. Hoyle and C. Wickramasinghe, *Lifecloud: The Origin of Life in the Universe* (New York: Harper & Row, 1978).

[16] See F. Crick, *Life Itself: Its Origin and Nature* (New York: Simon and Schuster, 1981).

Contrary to the supernatural explanations, the Hoyle–Wickramasinghe hypothesis (but not Crick's, for the same reasons as we saw earlier) is at least in principle open to experimental verification, in that it makes some relatively precise predictions. For one thing, if it is correct, we should find many organic compounds in interstellar clouds and/or inside comets. Both of these expectations have superficially been verified. I say "superficially" because the kinds of compounds found by astronomers in these media are very simple—much too simple to provide any meaningful "seed" for the origin of carbon-based life forms on Earth.

Furthermore, extraterrestrial organic compounds have random chirality, unlike the organic compounds typical of living organisms. Chirality is a property of any chemical structure characterized by the three-dimensional arrangement of that structure's atoms and molecules. For example, all amino acids, the building blocks of proteins, can in theory come in two versions, which are mirror images of each other. These are called *left-handed* and *right-handed* forms, and they are characterized by exactly the same chemical properties, so there is no physical–chemical reason for one form to be more abundant than the other. Accordingly, the organic compounds (a general term for carbon-based compounds and, contrary to the misleading name, not necessarily the result of an organism's metabolism) found in space or in meteorites come in equal proportions of right- and left-handed forms.[17] This is not true of the compounds that are actually used by living organisms on Earth, which are found in only one version. If indeed life had come from space, one would expect to find some sort of chiral asymmetry in space organic matter as well (alternatively, we need a mechanism to turn the symmetry of space molecules into the asymmetry of living matter on Earth). What is possible is that comets have brought to Earth—and perhaps to other planets around the cosmos—some of the basic elements on which life is based, starting with water. But this is not the same as saying that life originated in space or was transported through space.

A second crucial objection to the hypothesis of life from space is the solution of continuity problem. If comets and meteors brought life forms—literally—to Earth a few billion years ago, why are they not doing it now? Meteors continue to bombard our planet and our neighbors in the solar system on a regular basis, yet so far not a single living organism or complex organic molecule has been found inside any of them. There is no reason to think that the primordial "shower of life" has ceased. Even though the conditions for the persistence of primordial life on our planet probably do not hold any longer (because of dramatic changes in the composition of the atmosphere, or because of competition from "resident aliens"—i.e., from currently living organisms), presum-

[17] This particular statement may need to be modified because of recent research showing that cosmic radiation can cause asymmetric chirality without the intervention of living organisms. See W. A. Bonner and B. D. Bean, "Asymmetric Photolysis with Elliptically Polarized Light," *Origins of Life and Evolution of the Biosphere* 30(6) (2000): 513–517.

ably the space surrounding our solar system has not changed that much, leaving Hoyle, Crick, and the like with a major hole in their argument.

Another thought about extraterrestrial theories of life's beginning is that, at a minimum, they violate one of the most venerable principles of natural philosophy: the already mentioned Occam's razor. Because all extraterrestrial hypotheses still rely on organic chemistry, and because they require further assumptions—for example, that the "seeds" found a safe passage through Earth's atmosphere without burning into nothingness as happens to most meteors—they violate Occam's rule. On the other hand, there is no real guarantee that the universe behaves as Occam's rule suggests, so the razor can be invoked only as a provisional way of favoring simpler explanations, not as a definite argument against more complex or less likely alternatives.

Finally, we must realize that even if we admitted that life originated outside Earth and was then imported here, we still would not have an answer to *how* life started. We would have simply shifted the question to a remote and very likely inaccessible location.

Having excluded gods and extraterrestrials from the likely—at least temporarily—we are left with plain old biochemistry and biology to give us clues to the origin of life on our planet. The history of scientific research in this field is indeed long and fascinating. It started in the 1920s with the Russian Alexander Oparin and his "coacervates," blobs of organic matter (mostly containing sugars and short polypeptides) that were supposedly the precursors of modern proteins. It was Oparin, together with the British biologist J. B. S. Haldane, who came up with the idea of a *primordial soup*[18]—that is, the possibility that the ancient oceans on Earth were filled with organic matter formed by the interaction between the atmospheric gases and energy provided by volcanic eruptions, powerful electric storms, and solar ultraviolet radiation.

We had to wait until the 1950s for Stanley Miller to attempt to reproduce the soup experimentally.[19] Miller started with what at the time seemed (but now no longer does) a reasonable composition of the ancient atmosphere: mostly methane and ammonia, with no oxygen—since atmospheric oxygen, together with the ozone that blocks ultraviolet radiation, was in fact produced by the organic process of photosynthesis in blue-green algae much, much later than the time of the primordial soup (which is fortunate, given that oxygen attacks and destroys—technically it *oxidizes*—organic compounds at a very fast rate). Miller put these ingredients in a glass ball, added an electric charge, and waited (Figure 6.2). He indeed found that amino acids and other fundamental complex organic molecules were accumulating at the bottom of his apparatus.

[18] On the primordial soup, see A. I. Oparin, *The Origin of Life,* translated by S. Morgulis (New York: Macmillan, 1938); and J. B. S. Haldane, "The Origin of Life," in *On Being the Right Size and Other Essays* (Oxford: Oxford University Press, 1985).

[19] The classic paper by Miller is S. L. Miller, "A Production of Amino Acids under Possible Primitive Earth Conditions," *Science* 117 (1953): 528–529.

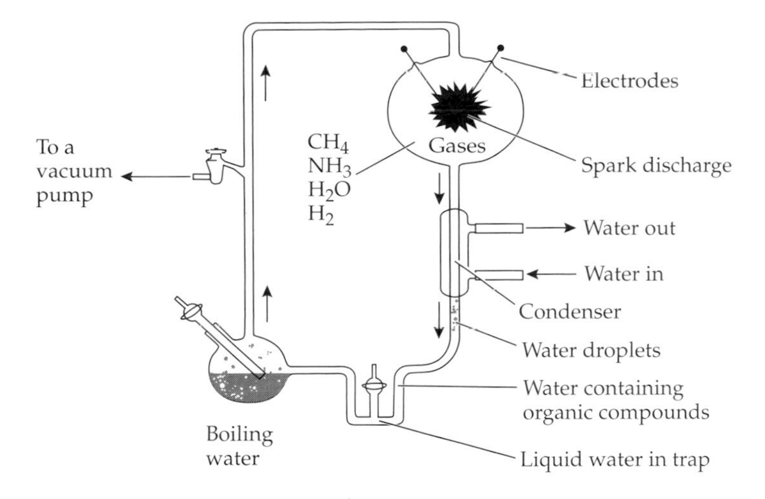

Figure 6.2 The basic idea behind Miller-type experiments to investigate the production of organic molecules under conditions simulating the prebiotic Earth. The ball of gases represents the primordial atmosphere, and the electrodes provide energy such as could have been furnished by lightning (or volcanic eruptions, or a variety of other possibilities). The boiling water provides moisture to create the "soup," which accumulates and is purified in the lower right-hand part of the apparatus. These experiments are now so easy to replicate that the apparatus can be used for high school projects.

Miller's discovery gave a huge boost to the scientific investigation of the origin of life. Indeed, for some time it seemed that the recreation of life in a test tube was within reach of experimental science. Unfortunately, Miller-type experiments have not progressed much farther than their original prototype, leaving us with a sour aftertaste from the primordial soup. Furthermore, recent research has questioned the notion of the complete absence of oxygen from the primordial atmosphere, although the jury is still out on exactly how chemically reducing (no oxygen) or oxidizing (significant oxygen) the primordial conditions were. One alternative to the soup, the "pizza" theory that I will discuss shortly, would solve the problem by making the atmospheric conditions pretty much irrelevant.

One important point concerning Miller's experiments and creationism must be made here. Creationists such as Duane Gish (e.g., during several of my debates with him) sneer at the idea that the biosphere is an open system and that therefore evolution is not made impossible by the second principle of thermodynamics, as discussed earlier. Gish claims that energy from outside does not do the trick and explains his point in his typically humorous way: You know what happens when energy (say, from lightning) hits you? You don't become more complex; you become dead! Except, of course, Gish—as a biochemist—should know that Miller's experiments showed him wrong on this point. When lightning, or ultraviolet radiation, hits simple gases in a mixture, more complex organic molecules do indeed form. Even if Miller's experiments turn out to have been a dead end at explaining the particular problem of the origin of life on Earth, they surely demonstrate that energy can build complexity, in accordance with both the second principle of thermodynamics and the theory of evolution.

Both Oparin and Miller, as well as other prominent researchers in the field up until the 1960s, thought that the problem was how to explain the appearance of proteins, because it was believed that they must have caused the initial spark of life.[20] Today any student of introductory biology knows that there are *two* major players inside every living cell: proteins and nucleic acids (such as DNA and RNA). The problem is that the structure of DNA was discovered only in 1953 (in fact, the same year of Miller's experiment), and the nature of DNA as the information carrier of the cell was little appreciated before Watson and Crick unveiled the double-helix nature of this remarkable molecule.

The origin-of-life debate after the 1950s therefore became decidedly slanted in favor of nucleic acids preceding proteins. The new discipline of molecular biology was making spectacular progress—first by uncovering the universal code by which the instructions for making proteins are embedded in the nucleic acids, then by finding ways to extract and compare that information from different and distantly related species, and finally by developing modern genetic engineering and the ability to directly modify the genetic information, thereby transforming the characteristics of species more or less at will. Scientists such as Leslie Orgel, Walter Gilbert, and others proposed that the egg, so to speak, came before the chicken. Some sort of primitive nucleic acid had appeared first, followed only later by proteins.

Now, in today's biochemically sophisticated cells, proteins and nucleic acids play very distinct roles. In fact there are four fundamental activities that we must discuss (Figure 6.3):

1. DNA (deoxyribonucleic acid) encodes the information that eventually will give rise to proteins.
2. *Messenger* RNA (mRNA) then carries the information to specialized structures known as ribosomes. (*RNA* stands for *ribonucleic acid,* which is the same as DNA, but with an extra oxygen atom and a few other chemical differences.)
3. Inside the ribosomes (which are made of both nucleic acids and proteins), the message is translated into proteins by virtue of a second type of RNA, known as *transfer* RNA (tRNA). Transfer RNA has the peculiar ability to attach itself to the mRNA on one side and to amino acids (the blocks that make

[20] Another student of the problem who thought along similar lines was Sidney Fox, who discovered the possibility of forming *proteinoids,* proteinlike structures that can be obtained from the heating of mixtures of amino acids in a dry state. But proteinoids are only a very distant cousin of actual biological proteins.

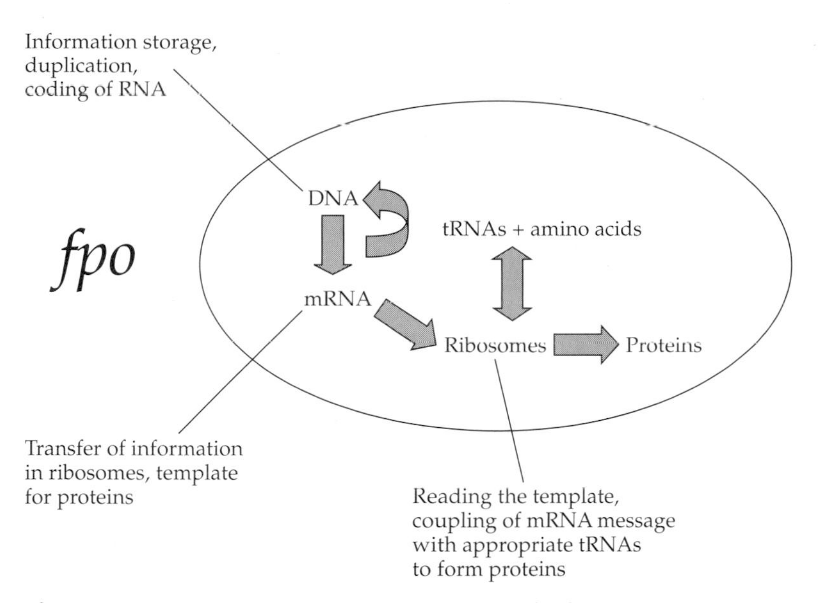

Figure 6.3 The most important functions of the basic constituents of life as we know it today inside a cell (large oval). The DNA template is copied into messenger RNA (and from time to time copies of it are made to prepare for cell duplication). The mRNA transfers the information that encodes proteins to the ribosomes, organelles made of nucleic acids and polypeptides that attract and align the correct transfer RNAs. The sequence of amino acids that results folds in three-dimensional space to yield a protein.

up proteins) on the other side. Thus a chain of mRNA is paralleled by the forming chain of amino acids, which in turn will eventually result in the final protein.

4. The proteins, most of which (but not all) are enzymes, are the "doers" of the cellular world, in that they are both the building blocks of cell structures and membranes, and the builders themselves, in the form of enzymes capable of catalyzing all sorts of chemical reactions, including the replication of DNA and the transcription of its message into RNA—which, of course, closes the circle!

It should be clear from the preceding extremely concise and greatly simplified description of what goes on in a cell that we are indeed facing a classic chicken–egg problem. If the proteins appeared first and thus could eventually catalyze the formation of nucleic acids, how was the information necessary to produce the proteins themselves coded? On the other hand, if nucleic acids came first, thereby embodying the information necessary to obtain proteins, by what means were the acids replicated and translated into proteins? It seems pretty clear that the answer, still very nebulous at the moment, must lie in the proverbial middle. In fact, the existence of tRNAs points to the distinct possibility of dual structures, containing both RNA and amino acids. On a slightly different take, the discovery by Sydney Altman, Thomas Cech, and others that some RNAs are at least partially self-catalytic (i.e., they can catalyze chemical reactions in which they themselves are reactants), lends support to the idea of a mixed origin of life in which the early molecules were both replicators and enzymes, with the two functions slowly diverging through evolutionary time and assigned to distinct classes of molecules. Moreover, this solution is certainly more parsimonious (and has some aesthetic appeal as well, something that in science is often synonymous with parsimony).

Although a hybrid protein–nucleic acid origin is a likely possibility, there is one major problem with the Haldane–Oparin soup scenario: It could get too watery. Because the organic compounds would be freely bumping into each other within the ocean, unless their concentration was extremely large, it is difficult to see how dense enough pockets of organic molecules could have formed often enough to allow some significant prebiotic chemistry to occur.

This is even more of a problem when we consider the question of the origin of the first metabolic pathways. Metabolism requires proto-enzymes to interact with their substrates so that a given reaction can take place. However, it is unlikely that enzymes and substrates could come close enough together in a three-dimensional space with no enclosing barriers (hence there are several hypotheses about the formation of proto-cells with a lipid membrane). Furthermore, most of the necessary reactions for prebiotic chemistry, such as the formation of polypeptides by aggregating individual amino acids, produce water. This kind of reaction is difficult to have in an aqueous environment because of thermodynamic considerations (it requires energy, and the products are unstable and can be hydrolyzed back to their component parts).

An alternative to the primordial soup has therefore been proposed, and it has become known as the *primordial pizza*.[21] The idea is that early organic chemistry occurred in dry environments, most likely on the surface of minerals with physical properties conducive to accumulating and retaining organic molecules in place. Perhaps the best candidate for such a role is pyrite, "fool's gold." On a two-dimensional surface, enzymes and substrates, or even simply different amino acids or nucleotides without enzymes, would find themselves constrained by much less freedom of movement, which of course would increase the chance of reciprocal encounters. Furthermore, because the pyrite surface would have no water on it, the occurrence of water-producing reactions would be much facilitated. In fact, thermodynamic calculations show that these reactions would increase entropy and would therefore occur spontaneously.

Very little empirical research has been done on the rather novel concept of a primordial pizza, and other candidates besides pyrite are possible as material substrates, but the concept is appealing in its elegant solution of two major problems

---

[21] See J. Maynard-Smith and E. Szathmáry, *The Major Transitions in Evolution* (Oxford: W. H. Freeman Spektrum, 1995).

facing prebiotic chemistry. We should see some progress in this area in the next few years. But an even bigger question is, "What happened after the primordial pizza?" There is still a very large gap between a semicatalytic, semireplicating nucleoprotein and the first living "organism," whatever that may have been. Moreover, the uncertainty about what the original organism was (from a biophysical standpoint) is an important part of the problem: What exactly *is* life?

That question was asked precisely in such fashion by physicist Erwin Schrödinger in 1947. Although Schrödinger's thinking led him to predict some of the properties of DNA as a necessary component of a living organism, we still have only a vague notion of the boundary between life and inert matter. And so it should be, if we accept the idea that living organisms are made of inert matter that happens to acquire some "emergent properties" when it is assembled in particular ways. To put it another way, living beings are not separated from the rest of the universe by the possession of a mysterious force or vital energy. At least, we have no reason to believe so, as much as this may be maddening to creationists such as Gish (who—as you might remember—claims that evolution tells us that hydrogen gas becomes people when given enough time) or Kent Hovind (who maintains that evolution is the theory that humans come from bananas and ultimately from rocks).

How, then, do we distinguish life from nonlife? Well, we can come up with a list of attributes, some of which may be properties also of nonliving systems, but the ensemble of which defines a living organism.[22] Here is my list:

- The ability to replicate, giving origin to similar kinds (reproduction)
- The ability to react to changes in the environment (behavior, not limited just to the special meaning that the word has in animals)
- Growth (i.e., reduction of internal entropy at the expense of environmental entropy; note that even single cells grow immediately after reproduction, so this is not a property restricted to multicellular life)
- Metabolism (i.e., the capacity of maintaining lower internal entropy, including the ability to self-repair)

How did we get from a nucleoprotein to an entity capable of all of the above? And what did this entity (sometimes referred to as the *progenote*) look like? There are a few tentative answers to these questions, and this—I think—is

[22] Philosopher Ludwig Wittgenstein would have called this a *family resemblance* concept of life. According to Wittgenstein, in his *Philosophical Investigations* (see G. P. Baker and P. M. S. Hacker, *An Analytic Commentary on the Philosophical Investigations*, Vol. 1: *Wittgenstein: Understanding and Meaning* [Oxford: Oxford University Press, 1980]), complex concepts have a fuzzy edge, so no individual property captures the essence of each one. Rather, there is a sliding scale and a "group" quality to them. Wittgenstein's example was the concept of a game. It is actually rather difficult to define what a game is. For example, the *Oxford Dictionary* defines *game* as a "form of contest played according to rules and decided by skill, strength, or luck." But according to such a definition, some things that we do not normally consider games (such as career advancement or war) would fall under the definition because there is no single characteristic that neatly divides games from nongames. A similar problem (and the Wittgensteinian solution) applies to the definition of life.

where the real problem of the origin of life lies. The German scientist Manfred Eigen has come up with one possible scenario that invokes what he called "hypercycles." We can think of a hypercycle as a primitive biochemical pathway, made up of self-replicating nucleic acids and semicatalytic proteins that happen to be found together in pockets within the primordial soup or on the primordial pizza. It is possible to imagine that some of these hypercycles are made of elements that "cooperate" with each other; that is, the product of one component of the cycle can be the substrate for another component. Different hypercycles could have coexisted before the origin of life, and they would have competed for the ever decreasing resources within the soup or pizza (the resources were decreasing because the hypercycles were using up some organic compounds at a higher rate than they were being formed by comparatively inefficient inorganic processes). Eventually, this competition would have favored more and more efficient hypercycles, where the "efficiency" would be measured by the ability of these entities to survive and reproduce—that is, by the parameters of Darwinian evolution. Life as we know it (sort of) would have begun.

Eigen and modern followers of complexity theory also expected these systems to become more complicated by the addition of new components to the cycle. From time to time the addition of one component would modify the whole system dramatically, giving it properties that the previous group did not possess (sort of like adding an atom of oxygen to two of hydrogen and suddenly getting something completely distinct and more complex: water). Complexity theorists such as Stuart Kauffman have indeed demonstrated on the basis of mathematical models that some self-replicating systems can display unexpectedly complex patterns of behavior (see the "Additional Reading" section at the end of this chapter). The textbook examples of this phenomenon are the so-called cellular automata, mathematical entities that were first imagined by John von Neumann in 1940 and that can now be studied at leisure by anybody who has a personal computer and a copy of a game aptly called "Life."

To summarize, the general path leading to the origin of life seems to have been something like this:

1. **Primordial soup or pizza** (simple organic compounds formed by atmospheric gases and various sources of energy)
2. **Nucleoproteins** (similar to modern tRNAs)
3. **Hypercycles** (primitive and inefficient biochemical pathways, characterized by emergent properties)
4. **Cellular hypercycles** (more complex cycles, eventually enclosed in a primitive cell made of lipids)
5. **Progenote** (first self-replicating, metabolizing cell, possibly made of RNA and proteins, with DNA entering the picture later)

How plausible is all this? It is fairly conceivable, as far as modern biology can tell. The problem is that each step is really difficult to describe in detail from a theoretical standpoint, and so far (with the exception of the formation of organic molecules in the soup and of some simple hypercycles) has proven remarkably elusive from an empirical perspective. It looks like we have several clues, but the puzzle may very well prove one of the most difficult for scientific analysis to solve. The reason for such difficulty could be—as pointed out at the beginning of this section—that, after all, we have only one example to go by: life on Earth. Or it may simply be that the events in question are so remote in time that there is very little about which we can be certain. Consider that the fossil record shows "modern-looking" bacterial cells a few hundred million years after the formation of Earth—that is, about 3.8 billion years ago. This tells us that whatever happened before the formation of these bacterial cells happened fast (by geological standards, nothing creationists can exploit), but there is no record of it. Finally, it could very well be that we are missing something fundamental here. It may be that the origin-of-life field has not had its Einstein or Darwin just yet, and that things are going to change possibly very soon, or maybe never.

Be that as it may, a contemporary discussion of the question of the origin of life cannot be complete without inclusion of A. G. Cairns-Smith's theory of clay crystals.[23] Briefly, the idea is that life did not originate with *either* nucleic acids or proteins (and for that matter, not with a combination of the two). No, the original replicator and/or catalyzing agents were actually crystals to be found everywhere in the clay that was scattered around the primitive Earth. The cardinal points of the Cairns-Smith hypothesis are four:

1. Crystals are structurally much simpler than any biologically relevant organic molecule.
2. Crystals grow and reproduce (i.e., they can break because of mechanical forces, and each resulting part continues to "grow").
3. Crystals carry information, and this information can be modified. A crystal is a highly regular structure, which tends to propagate itself (therefore it carries information). Furthermore, the crystal can incorporate impurities while growing. These impurities alter the crystal's structure and can be "inherited" when the original piece breaks (hence, the information can be modified).
4. Crystals have a minimum capacity of catalyzing (i.e., accelerating) chemical reactions.

Cairns-Smith then proposed that these very primitive "organisms" started incorporating short polypeptides (proto-proteins) found in the environment—presumably in the soup or on the pizza—because polypeptides enhanced the crystals' catalyzing abilities. The road was then open for a gradual increase in the importance of proteins first, and then eventually of nucleic acids, until these two new arrivals on the evolutionary scene completely supplanted their "low-tech" progenitor and became the living organisms that we know today.

---

[23] See A. G. Cairns-Smith, *Seven Clues to the Origin of Life: A Scientific Detective Story* (Cambridge: Cambridge University Press, 1985).

I am rather skeptical of the clay hypothesis for a variety of reasons. First, Cairns-Smith seems to completely ignore what a living organism is to begin with. Crystals don't really have a metabolism, at least not in a sense even remotely comparable to what we find in actual living organisms. This may have something to do with the fact that not only are crystals structurally much less complex than a protein or a nucleic acid, but also their non-carbon-based chemistry is recognizably much simpler than the chemistry utilized by living organisms on Earth. The lower complexity and simpler chemistry may be insurmountable "hardware" obstacles to the origin of a true metabolism in clay matter. Second, crystals don't really react to their environment either—another hallmark of every known living creature. Notice that this is a property distinct from metabolism, in that metabolism can be entirely internal, with no reference to the outside (except for some flux of energy that must come into the organism to maintain its metabolism). On the other hand, living organisms universally respond actively to changes in external conditions—for example, by seeking sources of energy or by avoiding dangers. Furthermore, an argument can be made that crystals are not actually capable of incorporating new information in their inherited "code," unlike what happens with mutations in living beings. True, they can assimilate impurities from the environment and transmit such information to their "descendants" for some time, but these impurities are not replicated, they need to be imported continually from the outside, and they do not become a permanent and heritable part of the crystal. Moreover, impurities do not create new types of crystals, the way mutations eventually produce entirely new kinds of organisms.

Another major hole in the clay theory is—of course—that we have no clue about how the "mutiny" of nucleic acids and proteins occurred; in fact, we are given only very faint hints about how a crystal could possibly co-opt a polypeptide to enhance its growth (which, by the way, should be something relatively easy to test in a modern biochemistry laboratory). It is true that the competing, more biologically traditional, hypotheses also are at a loss to provide detailed scenarios. But in the case of Cairns-Smith's suggestions, we don't simply miss the details; we literally have no idea about how such a transition would come to pass. So as much as creationists might like the flavor of a theory of the origin of life in which the first living beings came literally from dust (although Cairns-Smith is certainly no creationist), we are still left with ribonucleoproteins as our best, albeit fuzzy, option. This is one question that scientists will be pondering for some time to come.

As for creationists, I provided this detailed discussion of the problem to show how complex some scientific problems are and how difficult (sometimes even impossible) progress may be. However, I am also hoping that what I have written here will be accepted as an intellectually honest appraisal of the situation that does not lend any credibility to creationist "alternatives," for one thing because there are no plausible alternatives that creationists have offered. To say simply that "God did it" is a much more presumptuous and equally unhelpful way of admitting our ignorance. It is fascinating how creationists repeatedly miss this point, convinced as they are that supernatural intervention is a commonsensical explanation for everything we don't understand. It is not, and they need to do much more legwork before they can produce anything that looks like a positive alternative to the naturalistic research program. Of course, I think it is simply impossible for a creationist to do so, given the very assumption of supernaturalism embedded in the attempt. At most, the problem of the origin of life is a tie between the two paradigms of thought, but beware: All past ties of this sort have been resolved in favor of naturalism, and if history teaches us anything . . .

Conceptual Summary

Creationism	***Evolutionism***
Science cannot explain the origin of life; therefore it must have been the result of supernatural intervention.	This is an argument from ignorance, and it assumes that creationism wins by default. Either creationists need to make a positive case for their explanation, or they must adopt the most honest position, which is simply to conclude that we do not know what happened.
Because Miller conducted his experiments by simulating an unrealistic primordial atmosphere, they should be discounted as irrelevant to the problem of the origin of life.	This is another oversimplification of how science works. First, the importance of Miller's experiments lies in the fact that he showed that the problem could be attacked scientifically, something that was very much in doubt until that point. Second, Miller has shown that the formation of organic compounds is very easy to reproduce under a variety of conditions, some of which may be relevant to life on other planets, as well as to the formation of organic compounds in space, from which they (but likely not life itself) could have been imported to Earth. Third, geochemists are still debating what the composition of the early Earth atmosphere was, and until that debate is settled it seems more reasonable to suspend judgment on the significance of Miller's experiments.
God did it.	Needless to say, this is simply not an explanation.

The Cambrian Explosion and the Incompleteness of the Fossil Record

Creationist Claim: The sudden appearance of life forms at the beginning of the Cambrian period is proof of the special creation of all living organisms by God.

Evolutionist Claim: The Cambrian explosion is one of many instances of relatively rapid, but by no means miraculous, changes that characterize the history of life on Earth.

The controversy surrounding the Cambrian explosion requires a background almost as extensive as evolutionary biology itself because pertinent information comes from paleontology (the study of fossils), geology, developmental biology, molecular biology, ecology, and systematics—to name a few. But I will try to be brief and to the point. First, let me define what the Cambrian *is*. The term *Cambrian* refers to a period that marks the beginning of the Paleozoic era, currently estimated at 570 million years ago (if we accept the estimates of geologists and geochemists). Geologists divide the history of Earth into 11 eras. The most ancient era is termed the Hadean, and it started 4.56 billion years ago (the estimated age of our planet) and ended 3.8 billion years ago (BYA). The current era is the Cenozoic. It started at the time of the extinction of the dinosaurs, 65 million years ago (MYA), and is continuing today. The dinosaurs dominated most of the era preceding the Cenozoic, called the Mesozoic (65–245 MYA). Before the Mesozoic, the Paleozoic era spanned from 245 MYA all the way back to 570 MYA. Between the Paleozoic and the Hadean are seven more eras, whose characteristics are less important for the current discussion.[24]

Why is the Cambrian period of the Paleozoic era so important? Because until fairly recently, no fossil record at all was known from Precambrian rocks. Yet Cambrian sediments are rich in all sorts of fossils, including all fundamental types[25] of invertebrate animals living today (the vertebrates came slightly later, appearing for the first time in the fossil record in the Ordovician period of the Paleozoic era, about 510 MYA). The lack of Precambrian fossils (in Darwin's time) was recognized as a problem by Darwin in his *Origin of Species:*

> *If my theory be true, it is indisputable that before the lowest [Cambrian] stratum was deposited, long periods elapsed, as long as, or probably far longer than, the whole interval from the [Cambrian] age to the present day; and that during these vast, yet quite unknown periods of time, the world swarmed with living creatures.*

---

[24] For the compulsive among the readers, these are the Sinian, 570–800 MYA; Riphean, 800 MYA–1.65 BYA; Animikean, 1.65–2.2 BYA; Huronian, 2.2–2.45 BYA; Randian, 2.45–2.8 BYA; Swazian, 2.8–3.5 BYA; and Isuan, 3.5–3.8 BYA.

[25] *Type* in this sense means a very specific rank of the taxonomic hierarchy, formally known as *phylum.* The phylum is a very broad category. For example, all vertebrates (fish, amphibians, reptiles, birds, and mammals) are a *class* within the larger phylum known as Chordata.

Darwin honestly recognized that the state of knowledge current at that time about the fossil record was a problem for his theory of evolution. The problem was that a core component of the Darwinian view of *descent with modification* (as Darwin referred to the evolutionary process) is that organisms evolve relatively gradually, with many intermediate stages linking one species to another (but see Chapter 5 on the theory of punctuated equilibria). If the fossil record as it was known in 1859 was to be read literally, scientists were faced with the appearance of relatively complex forms of life literally out of nothing, in direct contradiction to Darwin's theory.

In fact, such a contrary position had been stated a few years before the publication of *The Origin* by a famous geologist of the time, Roderick Impey Murchison. As cited in Stephen Gould's *Wonderful Life*,[26] Murchison wrote in 1854:

> *The first fiat of Creation which went forth, doubtlessly ensured the perfect adaptation of animals to the surrounding media; and thus, whilst the geologist recognizes a beginning, he can see in the innumerable facets of the eye of the earliest crustacean, the same evidences of Omniscience as in the completion of the vertebrate form.*

Darwin clearly saw it differently. But he was faced with the then incontrovertible absence of ancient fossils. He advanced two predictions to reconcile the evidence with the theory:

1. The time that had elapsed between the formation of Earth and the beginning of the Cambrian must have been enormous, probably longer than the time that had elapsed from the Cambrian to the current day.
2. This Precambrian time must have seen a panoply of life forms originating and evolving gradually over millions of years.

If correct, these two predictions would be a confirmation of Darwin's theory of evolution because they were made before anybody knew anything about the Precambrian, let alone about the living organisms that existed during that period.

The evidence came long after Darwin's time, in the form of a series of rare fossil traces of early life, which must be understood in order to comprehend the creationist diatribe about the Cambrian explosion. The most ancient evidence of life on the planet is almost as old as the oldest rocks currently available to the paleontologist: 3.75 BYA. These fossils do not come in the form of actual organisms, nor even of single cells. All we have is an altered ratio between two forms of carbon, ^{12}C and ^{13}C. These two atomic forms (called *isotopes*) are found in nature in a certain ratio with respect to each other. Plants and other photosynthetic organisms, however, preferentially use ^{12}C, which accumulates in their tissues. When the organism dies, its entire body may be destroyed with time,

[26] S. J. Gould, *Wonderful Life: The Burgess Shale and the Nature of History* (New York: Norton, 1989).

but the atoms of carbon that it collected will be embedded in the sediments that will later become solid rocks. Scientists can then compare the ratio of the two isotopes of carbon in any rock with those known to be typical of photosynthesis (and hence the hallmark of organic processes). If the two match, that is a good indication that there were photosynthetic organisms embedded in the rocks as fossils, even though any clue as to their physical appearance may have been destroyed in the meantime. Therefore, we know that there were photosynthetic organisms on the planet as early as 3.75 BYA, and possibly earlier.[27]

A little later in the geological column we see the appearance of the first fossil evidence of organisms as we think of them today: Rocks dating from 3.5 to 3.6 BYA contain bacterial cells and stromatolites. (The latter are colonies of photosynthetic bacteria, which still form today in warm seas.) We also have fossil evidence of eukaryotic cells going back about 1.4 billion years. These cells are more complex than bacteria and represent the building blocks of all modern animal and plant life. The earliest known multicellular organisms appear in the fossil record about 700 MYA—that is, slightly earlier (in geological terms) than the beginning of the Cambrian, which is dated at 570 MYA.

The three most important sets of fossils relevant to the Cambrian debate belong to three distinct *faunas,* which are assemblages of animals found worldwide. The *Ediacaran* fauna (named for the locality in Australia in which it was first described) dates to about 590 MYA. It consisted entirely of soft-bodied (and therefore rarely fossilized) organisms. To date, paleontologists are divided about exactly what kind of organisms lived in Ediacaran times. M. F. Glaessner, the discoverer of the first such group of fossils, has interpreted them as direct ancestors of modern, post-Cambrian forms such as corals and roundworms. A. Seichaler, however, has questioned such an interpretation, suggesting that most organisms of the Ediacaran period share a common body plan (i.e., the way in which their bodies are constructed) which is different from the body plans of later animals. Basically, Seichaler thinks of the Ediacaran fauna as consisting of organisms with flattened bodies, made of two equal, symmetrical, juxtaposed parts. If this interpretation is correct, then the Ediacaran fauna was a dead end in the history of life—sort of a failed natural experiment—since afterward no organism with that body plan survived. The controversy, however, has not yet been settled.

The second fauna relevant to our discussion is the Tommotian fauna (named for the locality in Russia were the first fossils of this type were discovered), dating from the very beginning of the Cambrian. Hard bodies, which fossilize more easily, characterize Tommotian animals. Many (but not all) of these are clear relatives and precursors of modern forms. Finally, immediately after the beginning of the Cambrian we have the Burgess Shale (named after a locality in British Columbia). The Burgess forms are extremely

---

[27] Obviously, those organisms were not plants, which appeared much later; it is more likely that they were similar to modern photosynthetic bacteria.

Figure 6.4 An artistic representation of the Burgess Shale fauna (500–600 MYA). Some of these animals do not have any known subsequent or modern relative, and some may have been the ancestors of arthropods (insects and crustaceans), worms, and corals. (From P. W. Price, *Biological Evolution* [Fort Worth, TX: Saunders College Publishing, 1996])

diverse (Figure 6.4)—some clearly related to animals that survived and diversified later, such as snails, clams, and some types of worms. Other lineages, however, went extinct quickly, without leaving any relative to inherit their fundamental characteristics. This in and of itself, of course, is something that happened again after the Cambrian: Many kinds of organisms disappeared completely after periods of mass extinction without leaving close relatives, including trilobites at the end of the Paleozoic, and ammonites and dinosaurs at the end of the Mesozoic.

Another piece of the puzzle that we must understand before getting into the heat of the evolution–creation controversy on the Cambrian explosion is how scientists attempt to determine relationships among organisms. In other words, how do we know that humans are more closely related to chimpanzees than they are to New World monkeys (or to elephants for that matter)? The ability to construct verifiable hypotheses about ancestor–descendant relationships is relevant here because the debate about the Cambrian explosion hinges in part on which—if any—of the Precambrian forms were ancestors of modern groups of animals.

The history and theory of systematics, the branch of biology that deals with *family trees* (technically known as phylogenies), is long and fascinating. Essentially, however, biologists have devised three ways of achieving the goal of reconstructing relationships among groups of animals or plants (or between the two and other kinds of organisms). First, consider what is now known as *classical* systematics, which has been practiced at least since the publication of *Systema Naturae* by the Swedish biologist Carolus Linnaeus in 1735. Simply put, classical systematists group together organisms that share similarities in major characteristics. This amounts to saying that if two organisms look alike in fundamental ways, they probably derived from a recent common ancestor. In the same way, we may recognize that someone is the daughter or son of a given couple because he or she "looks like" the parents. What defines a "major" similarity, however, is mostly the experience of the systematist and her knowledge of the group under study. Because of this element of subjectivity, two more recent approaches have been proposed, one of which is currently the standard way of reconstructing phylogenies.

The first modern approach is called *phenetics,* or sometimes *numerical taxonomy.* The idea is simply to extend the approach used by classical taxonomy to its logical extreme. Instead of considering the similarities of organisms according to *some* characteristics, numerical taxonomists collect a large database of information concerning *every* characteristic that is possible to measure reliably, feed the information into a computer, and let the machine determine the *overall* similarity among different species or groups of species. Even though this method is indubitably less subjective than classical systematics, it runs into several problems. First, it is simply not possible to collect data on every conceivable characteristic, so choices based on logistics and judgments of relative importance still have to be made. Second, in phenetics all characteristics are considered equivalent, regardless of how important they have been in the actual history of a given group of animals or plants. It makes sense that some traits are more "characteristic" of a particular species or group of species, and they should be given a higher priority. For example, among mammals the ability to walk on two limbs (bipedalism) is unique to humans and closely related hominids, and it sets them apart from all the other primates, no matter how similar or dissimilar they are with respect to other characteristics.

But how is it possible to weigh the intuitively unequal importance of some traits, without falling back into the subjective judgments typical of classical systematics? A useful method—known as *cladistics*—was developed in the 1960s by Willie Henning. Fundamentally, cladists divide characteristics into two categories: those that are shared by a large number of species within a group, and those that are shared by only a few members of the group. For example, all vertebrates share the common trait of having a spinal cord. Although this is probably the case because they all inherited it from a common ancestor, the spinal cord does not tell us which vertebrates are more closely related to which others. Only two types of vertebrates, however, share the presence of a wishbone: birds and dinosaurs. This fact (and the existence of other features shared only by these two groups) is what led biologists to the conclusion that birds and dinosaurs are closely related within the "family tree" of the vertebrates. In other words, cladistics combines the objectivity of a method in which characteristics are selected by criteria established a priori, with the intuition of classical systematics that some traits are more informative than others (and therefore should be given more attention).

Although cladistic methods can be applied to fossils (as long as one can measure enough features), how do we know which major groups of organisms are related to which others when most of the fossil record is (at least for now) lost in the mists of time? To some extent, of course, systematists rely on the morphological characteristics (i.e., the physical appearance) of living organisms, subjecting them to cladistic analyses to reconstruct as many branches of the tree of life as possible.

However, often living organisms are so different from each other morphologically that one cannot even find comparable traits to measure. For example, although many animals share at least some characteristics, what would we measure that is common to plants, fungi, algae, and animals to determine which is more closely related to which? Fortunately, all organisms do share a very high number of traits that are phylogenetically informative: the sequences of their DNA. All known species have DNA inside their cells (viruses and a few others are exceptions that carry the chemically similar RNA), and each molecule of DNA is made of millions of combinations of four chemicals: adenine, thymine, cytosine, and guanine. These four fundamental types of chemicals are usually represented by the letters A, T, C, and G, respectively.

The "letter" in each position on a particular gene can be thought of as a characteristic to be used in phylogenetic analyses. The letters at corresponding positions can be compared among different types of organisms, and these can be grouped on the tree of life according to which letters they share at each position along the DNA molecule. Even though in practice it is not as easy as it sounds, modern computers can process the necessary data, explore millions of possible alternative phylogenies, and provide researchers with the most likely scenarios. Molecular phylogenies are not a direct way to study the fossil record before the Cambrian explosion, but they are the crucial piece of evidence that allows biologists to explore the "family tree" of ancient forms—including Precambrian forms—despite the incompleteness of the fossil record.

For combining paleontological and molecular data, the scenario proposed by biologists for the early development of life is of a series of simple organisms appearing at long intervals of time throughout the Precambrian. What, then, characterizes the beginning of the Cambrian as an "explosion" is the fact that the number of types of invertebrates present in the fossil record increased dramatically over a relatively short period of time (estimated at 15 or more million years—a short but certainly not instantaneous burst). What exactly caused the "explosion" is unclear, but several hypotheses have been proposed. Most of these revolve around the ecological concepts of niche, competition, and adaptive radiation, and therefore we need to understand these concepts as well before proceeding with examining the creationist and evolutionist claims.

It is very well known that organisms will avoid competition with other species, if at all possible. The reason is that direct and prolonged competition is likely to result in the extinction of one (and sometimes both) competitors. Two (or more) species of animals or plants enter into competition with each other if there is a substantial overlap in their niches. A *niche* is the range of environmental factors that play a significant role in the organism's life. For example, a fundamental aspect of the niche of some birds is determined by the size of the seeds that they can eat. If the seeds are too small, they will not provide enough nourishment. If they are too large, the bird will not be able to crush them in its beak. If the seeds in question are produced by only one species of plant, then this species also becomes part of the definition of the bird's niche because the right plant and the right size of seeds have to be present for the bird to survive. And so on.

Now, if another species of bird starts feeding on the seeds of the same species of plant, the niches of the two birds will overlap somewhat. If both species happen to be able to process only seeds of exactly the same size range, they are in direct conflict. Unless the populations of the two species of birds are small and the number of available seeds is very high, competition will become more and more intense, eventually causing the collapse of one (or both) species of birds. What happens in many cases is that the two species of birds diverge with respect to their favorite food targets. One species will start eating seeds that are slightly smaller, perhaps developing strategies to find more of them in the same amount of time; the other bird species might be able to start feeding on larger-than-average seeds, possibly as the result of developing larger and stronger beaks. In time, these changes in eating habits will reduce the overlap between the niches of the two bird species, increasing the likelihood of survival of both. Notice that a very important by-product of this process of niche differentiation is that now the two species of birds will have different morphologies or behaviors that better enable them to go after a specific type of food. Therefore, the degree of diversity among birds has increased as a result of competition avoidance. Far from being "red in tooth and claw," nature is more appropriately seen as an endless diplomat, accommodating as many species with the least overlap and reciprocal interference as possible.

We need one more piece of information before we can start understanding what probably happened during the Cambrian explosion. Ecologists now know that there are exceptional times when a single lineage of animals or plants becomes incredibly successful and rapidly diversifies, producing many related yet distinct species. This phenomenon is called *adaptive radiation.* It can occur for one of two reasons. On the one hand, a species might colonize a habitat where there is little or no competition, such as an island that emerges after a volcanic eruption in the middle of the ocean. The island is initially completely empty, and if enough time is allowed for plants to colonize it, the first herbivores arriving on the island will find much food and little competition. But these first arrivals will soon be followed by a population explosion and possibly by the arrival of other species. All are naturally going to look for places with plentiful resources. Among humans, this is what actually happened when the Europeans colonized the New World. The new crowd augments the level of competition, and by the principle explained earlier, it also spurs the origin of new forms that increasingly specialize on distinct subsets of resources. In nature, the whole process can occur quite rapidly, over the course of a few dozens or hundreds of generations, leading to what looks like an "explosion" of forms.

An adaptive radiation (or explosion) may occur also if some species happen to develop a new way to exploit the environment. These new ways are called *key adaptations* or *phenotypic novelties.* This is rather analogous to somebody inventing a new machine, or a new way to do things, like the idea of offering rapid delivery of packages through private enterprises rather than the post office. Initially, only the first company (species) that came up with the idea will benefit, and greatly for that matter. But soon other companies (species) will follow suit, until the level of competition again rises and the only way to survive is to differentiate (by, for example, offering ultrafast delivery but only locally, or delivering on Sundays, or providing another special service).

What scientists think happened during the Cambrian was a combination of key innovations and new spaces to colonize. Although the details are not known (because it is incredibly difficult to reconstruct fossil niches), perhaps the key innovation was the invention of hard body parts by some animals. A hard skeleton has innumerable advantages over a soft body, including the ability to defend oneself against predation (or to be a more effective predator), to develop more complex and efficient internal organs, to experiment with new and more efficient ways of locomotion, to grow larger (and therefore more powerful), and so on. Animals with hard bodies also found themselves in the situation of being able to exploit resources and colonize places that were simply not available before that, thereby increasing biodiversity even more by a rapid series of adaptive radiations.

This combination was in all likelihood pivotal to the Cambrian explosion. There are other possible candidates as key innovations in the Precambrian, including the invention of multicellularity itself, which had occurred much earlier, but whose effects may have taken some time to be compounded to the point of generating a major adaptive radiation. More importantly, adaptive radiations are known from other periods of Earth's history (e.g., the origin and spread of dinosaurs and, more recently, of birds and mammals). Similar events can even be observed today—on a smaller scale—every time a new, sufficiently large island or group of islands is colonized (the Hawaiian Islands are a prime example).

In his book *Wonderful Life*,[28] Stephen Gould has suggested that adaptive radiations are also characterized by a special dynamic that unfolds in time. I have already described the first phase, marked by one or a few forms exploiting a new environment or a new key innovation. Then follows the "explosive" portion of the radiation, with the panoply of species and forms typical of events such as the beginning of the Cambrian. But afterward, Gould maintains, there is a period of stabilization during which things cool down and some forms that were initially successful because of the special circumstances are eliminated. Such elimination may result because certain characteristics of these forms do not stand up to long-term competition, or perhaps simply because of random events, for there is not enough space for everybody and somebody inevitably succumbs. This limitation could explain why some of the unusual-looking animals found in the Burgess Shale did not survive, and why possibly entire phyla died out during the early stages of the Cambrian.

This whole scenario is rather analogous—again—to what happens in human affairs after the initial burgeoning of forms following the discovery of a new idea or invention. For example, after the bicycle was invented in Europe during the mid-seventeenth century, it underwent a dramatic series of changes and an explosion of models, which yielded first the French velocipede around 1855, and then the first high-wheel bicycle patent with that name in England in 1873. Later, safety considerations stabilized the market, resulting in pretty much the kind of bicycle we use today. Even though there are many variations on the same theme (mountain bikes, all-terrain bikes, lightweight touring bikes, and so on), they all reflect the fundamental, "winning" design. Analogously, once animals "invented," for example, the vertebrate body plan,[29] they stuck with it, even though it is found in a variety of subforms, from frogs to elephants. Perhaps some of the strange creatures living around the time of the Cambrian explosion went the way of the high-wheelers, only to be remembered as oddities in the history of life.

There is one more reason for caution that we must consider as background for this section. On one hand, the sequence of fossils from the earliest rocks known up to and after the Cambrian explosion is a fact. On the other hand, most of the reasoning advanced by biologists about the dynamics and causes

---

[28] See note 26.

[29] Of course, from a naturalistic viewpoint, natural "inventions" and "design" imply not a conscious designer, but a process such as natural selection that can produce complex forms adapted to their environments.

of the explosion itself, as well as concerning the systematic status of many of the known organisms of the time, remains hypothetical. This is a very active field of research in paleontology, but one in which progress is difficult because of the inherent rarity of the fossils and the impossibility of carrying out controlled experiments. Hypotheses can be tested, but only indirectly, and only as far as the painfully slow finding of new fossils allows.

Let us now examine the creationist's take on the Cambrian explosion, which can be summarized in the words of Frair and Davis:[30]

> *Significant from our standpoint is that at a certain time in the supposed geological calendar, popularly called the Cambrian era, are found a host of fossils which are virtually absent from older layers of rock. From a scientific standpoint alone it is evident that a spectacular event must have occurred at this time. It seems reasonable that the abrupt change at the period designated as Cambrian is a result of God's creative activity.*

More recently, David Buckna and Denis Laidlaw put it this way:[31]

> *Evolutionists believe the Cambrian explosion of new life began about 525-550 million years ago. . . . What is the approximate number of beneficial mutations which must have occurred per year during this 5-million-year period, given that billions and billions of information bits would have to be encoded?*

Two distinct though related arguments are presented here. The first one suggests that the Cambrian explosion is roughly equivalent to the moment in which God created life itself. The idea of a multitude of life forms all arising at once is indeed intuitively appealing from the point of view of the creation stories as recounted in the Bible. Creationists emphasize the fast pace of the Cambrian explosion, aided (unintentionally) in this endeavor by first-caliber evolutionists such as Stephen Gould:[32]

> *The history of life is not a continuum of development, but a record punctuated by brief, sometimes geologically instantaneous, episodes of mass extinction and subsequent diversification.*

The second argument is that even if the Cambrian explosion did not coincide with the Biblical episode of creation, it is certainly impossible to account for it within a naturalistic framework because of the high improbability of so many beneficial mutations occurring in such a brief period of time. Indeed, if billions of mutations were needed to produce the variety of organisms that made up the Cambrian explosion, and if all these mutations had to be benefi-

---

[30] W. Frair and P.W. Davis, *The Case for Creation,* 2nd ed. (Chicago: Moody Press, 1972).

[31] D. Buckna and D. Laidlaw, "Should Evolution Be Immune from Critical Analysis in the Science Classroom?" *Impact* no. 282 (1996), an Institute for Creation Research pamphlet.

[32] In *Wonderful Life* (see note 26).

cial, then biologists would have to explain how likely it is for something of the sort to happen on a scale of a few million years.[33]

A third point usually brought up by creationists is the peculiar observation that no new phyla originated after the Cambrian. As we have discussed, a phylum is a very fundamental "kind" of organism, characterized by a distinctly recognizable body plan. There are only a few such body plans today, and a few more were present at the beginning of the Cambrian. If evolution is a gradual, progressive process, why did so many major changes occur at one point in time and, for all effective purposes, not at all afterward? Is it not more logical to think that God put all the fundamental "kinds" of organisms on Earth at once, and that extinctions occurred only afterward?

Let us turn to what evolutionary biologists think of this matter. They are busy with, but not worried about, the Cambrian explosion. The reason they are busy is that the explosion does indeed represent one of the great challenges of modern biology, as I have already explained. It is a fascinating event, unique in some (though not many) respects—an event that is at the same time crucial for our understanding of the history of life and yet difficult to investigate because of its remoteness and the scarcity of available clues. The reason evolutionary biologists are not worried about creationist claims is that some of them have been answered, some do not hold up under further scrutiny, and some simply call for a fair admission of ignorance on our part but are not a fatal flaw in evolutionary theory.

The point implied in the statements by Murchison and by Frair and Davis quoted earlier is that the Cambrian explosion can be seen as synonymous with the creation of life as described in Genesis (though they do not actually make the link explicit, they do talk about a "creative burst," and there is only one burst of creation described in the Bible). This would be a unique case of the Bible's being more in accordance with the available data than the current opinions of thousands of paleontologists. Although superficially the parallel is indeed impressive, if we think for a moment about what Genesis actually says and what the fossil record tells us, there is no congruence whatsoever between the two. If the Cambrian explosion was the moment in which God created life on Earth, how do we explain the worldwide presence of fossils of various forms during the more than 3 billion years preceding the Cambrian events? Did someone else create those creatures, or did the Bible neglect to mention an earlier creation? Furthermore, what about life forms that did not arise until *after* the Cambrian?

Some creationists admit that some "microevolution" may have occurred, but they are firm in denying that any new "kind" has ever appeared after the initial

---

[33] The actual duration of the "explosion" is still under debate. The commonly reported average is 15 million years. Gould's estimate is the lowest so far: 5 million years. Recent research suggests that the Cambrian drama may have unfolded much more slowly, over several dozen million years or more. In any case, it still remains a very rapid pace of change when compared to the geological scale of life on Earth.

burst. Although they conveniently neglect to define accurately what a "kind" is,[34] surely a phylum, or even a class within a phylum, has to qualify as a new kind. After all, phyla and classes represent two of the highest levels of the taxonomic hierarchy. Well, we know for sure that several classes of organisms appeared after the Cambrian (e.g., the flowering plants and, most impressively, all vertebrates, including fish, amphibians, reptiles, birds, and mammals).

Whereas the appearance of life forms scattered over a long period of time definitely contradicts creationist expectations, it is what evolutionary theory expects. In fact, recall Darwin's own predictions that we would find Precambrian fossils, and that the span of the Precambrian would be found to be much longer than the time between the Cambrian and the current era. A century and a half of paleontological research has upheld his two expectations. First, current estimates of the length of the Precambrian are on the order of 4 billion years—a much longer time than the period that has elapsed from the beginning of the Cambrian until today (570 million years). Second, we now have evidence of fossils scattered throughout the Precambrian, and they do fit (to some extent) Darwin's prediction of simpler forms. For example, a clear member of the crustacean group has been found in the very early Cambrian, pushing the origin of the whole group farther back into Precambrian times.[35] The new fossil is exceedingly well preserved, down to its delicate limbs cast in calcium phosphate, so there is no doubt that this animal was closely related to the group to which crabs, shrimps, and lobsters belong. This particular finding shows one of the major weaknesses of the creationist position: It is based on our current ignorance, but science has a way of discovering new facts that fill in old gaps. Because of this crustacean and other fossils, it is no longer tenable to maintain, as creationists have done so far, that the Cambrian "explosion" was too sudden a burst of new forms, since some of these forms actually extend the time for the emergence of some new phyla by tens more millions of years.

Why did it take so long for paleontologists to uncover Precambrian fossils, and do these fossils fit the expectations of evolutionary biologists? Even though we now know of fossils older than the beginning of the Cambrian from geological formations throughout the world, they are still few compared to the richness of the fossil record after the Cambrian. Darwin himself proposed (in explanation of this phenomenon) that the Precambrian rocks are so old that they have been eroded to a higher degree than any other (younger) formation on Earth. A second reason for the scarcity of Precambrian fossils is the fact that Precambrian rocks suffered from a large amount of disintegration—larger than in successive periods of Earth's history. The amount of Precambrian rock disintegration is due mostly to the fact that the planet itself was more geologically active then, especially in the earliest times represented by the Hadean and

---

[34] Compare this with the thousands of papers and decades of research that evolutionists have devoted so far to refining the concept of species.

[35] See R. Fortey, "The Cambrian Explosion Exploded?" *Science* 293 (2001): 438.

immediately successive eras. In fact, no rock older than 3.8 billion years is known to have survived. So even if there were fossils of extremely primitive life forms embedded in such rocks, we will never have a chance to study them. A third reason for the paucity of Precambrian fossils is that "hard" body parts had not yet been produced by natural selection at the time.

The overwhelming majority of known fossils—including the post-Cambrian ones—preserve parts of the body of the organism that are made of strong, highly mineralized materials, such as shells and bones. These parts are much more likely to survive the many hazards of the fossilization process, and therefore to come to us almost intact. Before the Cambrian, no animal had developed such body parts, and all life forms were soft-bodied, very much like modern worms and jellyfish. Soft-bodied organisms (or soft parts of hard-bodied organisms, such as dinosaurs' skin and birds' feathers) sometimes escape destruction and come to us in a recognizable fossilized form. However, when one compounds the odds against soft-body fossilization with the extreme antiquity and high degree of disintegration of Precambrian rocks, it is easy to see why it has been so hard to find evidence of life before 570 million years ago.

Another major point raised by creationists (e.g., by Buckna and Laidlaw) is that the "explosion" was simply too rapid to be the result of a natural process. Unfortunately, this contention is aided by the somewhat careless writing of evolutionary biologists such as Stephen Gould. The earlier quote from Gould's *Wonderful Life* refers to "geologically" instantaneous events punctuating the history of life (see Chapter 5). Creationists usually ignore two considerations when they report such statements from evolutionists. First, *geologically* instantaneous does not translate into the *everyday* meaning of the word *instantaneous.* For something to be geologically instantaneous, it has to happen quickly relative to the normal pace of change we see in the fossil record. But that can still mean tens of millions of years, and consequently hundreds of thousands or millions of generations, depending on the duration of the life cycle of an organism. Although Gould should have been careful to specify this difference, because he was writing for a lay audience and not for technically savvy colleagues, the creationists' use of this and similar sentences is careless to say the least. It is all too easy for creationists to take something out of context, without providing any clarification, and uphold it as "proof" that even their opponents agree with them. This strategy implies that the opponent is somehow either so devious or so incompetent that he is not willing to recant in public, and it hardly constitutes a fair contribution to the debate.

Second, the reader has to consider that scientists are human beings, and as such they are subjected to all human passions and weaknesses (see Chapter 7). It is not inconceivable (although not to be condoned) that a scientist—especially when writing for a lay audience—will stretch a point if it fits with the overall scenario that he is trying to convene. Gould's major scientific achievement is the expansion of the neo-Darwinian theory termed *punctuated equilibria,* which I have already discussed. Claiming that the Cambrian events were "geologically instantaneous" fits Gould's view of evolution. He may not have been able to get away with such a casual statement in the technical literature, where peer review would have caught it and demanded a more precise formulation. But in the court of public opinion, one has a little more leeway, and good writing is of necessity coupled with enthusiasm for what one writes. We all know that enthusiasm can lead one to overstate a case in which one firmly believes, but that does not automatically make one a liar, nor does it disprove one's case.

Even if we admit that the Cambrian explosion did unfold over the course of a few tens of millions of years, is that enough time to generate by natural means the bewildering diversity of life forms that we can glimpse in the fossil record? We cannot say for sure because nobody can rewind the tape of life and play it back while taking notes. However, all the scientific evidence we have so far points in that direction. Creationists seem to imply that the only naturalistic possibility is for a mind-bogglingly high number of mutations to occur simultaneously, and somehow to generate a new type of organism in a single saltational event (such as a fully formed bird from a dinosaur egg). This concept has once again been involuntarily aided by the somewhat careless writings of biologists such as Gould, as we saw also in Chapter 5 in the discussion of Richard Goldschmidt's idea of "hopeful monsters."

We now know that the process described by Goldschmidt cannot occur. But we also know much more about genetics and developmental biology than Goldschmidt did (remember that at the time Goldschmidt's book was published, people had no idea what the structure of DNA was). Modern molecular developmental genetics has finally started to bridge the gap between evolutionary theory and developmental biology. We now know that not all mutations have equivalent effects, and that some are capable of significantly altering the morphology (or behavior) of an organism. Although no contemporary biologist maintains the likelihood of a Goldschmidt-type monster, changes in regulatory genes can produce many interesting forms that at least show us that it is genetically possible to bridge the gaps that separated new types of organisms at various times during the history of life on Earth.

Whatever the genetic mechanism underlying the origin of new types of plants and animals is, the question of sufficient time remains open. However, I truly think that most creationists do not really have a sense of the span of time that is included in a "few" dozen million years. Indeed, it is hard for any human to fully comprehend what that means. To focus our ideas, let us assume we are considering an organism whose generation time is 25 years, the likelihood of occurrence for any given mutation is one in a million (per generation per gene—an empirically reasonable estimate), and the total population size of the species is 1 million individuals (probably a gross underestimation for most species). Given these conditions, on average *each* gene will mutate within the population once every generation (in different individuals, of course). This means that during the 15 million years of the Cambrian explosion, each gene would have mutated 600,000 times! Certainly this is more than enough change to provide the necessary raw material for natural selection.

A milder interpretation of creationist claims could be that the Creator worked through natural means to produce the results He intended. Although there is no way ever to conclusively reject such a proposition, its consequences may be worth exploring in some detail. Consider the pattern that we can observe in the fossil record, particularly the slow, scattered appearance of simple forms over several billion years. This period was followed by a rapid diversification of forms, itself followed by a great number of extinctions when things settled down. If this pattern is indeed the result of an intelligent design, what does it tell us about the Designer? It seems that the whole process was very poorly planned, with billions of years of failed experiments. Then, when suddenly success was achieved, the Designer was sloppy enough to let millions of its creations disappear into nothingness in what—geologically speaking—can be considered the time immediately following the Cambrian. I know that this is not the kind of creator that most creationists are thinking of when they raise their objections against the backdrop of the Cambrian explosion. In fact, this kind of creator behaves in a suspiciously similar way to what is expected from a natural, undirected, and unsupervised process.

Let me close with a couple more considerations about the never-ending question of missing links and intermediate fossils, which raises its ugly head not just in the case of the Cambrian explosion, but throughout the creation–evolution debate. Skeptic Michael Shermer has pointed out that a typical creationist tactic goes like this: A scientist displays two forms—say, a modern whale and one of its most remote ancestors. She is then asked by the creationist to provide an intermediate fossil between the two—a perfectly reasonable request. As soon as the intermediate is introduced (after much fieldwork and searching), the creationist claims that there are now *two* gaps in the fossil record! One can easily see that this is an absurd form of reasoning: The more intermediate fossils are produced, the more gaps emerge. Instead of improving, the situation appears more and more desperate with the appearance of new data.

This is not only sheer nonsense, but it betrays once again a complete misunderstanding of how science works. It is *not* the goal of paleontology to find every single intermediate step along a line of descent (given that most individuals do not fossilize, this is simply impossible). In addition, evolutionary theory is confirmed by the appearance of intermediate fossils in the appropriate geological strata even if it is very difficult ever to be sure that those fossils are direct ancestors or descendants of each other. The situation is analogous to denying that one belongs to a family tree just because one cannot find pictures of a father and mother, but only of brothers and sisters. This is not a position that many people of "common sense" would readily take.

A second unbeatable (and therefore intellectually sterile) creationist strategy is often displayed by Duane Gish of the Institute for Creation Research.

When confronted with fossils belonging to prehuman species, he invariably declares them either entirely apish or completely human, thus eliminating a priori the very possibility of intermediates. He does this without ever having examined the fossils himself, in the complete conviction that such fossils simply *must* be mistakes. It is astounding to see the length to which the human brain can sometimes deny evidence that is right under its nose (more on this in Chapter 8).

Conceptual Summary

Creationism	***Evolutionism***
The Cambrian explosion is best explained as the moment in which God created life (or at least as a sudden burst of creative energy from Him).	The creationists' claim contradicts the fact that some fossils pre-date the explosion by billions of years. At a minimum, the fossil record suggests that there were multiple creations, some not mentioned in the Bible.
No new "kinds" of life appear after the Cambrian.	New classes of organisms, of both plants and animals—for example, flowering plants and vertebrates—have evolved after the Cambrian.
The Cambrian explosion was "instantaneous."	The Cambrian explosion was rapid in geological terms, but this still translates into millions of years and hundreds of thousands of generations.
Precambrian rocks contain no known ancestors of modern groups.	The creationists' claim is discovered to be less and less true, the more research advances. In general, however, there are so few Precambrian fossils that the likelihood that the necessarily soft-bodied organisms that pre-dated modern forms would fossilize is very slim.
Too many mutations would have had to occur naturalistically to explain the origin of so many new forms at the beginning of the Cambrian.	It is very difficult to calculate how many mutations needed to occur or did occur (and of course creationists do not supply us with the details). However, there certainly was ample time for hundreds of thousands of mutations to happen in each lineage. Furthermore, modern developmental genetics shows that significant changes in the appearance or behavior of organisms require few, not millions of, mutations.

Creationism *(continued)*

No explanation is provided for observed extinctions.

Evolutionism *(continued)*

Why did so many forms of life go extinct after the Cambrian? From an evolutionary standpoint, extinction is as natural a process as the formation of new species, and it is related to the ecology and genetics of each species.

Additional Reading

Fry, I. (2000) *The Emergence of Life on Earth: A Historical and Scientific Overview.* New Brunswick, NJ: Rutgers University Press. A complete treatment of the problem of the origin of life, from historical, philosophical, and scientific perspectives. It includes a discussion of creationism.

Kauffman, S. A. (1993) *The Origins of Order.* New York: Oxford University Press. A discussion of the application of complexity theory to explain the apparent order of the universe, with implications for both the question of the origin of life and evolution.

Maynard-Smith, J., and E. Szathmáry. (1995) *The Major Transitions in Evolution.* Oxford: W. H. Freeman Spektrum. A treatment of several of the major transitions that occurred during the history of life, from its origin to the evolution of societies and language.

Monod, J. (1971) *Chance and Necessity: An Essay on the Natural Philosophy of Modern Biology.* New York: Knopf. A classic book on the philosophy of evolutionary biology by a Nobel Prize–winning biologist. It includes a lucid discussion of why the second principle of thermodynamics does *not* contradict evolutionary theory.

Schrödinger, E. (1947) *What Is Life?* New York: Macmillan. Another classic, in which a physicist poses for the first time in scientific terms the question—until then purely philosophical—of what life is and how it may have originated

– 7 –

Scientific Fallacies

The only real mistake is the one from which we learn nothing.

– John Powell –

Like all other arts, the science of deduction and analysis is one which can only be acquired by long and patient study, nor is life long enough to allow any mortal to attain the highest possible perfection in it.

– Sherlock Holmes, in Arthur Conan Doyle's "A Study in Scarlet" –

This chapter is relatively brief, not because there are few fallacies that scientists and educators have committed while dealing with creationism, but because many of these problems have already been discussed in other chapters (and I will refer to them specifically in the following discussion). Nevertheless, I felt that scientific fallacies deserve a chapter of their own simply because it is time that scientists face what both creationists, and philosophers and sociologists of science, have been telling them for some time now: Science is a human activity, and as such it is fallible. Scientists not only make mistakes like all other humans, but they are prone to the same character flaws (and positive aspects) as other humans, though neither more nor less. A scientist, in my experience, is as likely as anybody else to be jealous, egocentric, or closed-minded. A few scientists have even been prone to unethical behavior and downright fraud while attempting to further their own careers.

But who is without blame? Can we think of any human profession that doesn't offer a similar cross-section of human traits, good and bad? What is different about science as a process (as distinct from the people who *do* science), however, is that it is inherently self-correcting because of two factors: (1) the very rules of the game that scientists have decided to engage in (and by which they are therefore judged) and (2) the fact that there is an independent arbiter that cannot be fooled: Nature is what it is, no matter how much we sometimes wish it were otherwise. No scientist can fake anything of any consequence for very long, simply because somebody will eventually compare his claims with nature and reach a verdict of rejection.

Perhaps the most important fallacy that scientists and educators often commit is what I refer to as the *rationalistic fallacy*. If you are among those who are stubbornly trying to improve critical-thinking skills around the world and feel a bit frustrated by the wave of nonsense that regularly hits the airwaves and the media (of which creationism is of course just one example), you are not alone. But if you insist on thinking that all you need to do is explain things just a little bit better and people will see the light, you are committing the rationalistic fallacy.

It is probably true that better knowledge and understanding of science improve one's ability to grasp the real world; if that were not the case, the entire education system should be thrown out, a step that only a minority of right-wingers are prepared to seriously suggest in the United States at this moment. But it is also undeniably true that explaining science to many people does not result in making them any more skeptical, distrusting, or disbelieving of pseudoscience. For example, John Moore reports in an article in *The Science Teacher*[1] that subjects were surveyed for their beliefs in the paranormal, UFOs, and astrology before taking a course that dissected the evidential bases for all these pseudosciences. Although skepticism had increased marginally toward the end of the course, credulity had returned with a vengeance only a year after the

[1] J. Moore, "Thought Patterns in Science and Creationism," *Science Teacher* May 2000: 37–40.

test! It seems to me that we should try to understand what causes the rationalistic fallacy if we hope to make any progress in fighting the rampant irrationalism that has always manifested itself in so many forms in human societies. It might save us a great many misdirected efforts and a trip or two to the psychotherapist when the depression hits.

The first thing to realize is that many people who believe in all sorts of weird things are not stupid, certainly not in the generally accepted sense of the term. Sure, if we define intelligence as the ability to grasp the intricacies of the real world, then anybody who does not understand quantum mechanics is an idiot. But remember the immortal words of physicist Richard Feynman: "If you think you understand quantum mechanics, you don't understand quantum mechanics." The fact is that many people who believe in pseudoscience live successful lives. Some are college graduates. They can understand very well the reality of everyday life; sometimes they even successfully make complex decisions such as investing their money or planning a career. The answer must therefore lie elsewhere.

I think the problem is in what we mean by "understanding reality." Thomas Henry Huxley, the nineteenth-century scientist known as "Darwin's bulldog," was very successful in lecturing to the general public, to an extent that neither Richard Dawkins nor Stephen Gould could dream of today. Huxley's fundamental philosophy was that science is common sense writ large. Because most people are equipped with both an innate curiosity and a moderate dose of common sense, he thought that if we explained things while appealing to people's already existing mental tools, they would understand. Indeed, this is the philosophy behind most science documentaries.

The problem is that most modern science is not a matter of common sense at all! On the contrary, from physics to cosmology, from evolutionary to molecular biology, our current scientific understanding of the world is extremely counterintuitive. The reason is that the realm of scientific investigation now literally spans the whole universe, from the beginning of time until now (roughly 20 billion years) and from the subatomic level to the largest aggregates of galaxies. Let us remember that in Huxley's time most scientists thought Earth was a few million years old, the existence of galaxies was yet to be discovered, and nobody had the foggiest idea of what either an atom or a gene was.

Evolutionary psychologists such as Steven Pinker suggest an explanation for this state of affairs. According to the standard Darwinian theory, our brains are at least in part the result of natural selection to improve our fitness. But to what kind of environment did the brain adapt? Obviously, the one that we have inhabited for most of our evolutionary existence: forests and savannas, where *reality* meant being able to procure food and mates while carefully avoiding predators. Is it any wonder, then, that we simply can't understand (in a deep sense) quantum mechanics?

If we add to this mix the fact that people still want answers to the fundamental questions of life (probably a by-product simply of our evolving as self-aware beings), it doesn't take much to understand why evolution and the Big Bang are discarded in favor of all-powerful and all-merciful supernatural entities who watch over every detail of our lives and help us through our daily tribulations. Even the much touted fact that Europeans accept evolution much more readily than Americans do, I would argue, is likely due to a far less flattering (for Europeans) explanation than is usually assumed. It is not that the French, Italians, British, and so on are smarter or know more science (the latter at least is demonstrably not so); rather, it is probably that, given their history, they have had their fill of religious wars and witch hunts and they are putting their current trust in another category of priests, the scientists (at least until these, too, screw things up in some major way, as they have already done on several occasions after Chernobyl, or during the mad cow disease epidemics).

So what do we do about the persistence of pseudoscience and other forms of irrationality? Unfortunately, identifying the causes doesn't necessarily cure the disease. We are in no position to reshape the human brain to bring it up to speed with the current human environment. We can, however, become more familiar with the large body of literature on human cognitive neurosciences; getting to know how the brain works has to be the first step toward designing better tools and arguments to educate people, and to that topic I devote the last chapter of this book. We can also be more understanding when we do confront an irrational position, rather than dismissing our interlocutor as a simpleton. Demonstrating sympathy and reaching out to the intuitive right brain may be a better way to get to the often hyper-rational (and rationalizing) left brain (see Chapter 8). A third thing to do is to better understand the nature of science itself, including its limitations and pitfalls—something which, astonishingly, neither scientists nor science educators are particularly well trained to do. This is the topic to which I will devote the rest of this chapter.

Of Whales, Bacterial Flagella, and the Big Bang

A good way to start is to appreciate that often evolutionary biologists dismiss creationist arguments out of hand because they are "obviously" wrong, without realizing that even a wrong argument can point the way to a legitimate question underlying the fabric of evolutionary theory. Let us consider a few examples to clarify. One of the creationists' persistent questions concerns the distinction between micro- and macroevolution. Scientists use these terms in a very different way from what creationists seem to imply, which is part of the problem. For a biologist, *microevolution* refers to changes in a species over short periods of time that can be explored with the conceptual and mathematical tools of population genetics and ecology. *Macroevolution,* on the other hand, refers to changes that occur over much longer periods of time; this is usually the research field of the paleontologist.

The relationship between micro- and macroevolution is a legitimate field of research within evolutionary biology, and one that has generated a large amount of controversy over at least the past 60 years. Indeed, the beginning of the debate is often traced back to correspondence between Darwin and Huxley, in which the latter pointed out to the author of *The Origin of Species* that there was no need to postulate very slow and gradual changes as the only pattern of descent with modification, and that sometimes things could move along significantly faster, thereby speeding up the appearance of novel structures or groups of organisms, such appearance being a typical macroevolutionary event.

The debate resumed during the first part of the twentieth century, during the so-called neo-Darwinian synthesis that has shaped much evolutionary thought until very recently. One of the major contributors to the synthesis was paleontologist George Ledyard Stebbins, who proposed the concept of "quantum evolution" to explain the sudden (in geological terms) appearance of certain groups of animals in the fossil record. According to Stebbins, evolution can proceed at different rates, depending on the environmental conditions and presumably on the amount of available genetic variation in the populations that are evolving (the larger the amount of variation, the faster evolution can proceed, other things being equal). But Stebbins, partly under pressure from his more "orthodox" (i.e., more gradualistically Darwinian) colleagues, such as systematists Ernst Mayr, eventually dropped the idea of quantum evolution from later editions of his book, a process that Carl Schlichting and I have referred to as the "hardening" of the synthesis.[2]

Still, during the 1940s a challenge to the recently established neo-Darwinian consensus came from a prominent geneticist Richard Goldschmidt. As I mentioned in Chapter 5, in his *The Material Basis of Evolution*,[3] Goldschmidt questioned the assumption that simply extrapolating from microevolutionary phenomena would yield a satisfactory explanation of macroevolution, and he explored some of the possible alternatives. With little knowledge of molecular genetics and developmental biology, Goldschmidt couldn't offer more than some bold speculations, the most famous of which is his idea of "hopeful monsters." We need to discuss this concept in a bit of detail because it is still used today by creationists to alternately ridicule evolutionary theory for proposing different hypotheses and accuse evolutionary biologists of ignoring important alternatives that some of their own colleagues have proposed.

Goldschmidt reckoned that one way macroevolutionary processes, such as the origin of a new body plan (say, the transition from reptiles to birds), could happen is by what he referred to as "systemic mutations." A systemic mutation, or genetic revolution, is a reorganization of an entire genome in which genes are duplicated and/or shuffled around on the chromosomes. Goldschmidt was aware that when much smaller genomic rearrangements take place (such as the change of physical position of one gene in the fruit fly), major

[2] See C. Schlichting and M. Pigliucci, *Phenotypic Evolution: A Reaction Norm Perspective* (Sunderland, MA: Sinauer, 1998).

[3] R. Goldschmidt, *The Material Basis of Evolution* (New Haven, CT: Yale University Press, 1940).

phenotypic changes can occur. Indeed, he had done part of the fundamental work in this field and was therefore on solid empirical ground. Goldschmidt thought that if a genetic revolution were to occur in the offspring of an organism, the result would be, literally, a "monster"—that is, an animal that would look (and possibly behave) very differently from its relatives and conspecifics. Goldschmidt realized that most such monsters would simply die, or at least would not be able to find mates and reproduce. However, occasionally some might survive to adulthood and even find mates among conspecifics that would not find them too repulsive. These "hopeful monsters" would then fuel the evolutionary change within a species and, if successful, in the long run serve as the origin of a different group of animals altogether.

Goldschmidt's ideas, although ridiculed by creationists as an example of how absurd evolutionary biologists can become to defend their "irrational" faith in naturalistic explanations, have actually been abandoned by professional biologists for more than 50 years (roughly since our increased understanding of molecular biology from the 1950s on). Interestingly, however, we do know of the existence of some very successful hopeful monsters: A large number of plant species (and almost all ferns) actually do originate through genomic revolutions caused by a spontaneous duplication of their entire set of genes—a phenomenon called *polyploidy*. Polyploidy can occur either because of the failure of meiosis (the cellular process that reduces the number of chromosomes to half before each generation begins and that allows sexual reproduction to restore no more than the original full complement) or because of the mating of gametes from two different species, which occasionally can yield a fertile hybrid offspring. Polyploidy does not occur frequently in animals for reasons that are still not well understood and that probably have to do with key differences in the developmental processes of animals and plants. Ironically, however, Goldschmidt did identify a major process leading to macroevolutionary change in many species, just not in the animals he had in mind.

Thirty-two years after the publication of Goldschmidt's book, two paleontologists—Niles Eldredge and Stephen J. Gould—reopened the macroevolutionary question with their theory of *punctuated equilibria,* which I discussed in some detail in Chapter 5. This is another topic that is very much misunderstood and used out of context by creationists, as we have seen before. Eldredge and Gould attempted to link a standard theory of the origin of new species proposed by biologist Ernst Mayr[4] with the observable fossil record—that is, to link micro- and macroevolution by means of an established theory and the available empirical evidence. They succeeded to a large extent, thereby taking the wind out of one more creationist objection to evolutionary theory (though Eldredge and Gould never saw themselves as responding to creationist critiques).

The publication of Eldredge and Gould's original paper in 1972[5] spurred a healthy series of research papers aimed at either poking holes in their theory or at supporting it. The current consensus seems to be that there is enough empirical evidence to grant punctuated equilibria real existence, although we do not know how often this mode of macroevolution occurs when compared with more traditional, gradual, evolutionary change. Even so, however, the theory of punctuated equilibria does not *solve* the problem of micro- versus macroevolution because it provides only part of the answer. Specifically, we now have a model of how biological phenomena at the level of population demography can explain the patterns observed in the fossil record, but we still don't know how these changes occur at the genetic and developmental levels, a question that Goldschmidt's theory of hopeful monsters tried (unsuccessfully) to answer.

Given all this, what, then, causes macroevolution in animals? By and large we still do not know. Before creationists start quoting the preceding sentence (out of context) as proof of their correctness on matters of evolution, let me clarify what I mean. Plenty of research has been done over the last few decades to study the developmental and genetic differences between different types of organisms, and this research is likely to lead eventually to a satisfactory answer to the question of micro- versus macroevolution. However, the fact that we do not have that answer yet implies absolutely nothing about the correctness of any other view, much less the nonanswer that creationists are likely to propose ("God did it"). As we have seen, this is a most pernicious and difficult fallacy to eradicate in creationist thinking: Their lack of understanding that it is normal in science not to have final answers is coupled with the non sequitur that therefore they must be right. Once again, all that follows from the fact that we do not know something is that we don't know it (yet). There is no shortcut to doing the hard work required to figure things out.

Where, then, is the scientists' fallacy in explaining macroevolution within the context of creationism–evolution discussions? It lies in the *pretense* that we have a full answer when at most we have a few (tantalizing) clues. As an example, remember the problem of the origin of modern whales discussed in Chapter 2. There is now more than enough paleontological evidence to conclude beyond reasonable doubt that the two kinds of whales existing today (baleen and toothed) share a series of common ancestors. Furthermore, it is very likely that such ancestors can be traced back to a lineage of terrestrial

---

[4] Mayr's theory was the so-called allopatric theory of speciation. *Allopatry* means simply "in different places," and the theory says that if a small population becomes geographically isolated from a larger population from which it originated, the small population may evolve independently from the mother group for some time. Population genetics theory (and empirical evidence) tells us that small populations often evolve faster than large ones (they have, literally, less "inertia") because their rare genes can spread relatively rapidly within small populations. If the geographical isolation between the new, small population and the original large one persists long enough, the new population might become different enough to be a new species. Once this happens, the new species, if successful, may expand its area of distribution and population size, and these demographic phenomena can occur very rapidly (during a few thousands of generations). Eldredge and Gould simply realized that allopatric speciation, seen in a geological time frame, would lead to long periods of no change (times of equilibrium, which they called *stasis*), "punctuated" by relatively rapid (geologically speaking) periods of change—hence their theory of punctuated equilibria.

[5] N. Eldredge and S. J. Gould, "Punctuated Equilibria: An Alternative to Phyletic Gradualism," in *Models in Paleobiology,* T. J. M. Schopf (ed.), pp. 82–115 (San Francisco, CA: Freeman, Cooper, and Co., 1972).

mammals that lived about 55 million years ago and belonged to the same group as modern hippos, camels, sheep, and pigs (the artiodactyls). Now, if that is all we say to a creationist asking about whales, we are on as firm an empirical ground as possible, and we are making not just reasonable, but very likely, statements. However, this is *not* the same as saying that we have *solved* the problem of the macroevolution of whales. Far from it.

To really solve that problem we would need (at least) a detailed idea of what sort of genetic and developmental processes can, over a period of 20 million years, produce an entirely marine mammal from a completely terrestrial ancestor. Furthermore, it would be nice to know what sort of ecological conditions catalyzed the shift from terrestrial to aquatic life in this lineage, as well as what sorts of behavioral changes had to occur for that change to be realized. Unfortunately, we don't know all of these things, so although it is almost certain that baleen and toothed whales descend from a terrestrial ancestor (and we have good clues as to the group of mammals to which they are most closely related), this is far from completing our task of explaining that macroevolutionary process. This incomplete knowlege, of course, offers scientists a challenge for renewed efforts, not a reason to abandon naturalistic explanations and embrace nonanswers such as those forthcoming from the Christian or any other religious tradition. It is this passion for the quest that scientists and educators for the most part fail to convey. I will discuss this problem in Chapter 8 and argue that doing something about it is the most urgent task for teachers.

Perhaps the mother of all macroevolutionary mysteries is the Cambrian explosion, which I discussed in Chapter 6. As we saw in that discussion, there is nothing to indicate that the relatively rapid origin of many new body plans in animals is a mystery requiring anything more than our current understanding of evolutionary and ecological problems. Yet the clues currently available are so few that it is not intellectually honest to say that we "understand" the Cambrian explosion. We are studying it, and we are making progress. One cannot ask more from science.

A similar problem arises when evolutionary biologists are challenged by intelligent design advocates like biochemist Michael Behe to explain the evolution of complex subcellular structures such as the bacterial flagellum or the biochemical cascade that leads to blood clotting. As I argued in Chapter 2, we actually know quite a bit more about both of these problems than Behe and his creationist colleagues are willing to admit. For example, we can show that the highly complex flagellum that Behe likes so much to describe as an impossible challenge to evolution (Figure 7.1a) is not the only flagellum that exists in nature. Much simpler versions of it are found in other bacteria (Figure 7.1b), which means that it is possible to evolve the complex model from simpler ones. Again, however, the fallacy of the typical evolutionary response to creationism is that often the mere existence of simpler bacterial flagella is presented as evidence of our complete understanding of the process of evolution at the subcellular level. It most definitely is not. We know very little about the phylogenetic (ancestor–descendant) rela-

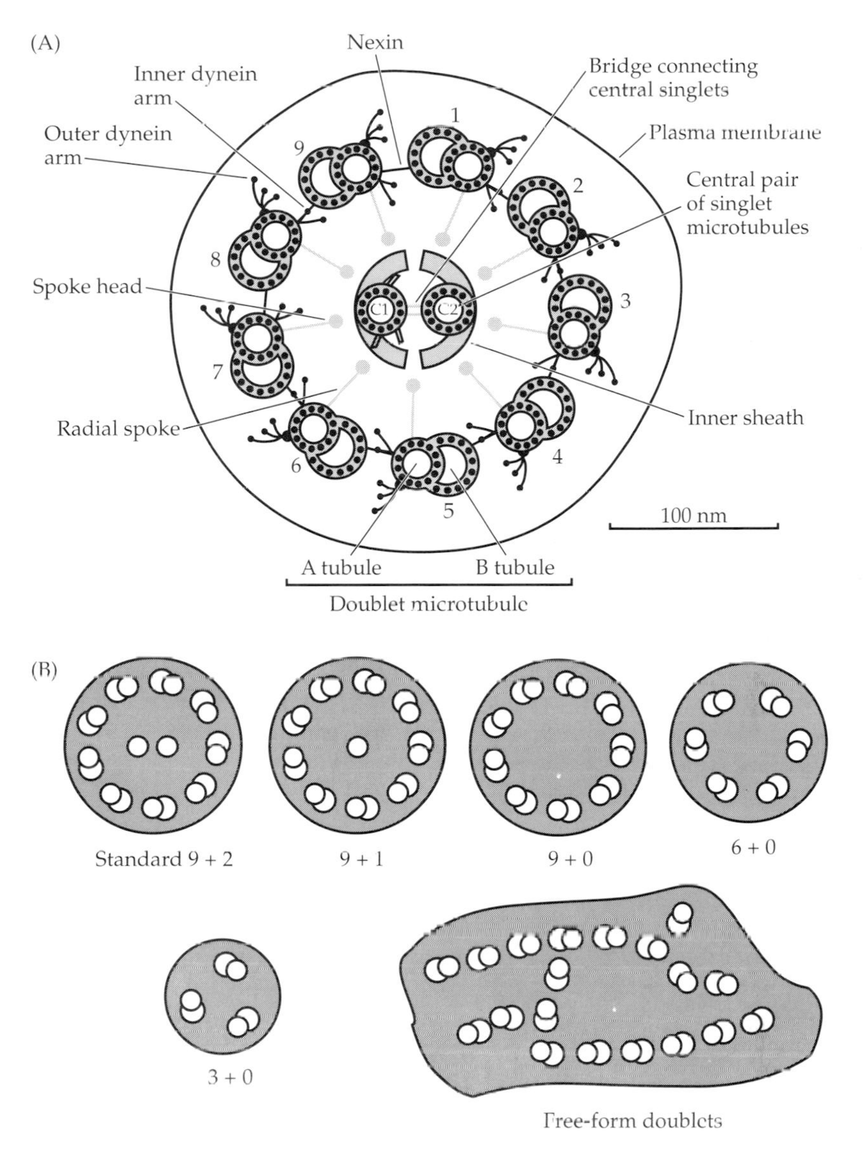

Figure 7.1 (a) The sort of complex bacterial flagellum (in cross section) that Michael Behe and fellow intelligent design proponents claim is impossible to explain with evolution. (b) The much simpler flagella that are found in related species of bacteria, exactly as evolutionary theory would predict.

tionships among different groups of bacteria. We suspect that flagella evolved independently on more than one occasion, but we don't know from which ancestors, and most importantly, we don't know what sort of mutations (in which genes and in what sequence) had to occur for the process to unfold.

This level of ignorance is not surprising, considering that very few people actually work on bacterial evolution and that our state of understanding of deep molecular processes is still very limited. Furthermore, evolution is by definition a historical process, and we might never be able to answer specific questions, simply because enough of the clues have been lost to time. Imagine what a historian would be unable to tell about Napoleon if most of the historical documents we had about him were missing. Imagine what our reading of the Bible would be if we had only a few fragments available today, as is the case for so many other historically interesting documents or people. Bacteria, unfortunately, very rarely fossilize. But rather than despair at our ignorance, we should be glad that we do have some clues to how flagella evolve. These clues are consistent with the theory of evolution and in no way indicate the intervention of a supernatural entity in the process.

Speaking of origins, the evolution–creationism debate is often cast in terms of *origins* (of species, life, the universe), which are some of the toughest problems for science to solve precisely because these are unique events that happened in the distant past. That, of course, doesn't mean we are clueless about them. I dealt with the question of the origin of life in Chapter 6 and explained why Miller's classic experiments—as historically important for the field as they are—are *not* the solution (or even a valuable starting point) to understanding the origin of life on Earth. An intellectually honest and well-informed science educator (they are usually the former but only more rarely the latter) should therefore point to the amount of progress that has been made in this field, describe some of the ongoing research, and stop far short of saying that the problem has been solved. This sort of approach to explaining science as a process is, as we shall see in Chapter 8, very labor-intensive, and it is hardly ever carried out either by professional scientists (who usually focus on their own specialty and research problems) or by educators (who normally are overwhelmed by the amount of material to cover in their classrooms and by their own lack of continuous training in the disciplines they have to teach). That is why easy dismissal of creationist questions is so common. But it is wrong, both ethically and educationally.

I have not written much in this book about the *other* origins question that is commonly raised by creationists: The origin of the universe. One reason for me not to dwell on this topic is simply that I am a biologist, not a physicist, and I do not therefore feel qualified to expound on the matter. However, it also represents a perfect example of scientific fallacy within the context of our discussion, so it deserves at least a few words. Most educators confronted by creationists on this topic simply dismiss it with a vague reference to the Big Bang theory having been confirmed by scientists. Although this is (almost) true,[6] it does not even begin to settle the question. To start with, cosmologists are still

[6] I say "almost" true because there are many aspects of the Big Bang theory, several of which have indeed been confirmed empirically, but many are still under active investigation by astronomers. Furthermore, alternatives to the Big Bang are still theoretically possible and are occasionally discussed in the professional literature on cosmology.

debating what exactly happened immediately after the Big Bang, and the evidence currently points toward a period of sudden acceleration in the rate of expansion of the universe referred to as *inflation*. This is important because inflation is a major addition to our understanding of the origin of the cosmos, and it potentially explains important features of our universe as well as helps us predict its future evolution. Second, it is not currently clear *how* exactly the Big Bang happened, which is of course a crucial question for creationists (as well as for scientists). The two currently dominant theories in physics—general relativity and quantum mechanics—actually make conflicting predictions when applied to the problem of the origin and early behavior of the universe, so much so that some physicists are considering a more overarching theory—known as *superstring theory*—to help reconcile relativity and quantum mechanics (see Chapter 2).

The simple version of the narrative one gets from physicists today does sound almost as incredible as a divine intervention (though not quite), since they talk about a quantum fluctuation that literally gave origin to the universe out of nothingness. This contradicts every intuitive notion of cause and effect that we have, and yet it is an empirically verifiable component of quantum mechanical theory. The point is that simply saying, "It was the Big Bang," doesn't even begin to address the question being asked. It requires a great deal of time and energy on the part of the science educator to look up secondary sources (in turn based on the primary, technical literature) and translate what they are saying into understandable concepts for whoever asked the question to start with. Science education is not an easy task and shouldn't be trivialized as such.

To make things even more complicated, even if one were to go to the trouble of learning about relativity, quantum mechanics, and superstrings, these theories open up a Pandora's box of *philosophical* (not only scientific) questions. One might wonder in what sense an event can be uncaused, what was there—if anything—before the Big Bang, or whether it makes sense to search for *the* fundamental constituents (the essence) of the universe, as physicists do while seldom asking themselves why they think they will eventually get an answer. These are philosophical conundrums for which both scientists and science educators are even *less* prepared than they are for the more common scientific questions that are posed by creationists.

Hoaxes and Blunders

Toward the end of 1911, Charles Dawson was walking on Piltdown Common, in Sussex, England, when a group of workmen told him they had found a strange fossil that might interest him. Dawson was an amateur fossil hunter, but he was so highly respected by the professionals that he had gained the fellowship of the London Geological Society at the age of 21. Indeed, he had discovered three species of *Iguanodon*, one of which carried his name. Dawson realized immediately that what the workmen were handing him was a remarkable find. By the end of that day he had three fragments of a skull that looked proto-human, and he rushed to the Natural History Museum in London to show them to Dr. Arthur Smith Woodward.

Woodward was an expert on fish and a fellow of the Royal Society of London, and he encouraged Dawson to go back to the site and look for more remains. Because of persistent inclement weather, Dawson had to wait until May 1912, when he went to look for additional fossils together with the famed theologian and paleontologist Pierre Teilhard de Chardin. The two were able to uncover the remainder of the skull and part of the lower jaw (Figure 7.2). When they brought the relics back to Woodward, the latter concluded that they were human. Better yet, the animal to which they belonged must have been a hominid that looked like a young chimpanzee, which was exactly what was predicted by the then popular (and now discredited) theory that "ontogeny recapitulates phylogeny"[7] proposed by the German biologist Ernst Haeckel. The new finding was quickly named *Eoanthropus dawsoni*, or "Dawson's man of the dawn," popularly known as Piltdown man (which was actually a woman, according to Woodward).

Piltdown man is still considered an embarrassment by evolutionary biologists, and one that creationists bring up at every turn. The reason for the embarrassment is that it took more than 40 years to realize that not only was Piltdown man not an intermediate between humans and apes, but it was a pure and simple forgery. A close examina-

(a)

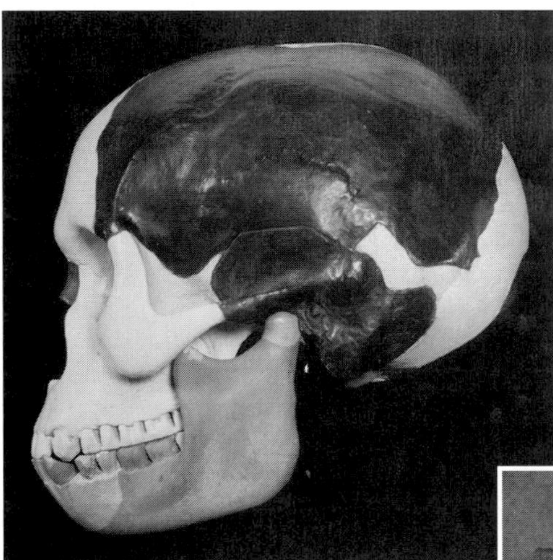

(b)

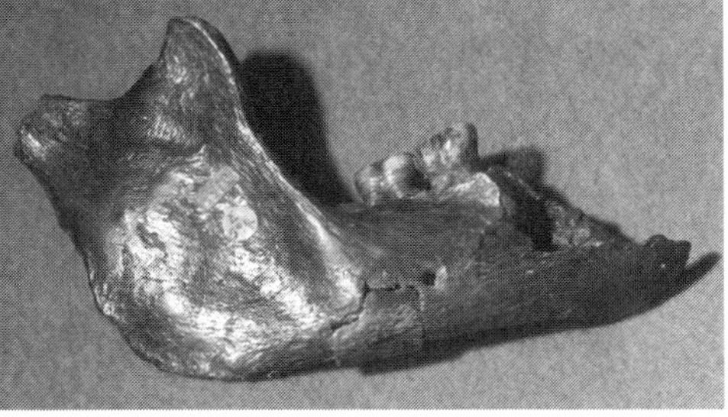

Figure 7.2 The reconstructed skull (a) and the outside and inside of the jaw (b) of the infamous Piltdown man, a hoax perpetrated on the scientific community in 1912. (Photos from the British Museum, Natural History)

[7] The phrase *ontogeny recapitulates phylogeny* means that the developmental stages of animals are parallel to the evolutionary stages of their ancestors. Although this is true in a few groups (such as some of the extinct ammonites' shells), it is generally not the case.

tion of the remains showed that the skull was indeed human, but only 50,000 years old, while the jaw and teeth were those of an orangutan, and an additional single tooth had been taken from a chimpanzee. These analyses also showed beyond doubt that the "fossil" carried stains of iron sulfate and chromium compounds, neither of which is found naturally at Piltdown and that were obviously used to make the specimen look older.

To this day it is not clear who was responsible for the fraud and why it was perpetrated. Dawson might have been aiming for an even higher level of recognition than he already had, or somebody simply wanted to make some quick money off of Dawson's credulity (he was famous for paying handsomely for new fossil finds in the area). Perhaps Woodward wished to make a jump in his career, or Teilhard de Chardin organized the hoax as a joke that got out of hand. And maybe even Sir Arthur Conan Doyle, the author of Sherlock Holmes, was involved, since he lived nearby, knew Dawson, and was interested in paleontology. Or, if one goes for really big conspiracy theories, perhaps all four men concocted the fraud together and also involved Sir Arthur Keith, an anatomist who provided "expert" testimony on the finding.

Whatever happened is of interest to historians, but what has always baffled me is why biologists are embarrassed by this episode. True, they were fooled for a long time, which is not a pleasant situation to be in. However, it is important to understand *how* exactly we came to discover the fraud. There was no confession, and all the people involved in the episode died with the apparent conviction of having reached fame and glory because of their role in it. What happened is that human paleontology (which was in its infancy at the time of the forgery) slowly but surely progressed. First, several more findings of Neandertals were unearthed, and then other human ancestors came to light in different parts of the globe, most crucially the australopithecines in Africa. It became more and more clear that Piltdown man *did not fit* into the emerging picture of the proto-human fossil record. In other words, the reason that biologists realized there was a problem was that science sooner or later has to face the objective reality of the world. You simply can't fool nature. Equally importantly, it was *scientists*, not creationists, who discovered the hoax and corrected the record—a perfect example of the self-correcting mechanism by which science makes progress *despite* the weaknesses and mistakes of individual scientists. Piltdown man should be taught in all introductory science classes as a paradigm of science at work, not an embarrassment to be skirted at all costs.

As an appropriate contrast, let us take a look at one of the creationists' greatest embarrassments and how *they* dealt with it. I am referring to the infamous Paluxy River (Texas) footprints (Figure 7.3) that allegedly show humans and dinosaurs walking next to each other (presumably not exactly at the same time, given the difference in size between the two and the fact that the dinosaur was likely a carnivore). The prints were discovered in the early 1900s, and creationists used them as major evidence of the flaws of evolutionary theory throughout the 1960s and through the 1980s. Geologists then clearly demon-

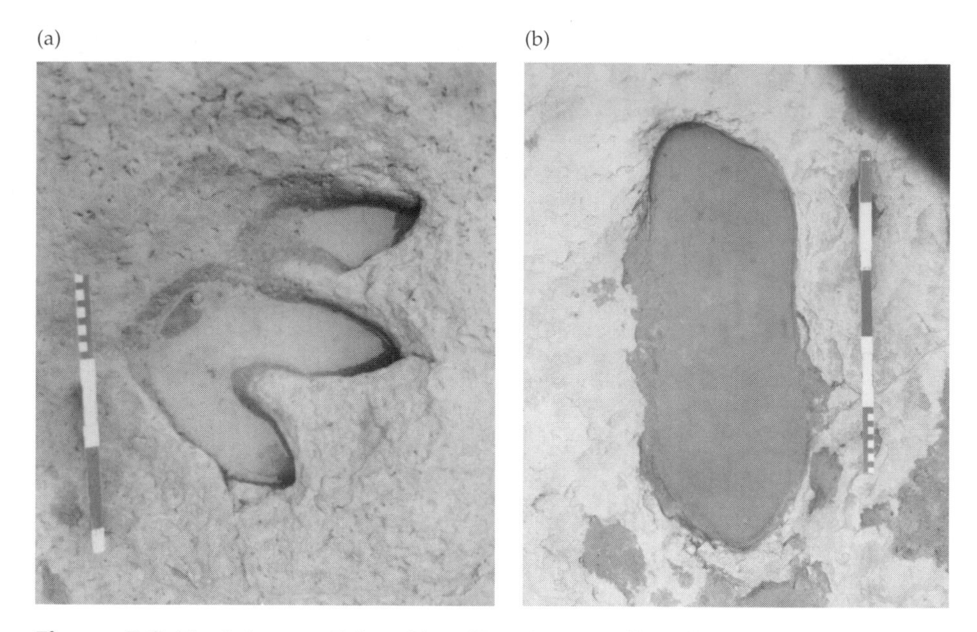

Figure 7.3 The infamous Paluxy River (Texas) prints allegedly showing dinosaurs and humans walking side by side. (a) A typical three-toed dinosaur track. (b) One of many pits that can be found on the same limestone surface as the dinosaur tracks. This is obviously a natural depression, not a human track (notice the absence of an arch, lack of toes, and large size [60 by 20 cm]). Scale: large divisions = 10 cm. (Photos by Ira Walters)

strated that no human being left those prints (they were in fact metatarsal dinosaur tracks, together with a few pure and simple fakes). But did the creationists retract their claims in this embarrassing incident (which was uncovered by scientists, not creationists)? Not really.

John Morris, of the Institute for Creation Science, did admit *in 1996* that there was a problem: "I am of the opinion that the evidence is, at best, ambiguous and unusable as an anti-evolutionary argument at the present time." Notice the extreme haziness in his words; the man just would not admit to being wrong! But Duane Gish, also of the Institute for Creation Research, wrote as recently as 2000 (in a booklet ironically entitled "Have You Been Brainwashed?"), "Undoubted dinosaur tracks have been found. In the same strata and very close by, other tracks have been found that some claim are human, although such claims are controversial." There is no controversy surrounding the prints, only the creationists' stubborn refusal to bow to the evidence.

A consideration of Piltdown and Paluxy River most clearly highlights the difference between science and creationism: The former is at times faulty, but always, if slowly, self-correcting; the latter is simply ideology in search of evidence, often trespassing the boundaries of intellectual honesty and integrity.

Science as a Social Activity

There is one issue about which creationists are completely right: Science is not above criticism. On the contrary, because of its influence on modern society, science and scientists need careful scrutiny as much as they deserve admiration and support. As Helen Longino eloquently puts it,[8] science is a social process, and one that is far too important to be left in the hands of scientists alone. Perhaps the most dangerous fallacy a scientist can commit, often subconsciously, is to only *do* science and never think about it. Yet many scientists who I know are not aware of the broad discussion about how science is done (or shouldn't be done) that permeates the literature in philosophy and sociology of science. Worse yet, when asked, they positively sneer at the idea of doing philosophy or sociology of science.

This lack of understanding of philosophy and sociology of science by scientists is, of course, at the root of the scientism I discussed in Chapter 4, and therefore indirectly helps cause the creation–evolution problem. As I mentioned in that discussion, a scientist of the caliber of Nobel Prize–winning physicist Steven Weinberg can even go so far as writing a book chapter entitled "Against Philosophy," in which he argues that philosophy is not only useless, but positively harmful to the scientific enterprise. It is this sort of hubris that offends many creationists (not to mention philosophers), and they have every right to be offended.

But in what sense is science a social activity? Certainly not in the sense advocated by some postmodernists and deconstructionists, such as those discussed in Chapter 3, as one of the main forms of anti-intellectualism. The conclusion that Earth rotates around the sun and not vice versa is not and cannot possibly be construed as a "social construction," and neither is evolution by natural selection. To insist otherwise is to engage in an intellectual game that is of no positive consequence and that can actually worsen the already bad state of public understanding and support for both the sciences and the humanities.

On the other hand, we have seen that the opposite—and equally naïve—concept of science as absolutely objective, as a steady, unimpeded march toward progress, as unqualifiedly good, can no longer be maintained after half a century of careful analysis of how science is actually done, not to mention after major blunders such as the invention of the atomic bomb, the hydrogen bomb, and biological warfare. As Enrico Fermi, who helped build the first atom bombs, put it, scientists have known sin, and they can no longer hide in their ivory towers and pretend complete detachment from the rest of the world.

A realistic way of looking at science is situated between these two extremes, and we would all benefit if both scientists and creationists would come to terms with it. Science is *both* objective and subjective. It is subjective because

[8] In H. Longino, *Science as Social Knowledge: Values and Objectivity in Scientific Inquiry* (Princeton, NJ: Princeton University Press, 1990).

it is done by human beings, and we cannot avoid being influenced by the times and cultural milieu in which we find ourselves. It is very reasonable to think that Darwin's book found such an astounding positive reception by the British public when it came out in part because it resonated with some of the fundamental ideas permeating Victorian England, as was made immediately clear by Herbert Spencer's invention of social Darwinism as a way to justify the newly emerging economic classes and the idea of capitalism itself. Some of the greatest American tycoons of the time used Darwinian metaphors to justify their exploitation of the masses. When the divine right is seen as a thing of the past, natural selection is a handy substitute.

This human subjectivity explains the blunders and dead ends in which science has found itself so many times and, no doubt, in which it will find itself in the future. But science is also objective, progressive in a cumulative sense, and self-correcting, and it is important to understand how this is possible. Essentially, there are two reasons: First, there really seems to be an objective world against which science is measuring its theories. That ultimate arbiter is often only partially and imperfectly accessible to us, but its verdicts cannot be distorted too much or for too long. Eventually, if they are about important findings, both mistakes and frauds are uncovered and corrected.

From the 1940s through the 1960s, a fool named Trofim Denisovich Lysenko managed to take control of Soviet biological research because his misguided theories about genetics fit Stalin's ideology better than the real thing. For decades, the best Russian biologists were driven into exile, imprisoned, or put to death. The result was a near collapse of the Soviet Union's agriculture and quite possibly a setback that hastened its demise at the end of the Cold War. Lysenko was able to fool politicians and some of his colleagues for a while, but nature caught up with him. His experiments failed miserably, causing incalculable damage to an entire society (in terms of both agricultural products and lost intellectual talents and opportunities). I have argued elsewhere[9] that the Wedge movement started by creationist Phillip Johnson (see Chapter 2), if it actually achieves the objectives that it set forth, will have a similar effect on American science, and Johnson and his collaborators will be remembered in history as the people who wasted the time and effort of a generation of American scientists and turned the clock back perilously close to the Middle Ages.

The second reason science is "objective" was identified by Longino:[10] precisely because it is a social, not an individual activity. Although we can all envision the stereotype of the lone scientist working hard at night in his laboratory until he comes up with a "formula" to solve problem X, that is not the way science works at all. In academia, every graduate student learns early on the saying "publish or perish," meaning that if she doesn't get her work out in print she will not have a career to look forward to. This publishing requirement is not just to feed the egos of people who like to see their names in print be-

[9] See http://fp.bio.utk.edu/skeptic/Rationally_Speaking/01-07-wedge.htm.

[10] See note 8.

neath obscure titles of papers that few are ever likely to read. Rather, it means that science is social: Scientists are a diverse group of people who communicate via publications (and other means, such as meetings) and whose activities are regulated by precise internal rules (which include "don't cheat") that are tightly enforced (through peer review).

The process of scrutinizing a scientist's work is, arguably, never over. Although most people (including creationists) seem to think that peer review ends at the moment of publication, that is in fact only the beginning of the process. Even if an author has passed the checkpoint of two or three anonymous colleagues before a work goes to press, once the article is out it is under constant scrutiny by any colleague who happens to read it or to work in the same field. That is why it was a foregone conclusion that the Piltdown hoax would eventually be uncovered. It was only a matter of how long it would take, which in turn depended on how many people were working on that topic and how much progress they were making.

Most importantly as far as creationism is concerned, scientists as a group are characterized by incredibly varied social and cultural backgrounds. Among their ranks there are women and men; Caucasians, Blacks, Hispanics, and Asians; gays and heterosexuals; young people and old; religious believers and atheists; and political conservatives and liberals. Some fall in love with every new idea; others despise the very idea of novelty and fight hard to keep things as they are. It is this variety, according to Longino, that ensures a high degree of objectivity in science. And it is the dearth of this sort of variety among creationists (who are mostly fundamentalist Christians, with a few fundamentalist Muslims, but very few people from more moderate denominations) that marks creationism as inherently nonobjective. Creationists really do start with a conclusion and select only the facts that best fit their preconception. Scientists can't succeed in doing that even when they try, because somebody among their peers will attempt to fit the data to a different conclusion and will catalyze a clash.

As I mentioned in Chapter 3, when creationists are confronted with the obvious paucity of ideological diversity among their ranks, they cry foul and accuse their opponents of committing what in philosophy is called the *genetic fallacy*— that is, of judging somebody's ideas not on their merits but because of who proposes them. But that's inaccurate; creationism loses on *both* tests of objectivity where science succeeds: Its predictions are repeatedly falsified by the arbitration of nature, *and* its claim to universality is undermined by its cultural homogeneity. The genetic fallacy is a fallacy only if a person's opinions are irrelevant to the argument being advanced. But in this case the opinions *are* the argument.

According to authors Collins and Pinch,[11] science is a *golem*, a Jewish mythological figure created by a human and endowed with large powers. The golem

[11] See H. Collins and T. Pinch, *The Golem: What You Should Know about Science*, 2nd ed. (Cambridge: Cambridge University Press, 1998).

is created for a good purpose, but it can get out of control and do harm through clumsiness. Science is like a golem, and it is for this reason that criticism of its activities and analysis of the consequences of its findings are crucial to our society. But it is the serious intellectual discourse of philosophers and sociologists that must do the job, not the hysterical cries of creationists.

Additional Reading

Collins, H., and T. Pinch. 1998. *The Golem: What You Should Know about Science,* 2nd ed. Cambridge: Cambridge University Press. An intriguing look at the scientific process through some of its greatest recent mishaps.

Collins, H., and T. Pinch. 1998. *The Golem at Large: What You Should Know about Technology*. Cambridge: Cambridge University Press. A follow-up to *The Golem,* focusing on technological mishaps.

Rothman, T. 1989. *Science a la Mode.* Princeton, NJ: Princeton University Press. A good look at how research in physics is directed as much by fashion and infatuation with certain ideas as by any objective criteria.

Youngson, R. 1998. *Scientific Blunders: A Brief History of How Wrong Scientists Can Sometimes Be*. New York: Carroll & Graf. A delightful booklet recounting many of the major blunders made by scientists, complete with a series of direct quotations showing that even the most intelligent human beings can be shortsighted and foolish.

– 8 –

What Do We Do About It?

The whole art of teaching is only the art of awakening the natural curiosity of young minds for the purpose of satisfying it afterwards.

– *Anatole France* –

Education never ends, Watson. It is a series of lessons, with the greatest for the last.

– *Sherlock Holmes, in Arthur Conan Doyle's "The Adventure of the Red Circle"* –

According to Jonathan Wells, an intelligent design creationist at the Discovery Institute in Seattle, the proof that evolutionary biology is a theory in crisis is that some high school and college biology textbooks contain errors in their descriptions of some major examples of evolution that students learn. This is, of course, sheer nonsense. It simply doesn't follow that because textbooks reprint mistakes, the intellectual foundations of a field are bankrupt. It is even less plausible—contrary to what Wells maintains—that scientists are engaged in a worldwide conspiracy to prop up the failing ghost of Darwinism. For one thing, nobody has yet come up with a reason why scientists would engage in any conspiracy to support a given theory, especially considering that their main hope, to make a rapid and glorious career, is to knock *down* somebody else's theory. Second, errors in textbooks are sadly present in any discipline, from history and philosophy to physics and chemistry, yet nobody in his right mind would suggest that quantum mechanics is a dead theory because it is not well explained in several high school (or even college) level books.

And yet Wells is onto something that is even more important than the evolution–creation controversy per se—something that deals with the whole way in which science education is done and that we need to change radically if we wish to produce a more rational population that will be naturally immune to creationist and other pseudoscientific nonsense. First, however, let's take a quick look at Wells's ten "icons" of evolution, as he calls them, and why they pose a serious threat not for evolutionary theory, but for the high school teachers who have to teach evolution without adequate preparation.

The Ten Icons of Evolution

Wells alleges that there are at least ten major pieces of factual evidence or crucial concepts in evolutionary theory that are simply not valid because their explanation in textbooks is incorrect, incomplete, or otherwise flawed. For each of these icons, I summarize Wells's argument and then provide a brief explanation of what the state of both textbooks and research in each of those fields actually is. This exercise is important not only because it provides us with a good example of how creationists exploit weaknesses in science education, but because the Discovery Institute has distributed tens of thousands of bookmarks to students across the United States with the explicit purpose of "embarrassing" their teachers by having students confront them with one of Wells's questions.[1] So then, here are the ten icons:

Wells's Argument: Some textbooks report an experiment done by Stanley Miller in 1953 as a milestone in the quest for the origin of life. But the experiment was based on an incorrect hypothesis concerning the chemical composition of the primordial Earth.

The Real Story: I explained the real status of Miller's experiment in Chapter 6, so here is just a brief summary: (A) The origin of life is not a field of research within evolutionary biology. Life began to evolve only *after* its origin, and the latter is the proper domain of biophysics. (B) Scientists still disagree on the composition of the early atmosphere (which is a very difficult empirical problem), so we don't know whether Miller's experiments have only historical or actual scientific interest. They almost certainly simulate the processes that originate organic materials inside meteorites, which are thought to be relevant to the origin-of-life question. (C) The origin-of-life field is not in disarray as Wells implies, but on the contrary is a very active and exciting research area, with new hypotheses and experiments being produced at a rapid pace.

Wells's Argument: Darwin's theory predicts that ancient forms of life in the fossil record should be more similar to each other than more recent ones (because they have slowly diverged). On the contrary, the Cambrian explosion shows that all the major phyla of animals appeared simultaneously and well distinct from each other. Furthermore, evidence of horizontal gene transfer (from one species to another) invalidates the idea of a "tree" of life.

The Real Story: Again, the reader is invited to go back to Chapter 6 for more details, but the gist of the story is this: (A) The Cambrian "explosion" was no explosion at all, but a series of changes that occurred over tens of millions of years, very likely more than that (the record is too incomplete to tell). (B) We now know of Precambrian forms that are difficult to relate to modern phyla precisely because they show mixed characteristics, which is what one would expect from evolutionary theory. (C) The Precambrian fossil record is still too sparse, and all of the ancestral forms were soft-bodied animals, which very rarely fossilize. The search continues, which is good science in progress. (D) Although horizontal gene transfer is a reality, it applies almost exclusively to bacteria and it complicates, but does not negate, evolution. This means that a phylogenetic tree is still a good depiction of the evolution of plants, animals, and other complex forms, but that its base actually looks more reticulate (meaning that various branches may split and later re-join). Darwin could not possibly have

---

[1] On October 4, 2001, Discovery Institute creationists William Dembski and Mark Hartwig were interviewed by James Dobson on his *Focus on the Family* daily radio broadcast. The interview was done as a follow-up to Dobson's discussion the previous week of the PBS series *Evolution,* which he termed "evolutionary propaganda" that "will be seen in public schools all across the country and probably throughout the Western world." During a part of the interview dealing with the concerns of Christian parents about what their children are taught in school, Hartwig provided a personal example of how such confrontations are being encouraged: "Well, we've been training our kid all along. I mean, this year my daughter confronted her eighth-grade teacher on evolution, on a couple of facts. She did it very politely. . . . We teach her what to watch for. We'll just say, 'Honey, you just keep doin' that in college!'" (From "God's Fingerprints on the Universe," Part II, *Focus on the Family* broadcast interview with William Dembski and Mark Hartwig, October 4, 2001.)

known about modern molecular genetic findings, and current evolutionary theory has progressed well beyond 1859 (the date of publication of *The Origin of Species*).

Wells's Argument: Homology (e.g., the similarity in the structure of the forelimbs of all vertebrates) cannot simultaneously be defined as an *outcome* of evolution and taken as *proof* of evolution. Molecular biology shows that some homologous structures are not controlled by the same genes.

The Real Story: (A) Wells does not understand basic concepts in the philosophy of science. Philosopher Karl Popper made a similar mistake in claiming that evolution by natural selection is a circular notion, only to retract the claim after scientists and others explained to him what evolutionary theory really is. Homology was recognized before the theory of evolution and is now interpreted as a logical outcome of the same. A perfectly noncircular way to think about homology is that similarities among organisms that cannot be explained by functional commonalities (i.e., same usage of similar organs) are the result of common descent. Alternatively, common descent can now be established by independent means (through the use of molecular data, for example), again avoiding circularity. (B) The molecular bases of homologous structures are not required by modern evolutionary theory to be the same because we know of instances in which a structure is maintained in a lineage even though its control is taken over by different genes (a phenomenon known in the technical literature as *genetic piracy*). (C) As in all other complex matters, the study of homology requires the convergence of information from a variety of disciplines, including comparative anatomy, developmental and molecular biology, genetics, and systematics. No magic bullet is available, and Wells's simplification is a naïve (or intentionally misleading) rendition of science.

Wells's Argument: The famous pictures of the embryos of various vertebrates at different stages of development produced by nineteenth-century embryologist Ernst Haeckel are a fake. Evolutionists are consciously using a fraud to fool the public.

The Real Story: (A) The drawings in questions were indeed fudged, but they are not complete fakes, and the fact of their being fudged was discovered by *biologists,* not creationists. Contrary to what Wells claims in his book, evolutionary biologist Stephen Gould was not "silent" about it, but published repeated essays[2] on the subject over an extended period of time. (B) Modern evolutionary and developmental biology do not rely on the work of a nineteenth-century scientist to make their case any more

[2] For example, in his book *Ontogeny and Phylogeny* (Cambridge, MA: Belknap Press of Harvard University Press, 1977).

than physicists still consider Newton's work the latest rage. (C) Current knowledge of developmental biology makes sense in the light of evolutionary theory because more closely related animals have more similar developmental systems. When exceptions occur, these are also predicted by evolution because they often turn out to be the result of adaptation to special environmental circumstances.

Wells's Argument: *Archaeopteryx,* the most famous missing link (between dinosaurs and birds) is disputed and now considered a side branch on the evolutionary tree. Cladistic analysis (a type of systematics) is based only on similarity among organisms.

The Real Story: (A) Wells completely misunderstands cladistics. It is based not on overall similarity (morphological or molecular), but only on shared, derived characteristics (a different kind of systematics, phenetics, was based on overall similarity; it is no longer used in modern biology). (B) The value of *Archaeopteryx,* as of any other fossil, is that it gives us a precious glimpse of the history of life on Earth and that it *does* fit the expectations of Darwinian theory. We might never be able to find intermediates between two forms along their *direct* line of descent, but that is a standard of proof that is not required in historical sciences because it is sufficient to show that related forms appeared in the right temporal sequence and that they were intermediate to each other. Fossils like *Archaeopteryx* do show us that such intermediate forms are possible and did in fact exist. More recently, many other intermediate forms between dinosaurs and birds have been found (indeed one of them was a hoax, again unmasked by scientists, as we'll see shortly). (C) Wells also does not seem to understand that ancestors can live simultaneously with their descendants. For example, you and your grandmother can be alive at the same time without negating the fact that you are on her direct line of descent. In fact, one of your ancestors could be alive after your death! Scientists know of several species (e.g., of plants and insects) in which the ancestor is still around today, simultaneously with its descendant. One of the best-studied examples is that of commercial corn and its wild progenitor, teosinte. (D) Wells has confused ideas about how historical sciences (such as evolutionary biology) work. The idea is to build a case for or against a theory that is based on accumulation of evidence, not on crucial, individual tests. The latter approach is now referred to as *naïve falsificationism* in philosophy of science and is only a caricature of how science actually proceeds.

Wells's Argument: The most famous textbook case of natural selection, the peppered moth, is a fake. Textbook pictures of dark and light moths on polluted and unpolluted tree trunks are staged. The moths don't actually rest on tree trunks. The original work on this case has been completely dismissed.

The Real Story: (A) Again, the peppered moth case has been revised by evolutionary biologists, not by creationists. This is another case of self-correction in science. (B) The current evidence still very strongly points toward natural selection in response to pollution, but the story is more complex than the original peppered moth research done by Bernard Kettlewell in the 1950s was able to show. (C) There is nothing wrong with staging pictures for illustrative or didactic purposes; it is done all the time to enable students to understand certain scenarios better by seeing them reenacted, and it is done in physics and other disciplines as well. The problem in this case is that the original story of what the experiment was about has been found to be incomplete and that textbooks have oversimplified what the experiments were used to show. The peppered moth experiments were not staged in an attempt to "prove" the theory of natural selection, but rather to illustrate the effect of coloration on the likelihood of predation. Unfortunately, textbook writers and publishers are in business for the money, and sometimes correcting mistakes in print takes years.

Wells's Argument: Observed instances of natural selection in Darwin's famous Galápagos finches have been improperly extrapolated to show that they could explain macroevolutionary differences among species. In reality, the environmental conditions reversed the direction of natural selection, leading to no net change in the current populations of finches.

The Real Story: (A) Extrapolation is a legitimate tool of scientific investigation, not an evil trick of corrupt evolutionary biologists. It simply means using known variables to draw a reasonably likely conclusion that takes one beyond the immediately available data; this is a valid variety of inductive reasoning. However, it does need to be used carefully, which was not the case in the textbooks cited. (B) The importance of the findings regarding the finches is to show that natural selection *can* cause meaningful morphological changes over a fairly brief period of time. That the changes in this specific example did not become fixed does not affect the example's support of this claim. The study of potential outcomes of a given phenomenon is an integral and legitimate part of science. To go beyond that, one would need additional information, such as knowing the sequence of environments for the past few millennia. We don't have that information for the Galápagos Islands where the finches are found, although paleoclimatology has established the fact that past environmental conditions on Earth were at times very different from the current ones. The fact that selection reversed itself in this case over a short period does not imply that it cannot be sustained for longer periods. (C) Even though some of the species of finches do hybridize, as Wells points out in his chapter, modern genetics has demonstrated the existence of *several* distinct species of finches (not just one, as Wells claims), all derived from a single ancestral stock. The origin of several species from a single ancestor in response to varying ecological pressures is exactly what the theory of evolution predicts. (D) Contrary to what Wells states, countless examples of natural selection have been measured in the field.[3] For example, my own laboratory at the University of Tennessee has repeatedly demonstrated selection for early flowering and increased production of leaves in the mouse-ear cress, a mustardlike plant. In that case, selection seems to work in the same direction regardless of the year or location of the study.

Wells's Argument: The discovery of mutations in fruit flies that cause doubling of the wings has been hailed as an example of beneficial morphological mutations, but in fact these flies are crippled and would not make for a viable new form of insect. There are no known beneficial *morphological* mutations, only beneficial biochemical ones.

The Real Story: (A) Only a few misguided molecular biologists (ignorant of evolution) have made such claims. Evolutionary biologists know that the significance of these morphological mutations in fruit flies lies in what they teach us about the developmental genetics, not the evolution, of organisms. Biologists are not just in the business of studying evolution. Another important field of inquiry is concerned with the effects of mutations on the basic developmental mechanisms that allow healthy organisms to grow into adults. The sort of mutation discussed by Wells is used by developmental geneticists for this purpose, not by evolutionary biologists to explain the appearance of new forms of life. (B) There are countless examples of beneficial morphological and behavioral mutations in the technical literature. If he would only stop and think for a second, Wells would realize that he carries many of these mutations himself, which is in part why he looks different from other human beings. Biologists know that every human being has a different genetic makeup (i.e., carries different mutations), and the overwhelming majority of these mutations are simply neutral (i.e., neither beneficial nor deleterious). A typical example of a morphological mutation that is beneficial (conditionally on the specific environment encountered by the organism) is the change in skin coloration that has happened several times during the course of human evolution. Light-skinned individuals are better adapted to northern climates (where they do not need to be shielded from the sun and would not produce enough vitamin D if they were dark-skinned). The reverse is true for the advantage of dark skin in tropical climates. Incidentally, this example

[3] Two recent surveys of studies of natural selection are H. E. Hoekstra, J. M. Hoekstra, D. Berrigan, S. N. Vignieri, A. Hoang, C. E. Hill, P. Beerli, and J. G. Kingsolver, "Strength and Tempo of Directional Selection in the Wild," *Proceedings of the National Academy of Sciences USA* 98 (2001): 9157–9160; and J. G. Kingsolver, H. E. Hoekstra, J. M. Hoekstra, D. Berrigan, S. N. Vignieri, C. E. Hill, A. Hoang, P. Gibert, and P. Beerli, "The Strength of Phenotypic Selection in Natural Populations," *American Naturalist* 157 (2001): 245–261.

also shows that the distinction that Wells makes between morphological and biochemical mutations is entirely artificial; the morphological mutations that affect the color of the skin are in fact biochemical mutations that change the rate of production of the skin's pigment, melanin.

Wells's Argument: The evolutionists' rejection of early hypotheses about a simple, linear evolution of the horse lineage shows their commitment to a preconceived ideology that rejects purpose and direction in the natural world. Evolution is simply materialistic philosophy in disguise.

The Real Story: (A) Wells himself acknowledges that the current view of the evolution of horses is very likely to be correct. One wonders, therefore, why he would attack an example of evolution that is so clearly and fully documented. It seems that he wants to have his cake and eat it too. His inconsistency only weakens his own position; it does not strengthen his arguments against natural selection. (B) General statements by evolutionary biologists that natural selection is a purposeless and directionless process are consistent with the evidence, since nobody has been able to demonstrate either a purpose or a direction in it (and many people have tried). There is simply no empirical evidence (which is what would be needed) to support either purpose or direction in nature. Although asserting that evolution is purposeless is indeed a philosophical conclusion,[4] it follows from everything we know and is consistent with the assumption in every science (not just evolutionary biology) that we can explain nature without recourse to the supernatural (gravitation seems to be as undirected and purposeless as evolution). This assumption has worked very well in practice, and Wells provides no reason to abandon it.

Wells's Argument: The Piltdown "man" was a hoax, showing how easily evolutionists can be tricked into accepting evidence because it fits with their theory. Paleoanthropology is not a science, but is really a form of storytelling.

The Real Story: I discussed the case of Piltdown man in some detail in Chapter 7, so here is just the gist of the story: (A) As Wells must acknowledge, the Piltdown man hoax was actually uncovered by evolutionary biologists, not creationists (and the hoax occurred back in 1912). More importantly, it was unmasked because it did not fit with the theory of evolution as applied to the new discoveries concerning the hominid fossil record. Although it did take several years to find out about the Piltdown hoax, part of the reason for the delay was that at the time we had very few other fossils for comparisons. A more recent hoax about a dinosaur-bird

---

[4] What is particularly irritating about this charge is that *every* conclusion is, in a sense, philosophical, simply because drawing any kind of conclusion requires the use of logic, which is an integral part of philosophy.

mixed form was exposed within a few weeks of its perpetration because of the current availability of more modern techniques of analysis of fossils and because we know quite a bit more about the reptile–bird evolutionary transition.[5] (B) As Wells also admits in his book, the hominid fossils discovered after Piltdown are genuine and do show intermediate stages of human evolution, exactly as predicted by the Darwinian theory. (C) Although it is true that there is a great deal of discussion on many important details of human evolution among paleontologists and paleoanthropologists, the reason is simply that it is difficult to find good fossil remains of creatures that lived in the savanna. This does not make the discipline just a matter of storytelling, and it certainly does not disprove the theory of evolution as a whole.[6]

Of course, Wells is perfectly right that biology textbooks should not contain mistakes and inaccuracies about their subject matter. Scientists and educators have always complained about the same thing, but most of them are not willing to undertake the tedious and difficult work that is necessary to correct the situation. A typical general biology textbook now has more than a thousand pages covering everything from molecular and cell biology to ecology and evolution. A single author (or even a team of coauthors) cannot possibly be directly knowledgeable about more than a tiny fraction of such a span of topics, which explains why most textbook authors simply copy from each other and perpetuate errors. It shouldn't be that way, but just complaining about it will not change the system. We have to realize that textbooks (unlike technical research books) are partially financial operations, so sometimes academic rigor is sacrificed in lieu of more pressing matters, such as publishing deadlines (although there are first-rate textbooks around, many of which Wells conveniently forgets to mention).

And yet the abysmal status of the textbook publishing industry is most worrisome not because of the few mistakes and inaccuracies that are of so much concern to Jonathan Wells, but for an entirely different reason that has a major impact not only on the creation–evolution debate, but more generally on how we teach science.

Teaching Science: Facts or Method?

The real problem reflected in current textbooks is that educators, especially in the nonphysical sciences, such as biology, geology, and psychology, are obsessed with the idea of "covering" the material and end up with no time to teach students how science is actually done. This is a terrible mistake because it does two things that we should avoid at all costs: First, it conveys the mes-

---

[5] See T. Rowe, R. A. Ketcham, C. Denison, M. Colbert, X. Xu, and P. J. Currie, "The *Archaeoraptor* forgery," *Nature* 410 (2001): 539–540.

[6] For a more extensive response to Wells's icons, see "Responses to Jonathan Wells's Ten Questions to Ask Your Biology Teacher," at http://www.ncseweb.org/resources/articles/7719_responses_to_jonathan_wells3_11_28_2001.asp.

sage that science is as boring as reading the yellow pages (which it is, if one simply enumerates all the factoids that are contained in the typical textbook); second, it reinforces the idea that science is about *results* that somehow—more or less magically—come out of people who are dressed in white coats and are involved in mysterious activities. No appreciation for the process of science is developed, and with each generation, we are directly responsible for rearing and educating citizens who will have no idea why global warming might be a threat and, most importantly, *how we know* that it is (or isn't).

Consider a simple fact: Although the typical introductory textbook in biology or psychology for college students is more than a thousand pages long, usually only about *ten* or so of these are devoted to the explanation of how all those facts were gathered and those conclusions reached.[7] A typical trend in the wrong direction is exemplified by the way some new textbooks in genetics are written and, consequently, how the discipline is taught (see Chapter 1). At one time these books followed a historical sequence, starting with Mendel and his experiments on heredity, progressing through their rediscovery at the beginning of the twentieth century, to the demonstration of the chromosomal theory of inheritance, up to the first evidence that DNA is the carrier of genetic information, and then on to the identification of the structure of DNA and the elucidation of the genetic code. Molecular genetics and recombinant DNA techniques were placed toward the end because they were the latest developments of a logical, if nonlinear, sequence of discoveries.

However, molecular biology has literally exploded during the past few decades, and teachers feel compelled to give "up-to-date" information about this rapidly expanding field, often skipping the historical sequence entirely, starting instead with the complexities of bacterial genetics. The result is that students are often confronted with a bewildering array of complex *facts* that they cannot link to each other conceptually, having no idea where they came from. For example, although every teacher of molecular biology knows what restriction enzymes are (because they are so useful in a variety of techniques), I doubt that most of them realize how they were discovered or what their natural function is.[8]

The point of teaching science in its historical context is not just that history is an important aspect of what we do and that we need to appreciate it (which is true), but that, most importantly, it is the best way to explain how and why scientific discoveries are made, to turn science from a barrage of meaningless and boring facts into a vibrant enterprise of discovery and human realization. Fortunately, some (few) universities and high school teachers around the country are coming to realize this and are changing their lesson plans accordingly,

---

[7] Also consider that many so-called introductory classes are actually *terminal,* in the sense that they are taken by students who have to fulfill a requirement and will probably never be exposed to a science class again. This is a sad waste of potential. One can only wonder what uninformed and even hostile citizens we are shaping in our classrooms, making sure that they will remember how much they hated science when they are in the voting booth.

but the overwhelming majority are turning students off from science by the hundreds of thousands every year. Is it any wonder that creationists find such fertile ground in high schools and even on university campuses?

Perhaps even more unfortunately, the major response so far to the sorts of concerns I am discussing here has been a shift in emphasis from traditional classroom lectures to simplistic "hands-on" activities in which students manipulate objects and perform experiments. As I shall argue in this chapter, moving away from lectures and getting students to do things is an excellent idea, but the way the hands-on approach is implemented in most classrooms I have seen actually produces *worse* results than the traditional approach. The problem with many hands-on classroom experiences is that the brain stays turned off. Kids just wonder about and giggle at whatever happens to be under the microscope without understanding what they are doing or why. The reason is that their teachers are often in no position to provide them with the conceptual background to derive the greatest benefit from the activities themselves.

Things are not much better in college, where teaching assistants who entirely lack training in teaching, and who are essentially slaves doing the job of professors for very little pay while trying to finish their Ph.D. dissertations, are unprepared to teach and simply don't care much about teaching (there are, of course, many exceptions). Even when college-level courses have a lab component, it is usually completely decoupled from the lecture class, so in effect students are taking two independent courses in the same discipline with little understanding of how to connect the experiments to the necessary concepts. Worse yet, most of these exercises are "prepackaged" labs designed to obtain a predetermined outcome, which often does not occur because of the carelessness of both students and teaching assistants. The latter then do the worst thing they could possibly do in teaching science: They tell the students that they should have gotten result X and to write up their reports *as if* they had. Is it a surprise, then, that the whole enterprise becomes meaningless and that most students think science is either too difficult for them to grasp or, worse, is actually done by cooking the results to come out according to a priori expectations (the perennial creationist paranoia)?

There is another pernicious myth that lingers in universities around the world and is a major cause of the problems we are considering: what Murray Sperber refers to as "the myth of the good researcher = good teacher" in his de-

---

[8] Restriction enzymes are proteins that cut DNA at particular locations and are therefore extremely useful to molecular biologists intent upon opening up DNA molecules and splicing new genes into them, or in doing a variety of other laboratory operations that are nowadays common in genetics laboratories throughout the world. Restriction enzymes, however, are not a gift from God to molecular biologists, but a result of natural selection that enables bacteria to defend themselves against viral infections. The enzymes cut any foreign DNA inside the bacterial cell to shreds, avoiding the bacterium's own DNA because of specific recognition signals known as methyl groups. Of course, some viruses have evolved a way to counter this potent biological weapon by mimicking the structure of their host's DNA, thereby avoiding being cut into pieces by the restriction enzymes. The evolutionary struggle continues under our very noses . . .

lightful—and worrisome—*Beer and Circus*.[9] University professors are officially paid to teach, especially at state schools. They are also expected to do research and possibly bring in some grant money, but the main reason taxpayers and parents subsidize their salaries is teaching. On the other hand, academics who wish to advance in their career at any research university realize immediately that the *real* trick is to invest as little time as possible in teaching and concentrate on research, publishing, and especially bringing in grants. Worse yet, this is the current trend even at primarily teaching universities, where the main mission is teaching but the road to promotion and tenure is publishing. The result is a schizophrenic role for the university professor, who literally has to serve two masters, one of which is (unofficially) much more powerful than the other. Where do you think his allegiances will go?

The standard response of university administrators (and of many faculty as well) is to cite the "good researcher = good teacher" myth. The assumption (which is all it is) is that anyone who is good at doing research in a certain field will also be good as a teacher in the general discipline in question. There is not an iota of empirical evidence to confirm this, and quite a bit of my own experience goes directly against it. Sure, there are a few exceptions who are both excellent teachers and excellent researchers, but that's what they are—exceptions. By and large, universities are filled with mediocre researchers and even more mediocre teachers. One of the statistics that gives away the game immediately is that faculty who are good at their research (as measured by the number of their publications and the amount of money they bring in with grants) are immediately *shielded* as much as possible from teaching! Some don't teach at all, while others confine themselves (happily) to teaching only graduate seminars, avoiding the most crucial of all courses from an educational standpoint: The introductory sequences. If these are supposed to be our best teachers, why not employ them where it is most crucial to have good teachers?

The reality, of course, is that human talents are varied and multifaceted, and they are not necessarily coupled in the way we would like them to be. One can be an excellent teacher and an excellent researcher, but that is the least likely outcome because the two things require quite different skills, which seldom develop in one individual. The fact that a researcher can think up cunning experiments at the frontier of molecular biology doesn't necessarily mean that he can explain meiosis to a class of college freshmen. Similarly, being capable of writing a good book for the general public does not mean you will produce superb grant proposals that will be regularly funded, and so on.

I am not here advocating the dismantling of the current university system, much less doing away with funding of scientific research in universities. What I am saying is that we know there is a problem and we also know why. The real obstacle is that taking the obvious steps to improve the situation (discussed

---

[9] M. Sperber, *Beer and Circus: How Big-Time College Sports Is Crippling Undergraduate Education* (New York: Holt, 2000).

at the end of this chapter) is painful and requires vision and courage—two qualities often sadly lacking in faculty, administrators, and politicians, the very people who have the power to change things.

If we broaden the horizon from academia to our culture at large, the view does not improve very much. Richard Dawkins, in his *Unweaving the Rainbow,*[10] complained about the fashion of making science "fun, fun, fun," as if it were a comedy show. Indeed, educators are often expected to be stand-up comedians to make their subject matter entertaining. Remember our discussion in Chapter 3 of Neil Postman's accusation that we live in a society in which the major objective is to "amuse ourselves to death." The particular case of science teaching is just one example of a cultural trend that will take a great deal of effort to slow down, let alone to turn around. The point, of course, is not that science has to be boring, but that there is a difference between something being *interesting* and simply entertaining. Human beings need entertainment, even of the mindless kind, but we are reduced to little more than brutes if *everything* in our lives is packaged in terms of light and funny sound bites, as we have been conditioned to expect by the TV programs we watch.

In the United States, the major TV networks are required by law to provide a certain amount of educational programming for children in exchange for being allowed to use the public airwaves for free and to make a mountain of dollars out of them. One of the networks in question submitted as part of its share of "cultural" programming a famous cartoon series featuring humans and dinosaurs living at the same time in an idealized version of suburban America! Even American public television, considered a paragon of cultural value (and hence very little watched or financially sustained by the public) is airing programs that are, at most, of dubious value in their attempt to educate children. Many of these programs feature a bewildering array of characters that talk about a particular subject matter for maybe one or two minutes, then jump off to something completely different, and do that continuously for the whole show. Now, it is true that young children don't have the attention span and capacity to concentrate that we expect (and rarely obtain) from an adult or an older child. But the point of education is to *help* children develop those abilities, not to set the expectation that they will be always entertained by colorful characters who can't stay on the same subject for longer than the span of a TV commercial. It is no wonder that elementary (and higher) school teachers have to contend with a chronic inability of their students to pay attention during class time.

The "problem" with education is so multifaceted and daunting that it is not surprising that progress has been slow. The type of students we get in our colleges is the result of a long chain of events that starts with little emphasis on education at home (or with parents unable to provide the necessary jump

---

[10] R. Dawkins, *Unweaving the Rainbow: Science, Delusion, and the Appetite for Wonder* (Boston: Houghton Mifflin, 1998).

start), and continues with years of exposure to teachers who are trained in *how* to teach (and not necessarily so well) but who rarely know much about the subject matter they are supposed to teach. And the whole problem is compounded by peer pressure and cultural stereotypes that favor (and reward) nonintellectual activities. As we shall see toward the end of this chapter, however, things are changing in the right direction, at least in some schools and universities, and these are the models that all others should follow in the quest for a more intellectually aware citizenry.

Not Just Creationism, Not Just Education

Creationism, as I have argued throughout this book, is not an isolated phenomenon, although it has its own peculiarities that deserve special attention. The general problem is really the widespread lack of critical thinking in the population at large and the fact that by and large we don't teach critical thinking in schools and universities. As a result, not only do 58 percent of Americans believe that "God created human beings pretty much in their present form at one time within the last 10,000 years or so,"[11] but large numbers believe in UFOs, alien abductions, astrology, haunted houses, the physical existence of the devil, telepathy, the ability to predict the future, and a host of other phenomena one would have thought ended up in the dustbin of history at the end of the Middle Ages.

Is more science education the retort to all these superstitions? The answer that I lean toward, as unsatisfying as it may sound, is, "Yes and no." I don't think that more science education of the standard variety will make any difference. In fact, it may positively do harm. Let me explain.

Scientists and science educators have always *assumed* that the reason so many people believe in pseudoscience is that they simply don't know enough science. However, although the latter is an accurate empirical observation (most people don't know much about science), it doesn't follow that scientific illiteracy is the cause of widespread belief in all sorts of paranormal phenomena; therefore, more science education will not necessarily solve, or even ameliorate, the problem.

In fact, the connection between education (science education in particular) and belief in paranormal phenomena or explanations is an empirical matter, and it has been investigated as such. The results are not very supportive of the standard view of the problem, at least not entirely. A survey by the Pew Research Center for the People and the Press found that belief in heaven as a real place did diminish according to increasing levels of education from 92 percent among people with less than a high school education to 73 percent among peo-

[11] The results of several surveys cited here are summarized in an article by Erich Goode: "Education, Scientific Knowledge, and Belief in the Paranormal," *Skeptical Inquirer* 26(1) (2002): 24–28.

ple with a postgraduate education. But three out of four people with a college-level education still believe in the physical existence of heaven! The same trend applies to other measures of belief: Hell is considered an actual place by 80 percent of the respondents in the first category and by 56 percent in the second. If one is less educated, one is 20 to 30 percent more likely to believe in angels, but the most astounding fact is that 22 percent of college-educated people in the United States think that "people on this earth are sometimes possessed by the devil"! That equals one in every four or five of the best-educated people in the most prosperous country in the world—one that is proud of possessing many of the best universities on the planet.

There is more. Although an inverse relationship does obviously exist between the level of education and the beliefs mentioned here (a finding mitigated by the fact that even many of the most educated people still hold on to such beliefs), notice that the Pew survey addressed beliefs with a strong religious component. What about beliefs in paranormal phenomena that are not based on a religious mythology? Anecdotal evidence suggests that religious fundamentalists tend actually *not* to believe in paranormal phenomena that are not directly mentioned in their sacred texts, such as alien abductions or astrology. But this lack of belief in the paranormal is hardly a result of their critical thinking.

Quantitative evidence is available: A Gallup poll found that the highest level of belief in UFOs visiting Earth is found among people with a college education (51 percent), a figure that is only slightly *less* for respondents who had not graduated from high school (48 percent). Similarly, a study by the Princeton Survey Research Association that asked about belief in the paranormal and supernatural found that high school dropouts responded more positively by only a very slim margin over their more educated colleagues (43 versus 39 percent). Several other polls have produced similar results, which amount to an inconsistent pattern of alleged correlation between education and some kinds of paranormal beliefs.

This and other published research on different sorts of nonscientific beliefs seem to point to two important conclusions that anybody seriously interested in science education should keep in mind. First, there is a difference between religious-based and non-religious-based beliefs: The former are inversely related to the degree of education; the latter are not. This means factors other than just the general degree of education are at play. Second, even when education makes a difference, it leaves a staggering number of people believing all sorts of nonsense. Why? Before delving into this question from the perspective of what we know about how the brain works, let me report and briefly discuss one more set of results, which I obtained while surveying one of my own classes at the University of Tennessee.

I teach a course on science and pseudoscience, which was offered for the first time to honors students, most definitely not a random subset of the student population at the university. These are among the best and brightest students on campus. They also come from disparate backgrounds, with fewer than half of those I interviewed pursuing a science major. I asked them to respond to questions aimed at evaluating their general knowledge of science as it is assessed in aspiring high school teachers. These are questions about matters of *fact*, not principles of science or critical thinking. Not surprisingly, science students knew (slightly) more science than nonscience students did. I then asked them to rate their belief in a series of paranormal phenomena, from voodoo to astrology, from water dowsing to haunted houses, and so on. The astounding result (Figure 8.1) was that, contrary to expectations, the science majors held *more* strongly to paranormal beliefs than the nonscience students! At the same time there was absolutely no difference between genders (so much for the stereotype that depicts men as more "rational" than women).

I do not wish to claim too much on the basis of this one survey of a small sample at a particular university, but it was interesting to follow up with a few questions to the students in order to generate causal hypotheses to be tested with additional surveys. The most revealing thing was that most of the nonscience students were philosophy or psychology majors, who actually take courses on the scientific method and on critical thinking. In contrast, science majors are *never* exposed to that sort of course and spend most of their initial scientific education in large classrooms where somebody who calls himself a professor, and whom they can barely see from a distance, inundates them with a flood of disconnected facts that they are supposed to remember in order to pass the test.

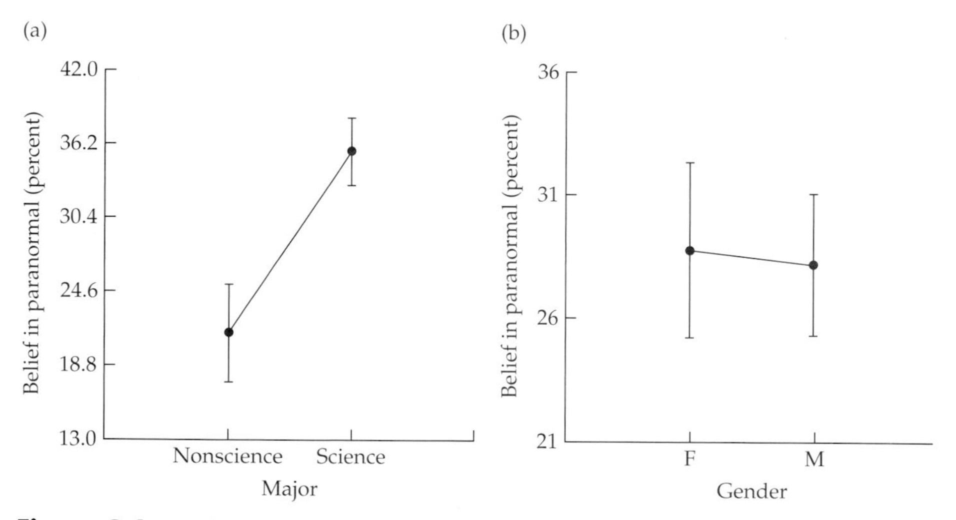

Figure 8.1 Results of a survey conducted by the author on the belief in a host of paranormal phenomena by honors students at the University of Tennessee. (a) Surprisingly, science majors actually show *more* belief in the paranormal than nonscience (mostly philosophy) majors. (b) In this sample there was no relationship between belief in the paranormal and gender. Brackets indicate standard errors.

Could it be that it is not just education (scientific or otherwise) that matters, but the *kind* of education we give to our students? If so, what must we do to improve the currently abysmal situation?

How the Brain Actually Works

Before proposing a series of steps to improve not only the state of the creation–evolution debate, but more generally the level of science education, it is useful to learn a bit more about the way the brain works, and in particular what neurobiology has found out over the last few years about how we learn (or fail to). This is obviously a huge topic, which by necessity I will have to treat very superficially. However, the suggested readings at the end of this chapter will provide the reader with fascinating introductions to this literature. The main point is that the brain is a crucial piece of our biological machinery, and yet we care little about learning how it works or how to improve its functionality. This is like having an expensive car and caring about replacing the tires, polishing the chrome, and making sure all the fluids are there, but completely ignoring the engine—not a good recipe for a long and happy ownership.

This lack of application of our knowledge about the brain to everyday life is especially astounding in the world of education (with a few exceptions of teachers or schools that are taking the first tentative steps toward making practical use of neurobiological research). It would seem that teachers and college professors should be mandated to take a course in how to help their students make the best of this wonderfully delicate organ that has been so crucial to human evolution and to our very identity as a species of primates.

Books like Eric Jensen's *Teaching with the Brain in Mind* and Pierce Howard's *The Owner's Manual for the Brain* are full of specific recommendations for teachers and students on how to use their brains. Some of the chief recommendations are summarized in Table 8.1. What I would like to discuss here instead are a few major characteristics of the functionality of the brain which explain why the rationalistic fallacy referred to in Chapter 7 is in fact a fallacy—that is, why simply explaining things to people won't solve the problem.

Perhaps the most astounding piece of evidence comes from experiments on split-brain patients (i.e., individuals with no connection between the two hemispheres of their brain) conducted by Michael Gazzaniga and other neurobiologists.[12] Neurobiological research on individuals with split brains has shown quite clearly that what we perceive as a unified self is really made of at least two distinct and largely independent components—one overseen by the right hemisphere, the other by the left hemisphere of the brain. The two hemispheres are normally connected by the corpus callosum, which contains

---

[12] For a summary of Gazzaniga's work and other material on split brains, see http://www.macalester.edu/~psych/whathap/UBNRP/Split_Brain/Split_Brain_Consciousness.

Table 8.1 *Suggestions for how to use your brain better (for more details see Eric Jensen's* **Teaching with the Brain In Mind***)*

1. About how the brain works:
   a. Learn the basics about how the brain is made and works. If you are an educator, share that with your students at the beginning of the term
   b. Learn about the literature on brain research (start at www.brain.com).
   c. If you are an educator, do your own research on what works and what doesn't, keeping track of results and sharing them with others.
2. About the physical environment:
   a. Learn (and practice) the basics of good nutrition and how it affects brain functionality. If you are an educator, share that with parents and those in charge of school meal programs.
   b. Drink water, not sodas. This may sound simplistic, but dehydration is a major impediment to good brain functionality.
   c. Pay attention to things like illumination (avoid fluorescent lights) and air circulation (more is better).
   d. Enrich your visual surroundings while you are learning. Lack of stimulation and boredom literally thin your cortex.
3. About changing conditions during the learning process:
   a. Allow yourself (or your students) to pace your learning and to introduce more variety in the way you learn about a particular topic (e.g., use more than one source and possibly more than one medium, including books, web-based resources, personal research, etc.).
   b. Vary the setting for your studies, move to a different classroom, studio, or go to the library or a coffee house.
   c. If you are an educator, the way to get attention from your students is to use contrast. Establish a balance between novelty and ritual.
4. About teaching methods:
   a. Consider that attention and retention cannot be performed simultaneously by the brain, so students need "down time" to absorb what they have been paying attention to. That is also why it takes a few seconds of silence before one can get a meaningful discussion going on a topic that was just presented.
   b. Avoid holding the students (or yourself) to unrealistic deadlines.
   c. If you are an educator, students see the connections between their actions and the outcomes. Do not assume that they know how to study, maybe they should, but often they don't.
   d. Organize (or engage in) as many interdisciplinary activities as possible. The brain is stimulated by novel connections and alternative ways to think about the same topic.
   e. Focus yourself (or your students) on implicit learning: how do you know X?
   f. Practice (or have your students practice) explicit learning by way of graphical techniques such as concept mapping (see http://158.132.100.221/CMWkshp_folder/CM.ResFolder.html).
   g. Ask yourself (or your students) as many "why" questions as possible and try to work out the answers.
   h. Practice (or have your students practice) the skill of finding patterns and using them to generate questions (more on teaching methods)
   i. Participate (or have your students participate) in group projects.
   j. Read a lot (or have your students read a lot), possibly primary sources.
   k. If you are an educator, provide a global overview for a topic before jumping into details. Roadmaps increase a sense of security and help keep pace.
   l. Before moving to a new topic, make sure that you (or your students) are capable of summarizing the most important concepts of the previous topic and their relations to each other. Better yet, try to explain them to someone else.

Table 8.1 *(continued)*

5. About stress:
   a. Practice (and teach your students) stress management techniques, such as time management, breathing, and the importance of down time. Physical exercise is an important component of a healthy brain, and so is a good amount of sleep.
   b. If you are an educator, work to reduce threats to the learning environment. These may come from outside, from other students, or yourself.
   c. If you are an educator, simply ask the students what is getting in the way of learning.
6. About motivation and rewards:
   a. Temporary demotivation is normal and should not be considered a crisis unless it persists.
   b. Increase your (or your student's) choices about how to learn.
   c. If you are an educator, eliminate any kind of embarrassment or sarcasm.
   d. Get (or provide) as much feedback as possible on how you (or somebody else) are progressing on a given course of studies.
7. About emotions and learning:
   a. Avoid the extremes of high emotions or complete emotional detachment, they are detrimental to the learning process.
   b. If you are an educator, be a role model. Show enthusiasm for your subject and what you bring to the classroom.
   c. Celebrate accomplishments (yours or your students) to create a positive feedback between achieving educational goals and the brain's production of "pleasure" neurotransmitters.
   d. Set up (or get involved) in a controversy, such as a debate, dialogue or argument. Research shows that the emotions generated by these venues help retain the material one is exposed to.
   e. In a classroom, the purposeful use of physical rituals (clapping, cheers, singing) to mark special moments of the day, if age-appropriate, can help build a positive emotional connection with the learning environment.
   f. Use (or encourage your students to use) journals, discussions and other informal communications with others. Learning is a social enterprise.
   g. Use current events (because of their interest) to learn or teach something about some fundamental aspect of a discipline. E.g., use an article on AIDS to learn about viruses and cell replication
8. About memory:
   a. While reading, stop often and take notes.
   b. Discuss what you read and try to explain it to someone else.
   c. Identify and repeat key ideas and how they are related to each other. The details will make more sense once that the big picture is absorbed.
   d. Divide the learning task into small, manageable chunks.
   e. Use techniques such as concept mapping whenever you need to understand and retain complex concepts.
   f. Use cliffhangers (especially if you are a teacher). Read or discuss a relevant problem to solve and leave the brainstorming until the following day. The brain is active at night and puts order in what you've learned during the day. That's why regular sleep is crucial for learning and why so often it happens that one wakes up the following morning with a solution to a problem that seemed unassailable the previous night.
   g. Retention is increased when it is connected to a particular event, such as a field trip, an event in the news, or a guest speaker.
   h. Create theme days or weeks (for yourself or your students), during which you concentrate on certain topics or certain approaches to those topics (skeptic day would be a good idea).
   i. Consider that bodily motion helps retaining concepts. Facilitate (or participate in) role playing, improvisational theatre, class presentation.

millions of neuronal bridges between the two sides. When the corpus callosum is surgically severed, the two hemispheres behave independently of each other, to the point that they can cause completely contradictory behaviors in an individual.

Experiments on split-brain patients show that the left hemisphere is dominant, meaning that it is in charge of unifying the different inputs from both hemispheres into one coherent narrative. The interesting thing is that we can show experimentally that this narrative is woven *a posteriori* (i.e., after the fact), as an explanation for what the individual has perceived or done. In a now classic example, a patient with a severed corpus callosum was shown a picture and then asked to pick a complementary picture from a given series (Figure 8.2).

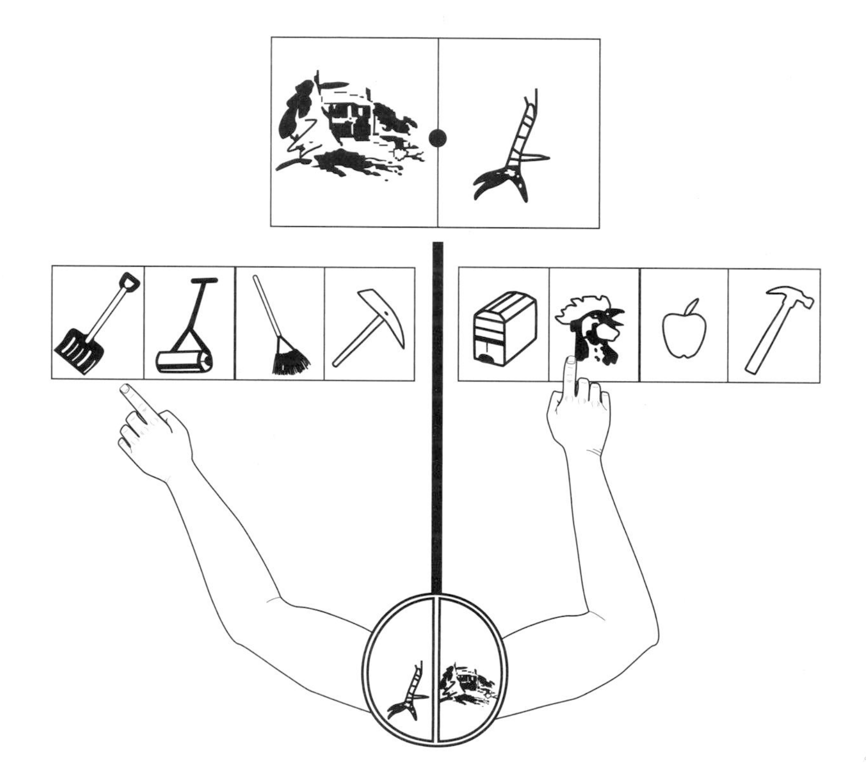

Figure 8.2 An experiment with a split-brain patient in which the two hemispheres of the subject are asked independent questions. Here, the right hemisphere was asked to pick a picture from the lower left series to match the image of a snow-covered house (top left). It picked the shovel. Analogously, the left hemisphere picked a chicken head to go with the chicken leg. However, when the left hemisphere (the only one that can communicate verbally) was asked why the subject picked the shovel and the chicken head, it answered by making up a story about having to clean the chicken's excrement. The left hemisphere did not know why the right did what it did (because of the lack of communication), but concocted the best explanation it could, given the available data.

The experimenter then directed a question separately to each of the two hemispheres (which is possible because the left one controls the right side of the body and its visual field, while the right one controls the left side of the body and its visual field; the experiment requires a computer that tracks the movement of the eyes and keeps the image in only one half of the field). The patient gave answers by hand gestures (because the right hemisphere does not have access to voice control). The right hemisphere was asked to pick out a picture that would go with that of a house covered with snow; the patient picked a shovel. The left hemisphere was asked to pick out a picture that would go with a chicken leg; the patient picked a chicken head.

Here is the interesting bit: When the left hemisphere (which is in charge of language) was verbally asked why the subject picked a chicken head and a shovel (notice that it was unaware of the house with the snow, because of the disconnection between the two hemispheres), the patient answered that he needed a shovel to clean up the excrement from the chicken! In other words, lacking complete information about the shovel (available only to the right hemisphere), the left hemisphere—our much-vaunted rational self—made up a story after the fact, the only characteristic of which was that it fit the available information, however awkwardly.

I think this sort of finding has profound implications for the evolution–creation debate and for teaching in general. Our brain is constructed in such a way that it comes up with an explanation (a story) for the world around it. How good an explanation the left hemisphere will create depends on how good the input is from the surrounding world: The better the information provided, the better the model the brain will create. Think of the brain as a virtual-reality device that literally creates not only your perception, but your understanding of the world. If the input is faulty, the device will construct something nonetheless, but as the data become more and more faulty, what is produced will bear less and less resemblance to the real thing.

Neurobiologist V. S. Ramachandran has suggested (in his delightful *Phantoms in the Brain*[13]) that one's views on the world or on a particular topic depend on a balance between the respective inputs of the two hemispheres. The right one plays the part of the devil's advocate, always feeding information that may at times be in dissonance with the currently held view. The left hemisphere filters this information in one of three ways: (1) It uses the information to reinforce the currently held view; (2) if the information does not quite fit the prevailing view, it can alter it slightly to accommodate it; or (3) it can temporarily just ignore any information that does not fit. Ramachandran suggests that when we change our mind about something, the reason is that the amount of information that doesn't fit has exceeded a certain threshold (which is probably variable within the human population) and has caused the left hemi-

[13] V. S. Ramachandran, *Phantoms in the Brain: Probing the Mysteries of the Human Mind* (New York: Morrow, 1998).

sphere to change its story in a radical fashion.[14] Whether you are very gullible, reasonably open-minded, or a die-hard skeptic may depend on the exact constitution of your corpus callosum and the way your left hemisphere handles cognitive dissonance![15]

Research on critical thinking has found yet another reason why that particular skill is so difficult to teach: The brain is apparently designed to jump to conclusions on the basis of little evidence. Figure 8.3a shows how we would ideally like our own and our students' brains to work on a problem: We would start with a preliminary study of the topic and form a provisional understanding of the problems involved; we would then defer judgment and move to collect new evidence, after which we would construct an informed belief (in the sense of an opinion based on a theoretical construct and empirical information, not in the sense of blind faith; see Chapter 4 on Bayesian inference).

Unfortunately, the human brain has an innate tendency to skip that all-important "defer judgment" phase, going instead from a preliminary under-

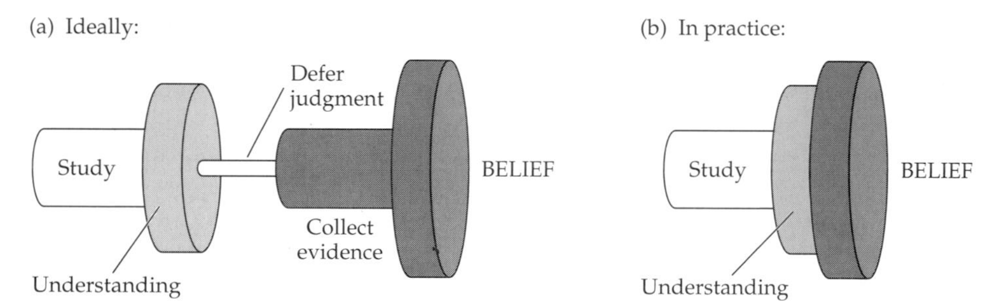

Figure 8.3 Diagrams showing how we would like the brain to work in order to maximize critical thinking (a) and how the brain actually works (b), probably as a result of natural selection for quick decision-making abilities. (After a suggestion from my colleague Donald Buckley)

---

[14] It is interesting to note that this continuous equilibrium between a certain view held by the left hemisphere in our brain and the dissonant information that doesn't fit, which eventually leads to a sudden shift in paradigm, closely resembles philosopher of science Thomas Kuhn's idea of periods of "normal science": Such periods are interrupted by "revolutions" (or paradigm shifts) when the number of "anomalies" (the data that don't fit the current paradigm) is too large to be ignored or bent to fit the current model. Of course, it should be no surprise that science, a human activity, works in a fashion parallel to the neural circuitry of the people who actually do it.

[15] I am not suggesting here that mental flexibility is therefore an innate characteristic. Indubitably it will depend in part on our genes, like everything else in our bodies, but the characteristics of the corpus callosum and of the left hemisphere are also shaped during the growth period of the brain and are therefore probably highly influenced by the environment. For example, experiments with rats show that visually stimulating environments during growth result in a significantly thicker cortex, whereas unusually dull environments actually produce a thinner cortex. I doubt things are very different for human beings.

standing to the formulation of a belief (Figure 8.3b). Once the belief is formed, it is very difficult to dislodge because of the innate conservativeness of the left hemisphere. It takes a great deal of cognitive dissonance over a long period of time for a belief to finally be replaced by another one. It is even possible, though this is highly speculative, that this "from understanding to quick judgment" mechanism was favored by natural selection during the ancestral history of humanity, when humans couldn't afford to wait too long before deciding if the spot at the horizon was a gazelle (worth hunting) or a lion (from which they were better off running in the opposite direction). All of this, of course, explains much of the difficulty encountered by the educators and scientists involved in the evolution–creation controversy. Ironically, it may even be that the major obstacle to understanding evolution in modern society has been put in place by natural selection during crucial phases of primate history!

Interestingly, the findings discussed here have some bearing on the controversy raging within the educational community concerning the wisdom of having scientists debate creationists (see Chapter 3). In the context of the current discussion, one has to realize that the point of a debate is *not* to convince people on the spot and carry the day. That would be committing the rationalistic fallacy in the highest degree (see Chapter 7). Rather, because of what we know about how the brain works, the goal is to plant seeds of doubt, to expose the often unethical tactics of creationists,[16] and to give pointers to interested members of the audience about where to go next. Let the right and left hemispheres of the attendees' brains battle it out over the following months and years and see what they come up with!

I have debated several creationists on many occasions, in front of crowds coming to places as varied as fundamentalist Christian churches and university auditoriums. The results are always positive. I receive plenty of e-mail and other correspondence, sometimes long after the actual debate, from people asking me for more information or sources. They often preface their letters with something like, "I still don't believe what you said, but . . . ," and it is that "but" that shows that the seed has been planted and is possibly about to germinate. People are unfailingly impressed by the spectacle of a scientist who is actually courteous and makes some sense, especially when they expect a blabbering, eggheaded intellectual or a devil incarnate, complete with horns and tail. Given that these are people who pay the taxes and tuition on which university professors live and conduct their research, it seems only ethically decent to spend a few evenings once in a while trying to explain to them why we are doing what we are doing.

Finally, I have talked to many people who, because of their upbringing, were hostile to evolutionary ideas and eventually overcame their belief systems. It is important to find out how they did it because that insight provides us with

[16] For an example of questionable creationist tactics, see Chapter 11 of my *Tales of the Rational: Skeptical Essays about Nature and Science* (Atlanta, GA: Freethought Press, 2000).

crucial clues as to what works and what doesn't. One thing that doesn't work is the "I'll explain it to you once, and you should be convinced" sort of attitude that underlies the rationalistic fallacy. Repetition, and especially repetition by different sources in different formats, is the key. For some people I know, the road to skepticism started after they read a book by Carl Sagan (or one of his articles in *Parade* magazine), watched a TV special, attended a debate or a lecture, or simply talked to friends or family. The process of change takes years, and the people who are responsible for this change rarely get feedback from those they have affected. Education may sometimes seem like a thankless, at times wasted, effort, but it isn't. It's just that the physiology of our brains is such that few educators will be around long enough to see the final outcome. Education is a work in progress that lasts a lifetime.

OK, but What Do We Do in Practice?

It's time to reflect over all I have said so far and come up with a few concrete suggestions that every reader can take home and use. This is obviously not going to be a magic bullet, but if the educational community will start adopting the steps I describe here on a consistent basis, change will be apparent within the next generation. Some universities and schools are doing it already, and occasionally individual faculty take the initiative where there is no institution-wide effort yet in place. For those who are interested, there is help. For example, the report of the Boyer Commission on Educating Undergraduates in the Research University (which can be found at http://naples.cc.sunysb.edu/Pres/boyer.nsf/) is an excellent starting point, as are some of the books in the "Additional Reading" section at the end of this chapter.

To make progress in the creation–evolution controversy, and more generally in education, the following steps are essential:

1. *Scientists must come down from the ivory tower!* It is high time for scientists to take seriously their role in their communities and give more back to them. This is not only for practical reasons (like the constant and very real threat to federal funding of certain areas of scientific research), but simply because it is the decent thing to do. Scientists who don't give back to the community in some tangible way should start thinking of themselves as social parasites—perhaps not of the worst sort, but parasites nonetheless. As much as it seems to some of us scientists that our particular field of research on aspect X of organism Y is so fundamentally important that of course we should be paid for life to carry it out, this is simply not so.
   - Conducting that study is a privilege that we enjoy because we serve the community. Our contributions take four forms: (1) the (remote) possibility that something we do will better the condition of humankind, (2) the fact that we are adding bricks to the edifice of knowledge (though often these bricks build dead ends or uninteresting side corridors), (3) the teach-

ing we do for graduate and especially undergraduate students, and (4) whatever contact we have with the community. Clearly, the last two are the most immediate contributions a scientist can make, and they are also those that receive respectively the lowest and no priority at all because of the way our system of rewards is set up. We don't get tenure because of how much community work we do or how good we are as teachers, but we are also the ones who set the criteria and standards for promotion and tenure, so we are the only ones to blame (together with shortsighted administrators, but that goes without saying).

- There are plenty of things scientists can do: We can contact our local schools or the university outreach program and volunteer to give occasional lectures, or better, provide demonstrations or facilitate discussions among students and community members. We can write both letters and occasional guest editorials to the local newspaper. We can engage in public debates on important issues (not just creationism, but the environment, ethical problems in science, the use of biotechnology, and so on—the possibilities are endless). Scientists with a particular talent for it can write articles for national magazines or work on an occasional book for the general public. It would not be necessary for a few people to do as much as a Carl Sagan, Stephen Gould, or Richard Dawkins if more people would bother writing for the public at least once in a while.

2. *Hiring practices in universities must be changed.* We should acknowledge the aforementioned myth of "good researcher = good teacher," and we should act accordingly. Let us hire researchers to do research, teachers to do teaching, and mixed breeds for what they are good at (e.g., there are people who can do research on how to teach). This means changing the current mentality in both faculty and administrators of colleges and universities, which is not something that is likely to happen anytime soon. But I really don't see any reason for perpetuating the horrors of the current system, with its abysmal rate of failure in producing thinking, well-informed citizens, which, let us not forget, is the *chief* goal of education.
   - Another good model is currently represented by Richard Dawkins at Oxford University. He chairs a position in the "public understanding of science." What a concept! Imagine somebody being paid in a university for the sole purpose of explaining to the public what that university is doing. Ideally, every major department in every university should have a position dedicated to the public understanding of X, where X can be any discipline that is actually the object of research in that department—from biology to physics, from philosophy to language, from law to psychology. It really would not cost much, it could probably be financed by special fund drives to which the public and several educational foundations are likely to be highly responsive, and it would pay enormously both in terms of public relations and—more importantly—in educational impact.

3. *There must be continuing education for teachers.* A teacher is really supposed to be a lifelong learner, but this ideal is seldom realized in practice. One of the major obstacles to achieving effective teaching of evolution in high schools, for example, is simply the fact that most of the teachers have never been trained in the discipline and do not have regular channels through which to update their knowledge of the subject matter. One of the experimental programs we have started at the University of Tennessee is to bring together faculty from biology, anthropology, and education to address this problem. The result has been a yearly workshop for teachers on how to teach evolution and deal with the surrounding issues, which has gradually crystallized into a full-fledged continuing-education course called "Evolution and Society," open to preservice teachers, as well as to students from other disciplines interested in the topic.
4. *Teaching training must be provided for university faculty and graduate assistants.* As I have already mentioned, not only is it simply not true that a good researcher is also necessarily a good teacher, but unfortunately, neither faculty nor teaching assistants get *any* training at all in teaching! The assumption is that, because they are professionals with the requisite professional expertise in their disciplines, they will be able to communicate what they know (or what they are learning) to an audience of undergraduates. How difficult could *that* be? Well, it is extremely difficult, and it is about time that universities started setting up support centers for faculty to learn the basics of classroom and laboratory teaching according to the most up-to-date research in pedagogy.[17] The same universities should require intensive training of their teaching assistants and make sure that they have more than a passing knowledge of the subject matters they are supposed to help the students understand. This is important; nothing less than the scientific literacy of the next generation and the continuing support for science as a government-funded activity are at stake.
5. *Schools and universities should institute truly interdisciplinary courses and curricula.* This has been done to some extent, of course, yet in many cases it is something that is true on paper but not in the reality of the classroom. Often interdisciplinarity simply means putting together two faculty from different departments who will split the teaching *load* of a course, without caring to coordinate their activities or even attending their colleague's classes.
   - Administrations tend to be favorable toward interdisciplinary courses in theory. In practice, however, they often revert to their bean-counting activities that penalize faculty who wish to create new courses or truly engage in coteaching experiences, failing to provide incentives and requir-

[17] Notice that I'm talking about *research* in pedagogy, not the fashionable trends that take our education departments by storm but are often based on nothing more than hunches and sound bites.

ing rigid accounting systems of the number of "contact hours" actually spent by faculty members with the class.

- Examples of interdisciplinary courses are easy to think up and not too difficult to implement. As far as the creation–evolution controversy is concerned, courses integrating philosophy of science and evolutionary biology come to mind, or multiscience courses on *origins* (of the universe, of the solar system, of life), or courses blending the history of scientific ideas with the current state of research in a given field.

6. *The textbooks must be rewritten.* The problem with most textbooks, both at the college level and earlier, is not that they carry the mistakes that so upset Jonathan Wells (although those should be corrected, of course), but that they tend to teach the sciences in a manner similar to a sequential reading of the yellow pages. A lot of facts are followed by a lot of other facts, which are followed by more facts and so on, with little if any attempt to put the major *ideas* in the foreground, to explain how they were arrived at and how they are connected. Most importantly, we need textbooks that clearly and extensively convey the message that science is a vibrant enterprise, very much alive, with a rich past of triumphs and mistakes, and hopefully a bright future of more discoveries and better understanding of nature. Let us remember that curiosity is innate in human beings and that most children are naturally drawn to ask questions about nature during their early school years. How is it that we succeed so systematically and almost completely in smothering that Promethean fire?

7. *The lecture format should be abandoned.* Lectures do have a place, but not in the classroom. Lectures are highly effective ways to convey a large amount of information in a small period of time. This is excellent when we are talking about professional meetings, or invited seminars by specialists, or when one is forced to present material for the general public to a very large audience (in the latter case, more for entertainment purposes than anything else). These are situations in which the listeners want to be there and are receptive to what the speaker is saying. That is hardly the situation in most classrooms, either in college or before.
   - What work much better than lectures are all sorts of activities in which students can exercise *active learning*—that is, where they can actually participate in their own learning process. Think about the difference between watching a play on TV and helping to put it together for a community theater, or actually performing in it. Which way do you think students will learn more about the play?
   - Pedagogical research has shown over and over what should be common sense:[18] More active and multimodal ways of learning produce better and longer-lasting results. Small discussion groups are ideal forums for brainstorming and active learning, but of course they are expensive because they require a much lower faculty:student ratio than most large universi-

ties aim for. But the currently widespread alternative is to provide the illusion of education at what is still a very stiff price, at least in the United States.

8. *"Canned" hands-on activities stipulating predetermined outcomes must be replaced with open-inquiry exercises.* Science is not about following the instructions found in a manual in order to obtain preordained results (if one is lucky and has followed the directions conscientiously). Science is an activity of open-ended inquiry. This does not have to be very complicated or high-tech. The point is to make students understand that sound reasoning and empirical evidence can help solve problems and find answers to questions. It doesn't matter what the question is. As I mentioned already, the currently popular variety of laboratory exercises is a terrible model of how science works—one that leaves students bored or, worse, suspicious of the methods employed to reach obviously predetermined conclusions. Several schools and universities have experimented with alternatives, and they have shown that a little effort goes a long way toward reaping large rewards for both students and faculty.

9. *More emphasis should be placed on the "how" of science, rather than merely on the "what."* Several of the sections in this chapter have touched on this issue, but I think the point needs to be stressed: It is not important merely to teach kids *what* exactly science has found on this or that subject. It is much more important to make them appreciate *how* it was done. I am not suggesting we do away with content altogether, of course. For one thing, one needs to know facts in order to study how scientists arrived at certain conclusions. Furthermore, we still need people who are knowledgeable in all fields, which requires *knowing* stuff. But more crucially, people must also develop a scientific habit of thinking. The details of the subcellular structures or of the classification systems of plants will soon be forgotten after the course, and at any rate they are available in books and articles if one knows what one is looking for and where to find it. The ability to approach problems in terms of rational thinking and empirical evidence, however—once devel-

[18] See, for example, J. Solomon, J. Duveen, and S. Hall, "What's Happened to Biology Investigations?" *Journal of Biological Education* 28 (1994): 261–266; M. D. Sundberg, M. L. Dini, and E. Li, "Decreasing Course Content Improves Student Comprehension of Science and Attitudes towards Science in Freshman Biology," *Journal of Research in Science Teaching* 31 (1994): 679–693; M. D. Sundberg and G. J. Moncada, "Creating Effective Investigative Laboratories for Undergraduates," *BioScience* 44: 698–704; J. D. Brasford, A. L. Brown, and R. R. Cocking, *How People Learn: Brain, Mind, Experience, and School* (Washington, DC: National Academy Press, 1999); G. Marbach-Ad and P. G. Sokolove, "Can Undergraduate Biology Students Learn to Ask Higher Level Questions? *Journal of Research in Science Teaching* 37 (2000): 854–870; R. L. Miller, W. J. Wozniak, M. R. Rust, B. R. Rust, B. R. Miller, and J. Slezak, "Counterattitudinal Advocacy as a Means of Enhancing Instructional Effectiveness: How to Teach Students What They Do Not Want to Know," *Teaching of Psychology* 23 (1996): 215–219; "Getting More Out of the Classroom" [Special issue], *Science* 293 (2001): 1607–1626; R. S. Velayo, "Retention of Content as a Function of Presentation Mode and Perceived Difficulty," *Reading Improvement* 30: 216–227.

oped—stays with students for life and can be applied to everything, from buying a car to making a career decision.

10. *We must teach critical thinking to all students.* The term *critical thinking* is now a very fashionable buzzword, the importance of which is in danger of being lost in the stampede to jump on the next educational bandwagon. It is rather amusing that we need to teach courses in critical thinking because one would think that critical thinking is what education is actually about. Ironically, critical-thinking textbooks are proliferating, yet most are not terribly useful. Nevertheless, good books on critical thinking do exist (see the "Additional Reading" section at the end of this chapter), and it is high time that at least one course of this type became mandatory for every student, no matter what the discipline. Are there any people who can seriously claim that they would be better off *without* knowledge of the basic kinds of reasoning and accompanying fallacies?

11. *Students' writing and communication skills must be improved.* It is astounding to me how many college-level students essentially don't know how to write and are barely capable of expressing themselves orally on complex subjects. Both oratorical and writing skills come with intense practice that should be started very early and continually required.
    - Again, research in pedagogy weighs heavily here (see note 18 for relevant references): The best way to learn something is to have to communicate it to others. If you don't know how to explain it, you don't understand it. The best way to learn something is to do it or to prepare a lecture or write a book about it. One of the chief motivations for me to teach and to write is that it is the surest way to really understand subjects in which I am interested. Everybody should try it at least occasionally. Surely it will not hurt if we have a citizenry that is more articulate and proficient in clear writing.

12. *The use of information technology must engage the student's brain, not bypass it.* There has been a concerted push to computerize classrooms and to put courses online—a push that will certainly continue in the near future, in part because of the novelty and in part because it is much cheaper to put computers in all classrooms than to actually train teachers to do what they need to do and to reduce the size of those classrooms so that real learning can occur.
    - Nevertheless, I am certainly not advocating that we discount the role of modern technology in education (I am writing this book on a laptop, not a typewriter or with pen and paper, and the first drafts of each chapter were available to my critical readers via a special Web page). What I am saying is that we need to use new technologies, like everything else, with our brains switched on. To repackage a traditional boring and ineffective lecture into a snazzy computer presentation will not alter the fact that the material is presented in a boring and ineffective way. Once students become more used to computer projection technology, they will be no more impressed by a PowerPoint slide than they were by the old-fash-

ioned slides or overhead transparencies. Nor should they be. There are plenty of positive applications of computer technology for classroom use, ranging from online interactive exercises to the exchange of electronic drafts of papers with the instructor, to simulation environments to help students understand complex quantitative concepts and how they might play out in realistic situations. These are good uses of the new technologies, but technology for technology's sake is a travesty, not education.

- Similar considerations apply to online courses. Again, the potential is high in the sense of being able to reach students who might not otherwise have the opportunity to attend a regular school or university, or to augment the classroom experience offered by traditional courses with ongoing discussion boards, test samples, and additional resources linked via Web pages. The danger is that many university administrators are looking at online courses as a cheap and "efficient" (financially, not educationally) way to solve the problem of ever-increasing enrollment in introductory courses. With Web-based courses, there is no physical limit to the number of students who can be "served," except, of course, that unless one also deploys a platoon of faculty with time to engage in online interactions with the users, the pedagogical effectiveness of these enterprises will soon plummet to a negligible level.

13. *Teachers should use controversial subject matter as a stimulus to thinking.* As I mentioned in Chapter 3, a school board in Kentucky recently claimed that certain subject matter (such as evolution, sexual education, and AIDS) should be kept out of the classroom because it is too "upsetting" to the students. I think that if students don't get upset at least once a week, they are not receiving a good education. Education is about challenging one's worldview and opening it up to external scrutiny. A student's worldview may or may not withstand such scrutiny, but either way the student will be better off for it.
    - In the case of the creation–evolution controversy, I take a position different from that of most of my colleagues. I think teachers should use the debate as a springboard to teach not only evolution, but science as a process. I am not talking about teaching creationism in the classroom; that would be not only illegal, but also simply wrong from a pedagogical standpoint. What I'm saying is that creationist tactics such as Jonathan Wells's "icons" book can be turned into critical-thinking exercises for a class. Teachers can direct the students to creationist and evolution Web sites, books, and articles. They can guide them through an understanding of how science works and why creationism is pseudoscience. The students may actually get excited about this more proactive approach, and the 50 percent or so who don't believe in evolution will not feel shut out of the learning experience.
    - I realize that there are many obstacles to implementing such a strategy (which, of course, would work with other controversial topics as well), starting with the fact that the teachers themselves might not be sufficiently prepared to understand the points being made by the two sides, not to

mention the potential resistance of parents and administrators. But if we are to change education, we have to take risks and try new things; taking a few risks is part of the job.

14. *Academics should organize community days.* Finally, I think it is of paramount importance that university departments start a series of "community days" during which they enlist their faculty and graduate students in an open house for the public so that people can appreciate what goes on inside the ivory tower of academia. These events are actually organized by some science and humanities departments scattered at a few universities, but they should become a regular feature of the academic experience, like homecoming days and football games.
    - My own experience at the University of Tennessee is with what we call *Darwin Day,* a day of learning about evolutionary biology and the nature of science that I helped start in 1997 (after the Tennessee legislature attempted to pass another bill curtailing the teaching of evolution in public schools). Darwin Day (which is normally celebrated on February 12, Darwin's birthday) includes a panoply of activities, ranging from an information booth where graduate students and faculty distribute literature and answer questions, to a series of videos followed by moderated discussions, from a book display to a keynote lecture by a biologist, a historian, a philosopher, or a civil liberties activist. Darwin Day became international in 2001, thanks to the effort of Amanda Chesworth, and dozens of groups in the United States, Canada, and Europe are now coordinated through a central Web site[19] which provides suggestions and materials to local organizers.
    - Darwin Day, "Socrates Day," "Geology Week," or whatever else a university may wish to organize, actually requires very little effort from a few people in a department and, once again, results in a huge impact not only on the image of a university, but more importantly on how people feel about education in their community. Universities that try this approach will not regret it.

The upshot of all this is, of course, that improving the quality of education is not easy or inexpensive. It requires energy, money, ideas, and enthusiasm, and it doesn't pay off in the immediate future, which means that only people with enough foresight and endurance are likely to engage in it. But the results last much longer than those generated by simplistic slogans and sound bites, and humanity has already paid an incalculable price for the ignorance of its ranks. Kurt Vonnegut once wrote that "it is embarrassing to be human."[20] It is up to educators to do their best to at least ameliorate that embarrassment for generations to come.

---

[19] At http://www.darwinday.org/.

[20] In *Hocus Pocus* (New York: Putnam's, 1990).

Additional Reading

Carey, S. S. 1998. *A Beginner's Guide to the Scientific Method.* Belmont, CA: Wadsworth. A delightful booklet on the basis of science as method.

Cogan, R. 1998. *Critical Thinking Step by Step.* Lanham, MD: University Press of America. One of the best books on critical thinking, with discussions on the distinction between science and pseudoscience.

Damasio, A. 1999. *The Feeling of What Happens: Body and Emotion in the Making of Consciousness.* New York: Harcourt Brace. Discussion by a neurobiologist of the concept of consciousness and some of the most interesting inner workings of the human brain.

Epstein, R. L. 1999. *Critical Thinking.* Belmont, CA: Wadsworth. Another good book on critical thinking.

Gazzaniga, M. S. 1998. *The Mind's Past.* Berkeley, CA: University of California Press. A brief but clear discussion of the pitfalls of the human brain by a first-rate neurobiologist.

Howard, P. J. 2000. *The Owner's Manual for the Brain: Everyday Applications from Mind-Brain Research,* 2nd ed. Austin, TX: Bard Press. An actual manual to handle your brain, full of practical advice and notes on critical thinking.

Jensen, E. 1998. *Teaching with the Brain in Mind.* Alexandria, VA: Association for Supervision and Curriculum Development. A resource for students and educators interested in making use of the findings of neurobiology in the classroom.

McGain, G., and E. M. Segal. 1988. *The Game of Science.* Pacific Grove, CA: Brooks/Cole. A very useful book on the scientific method.

Novak, J. D., and D. B. Gowin. 1984. *Learning How to Learn.* Cambridge: Cambridge University Press. A look at how to improve "active" learning, especially through the use of tools such as concept maps.

Pinker, S. 1997. *How the Mind Works.* New York: Norton. An evolutionary look at how the brain works and why.

Ramachandran, V. S., and S. Blakeslee. 1998. *Phantoms in the Brain: Probing the Mysteries of the Human Mind.* New York: William Morrow. An explanation of basic experiments one can do with the human mind and what can be learned from them, provided by "the Sherlock Holmes of neurobiology," V. S. Ramachandran.

Sperber, M. 2000. *Beer and Circus: How Big-Time College Sports is Crippling Undergraduate Education.* New York: Holt. A scathing critique of modern undergraduate education in the United States and its questionable ties to big sports and entertainment.

- CODA -

The Controversy That Never Ends

Evolution denial will not go away anytime soon. Indeed, if academics and educators don't take sustained and widespread action, it will at least temporarily increase its clout in the United States and perhaps in other countries. As I mentioned at the beginning of this book, this is to be expected because we are dealing with a cultural war, not a scientific dispute—a war that humanity has been fighting on ideological fronts throughout history.

At the risk of simplifying matters too much, let me say that some people tend to be guided by ideology and others by a passion for inquiry. The majority of the population falls in the middle and is constantly pulled toward these opposite extremes, leaning one way or the other depending on the historical circumstances.

Free inquiry saw a moment of triumph during the golden age of Athens in ancient Greece, but it had to wait through the very long Dark Ages to reemerge with a vengeance during the Enlightenment. It dipped again, but it is now alleged to be dominant in modern society, which is experiencing the "age of science."

And yet, the same century that saw a man on the moon and the start of widespread use of sophisticated computers to solve all sorts of practical problems (and to offer a good number of hours of mindless play) was also characterized by the extreme abuse of the power of science in the form of the development of weapons of mass destruction, as well as by a resurgence of irrationalism and superstition.

Copernicus, and especially Galilei, had to fight hard, at personal physical risk, to convince humans that their tiny planet is not the center of the universe. Darwin has made a compelling case for scientists, but not yet for the general public, that our species is neither the pinnacle of creation nor the direct handcrafted job of a god. The next battle has already started, though it has not been as prominent as the creation–evolution debate: Even though science is not even close to generating the first artificial example of consciousness, demagogues are already positioning themselves for the eventuality, claiming that that is yet another "sacred" topic not meant for humans to meddle with. Yet another line in the sand is about to be drawn. I am sure that if and when human ingenuity succeeds in creating artificial life and intelligence, the achievement will be denied in just the same way in which the cardinals of Rome refused to believe what they were seeing through Galilei's telescope and in which modern evolution deniers refuse to accept that evolution is happening under their very noses.

And yet, I am an incurable optimist (you have to be, if you wish to be in the business of science education). Even though the war might never end, we have made progress. Today, Socrates would not be killed for teaching critical thinking in most (though by all means not all) places on Earth. Nobody seriously questions that ours is one of many planets orbiting around an average star at the periphery of a run-of-the-mill galaxy. Perhaps in another century or two, few people will find it strange to be the cousins of chimpanzees and bonobos. Until then, we need to fight not in defense of a particular theory, but for the privilege of attempting to understand the universe. Dante Alighieri, in the *Divine Comedy*, said that humans are not made to live like brutes, but to seek virtue and knowledge. We help neither the search for virtue nor the search for knowledge by denying how the world really is.

– APPENDIX A –

Introduction to and Excerpts from David Hume's Dialogues Concerning Natural Religion, Where the Topic of Intelligent Design is Discussed Most Thoroughly

This appendix reprints parts IV through VII of David Hume's Dialogues Concerning Natural Religion, originally published in 1779. The Dialogues contains both a cogent presentation of the design argument and a sound refutation of it. Since most of Hume's reasoning still applies today, this text should be read by anybody seriously interested in the evolution–creation controversy.

Hume's Dialogues features three characters: Philo, Cleanthes, and Demea. Philo is the skeptic, who most commentators suggest speaks more closely for Hume himself (though other typically Humean arguments are used also by Demea). It is important to understand that Philo, as any good skeptic, does not argue for the nonexistence of God. He merely points out that reason is limited and cannot be used to elucidate such matters. Demea is a deist—that is, somebody who believes in the existence of a God who created the universe but did little or nothing after that. The French philosopher Voltaire and several of the founding fathers of the United States (most famously Thomas Jefferson) were also deists. It is relevant to note that the deist god does not interfere with nature, does not provide revelation, and has no control over one's personal life (and hence does not answer prayers). Cleanthes defends the idea that reason can be used to prove the existence of God, and he is the one who puts forward the design argument criticized by the other two.

In parts I through III (not reprinted here), the three characters debate the limits of the power of reason and agree that although there can be no doubt about the existence of God, the nature of such a god is very much debatable. Cleanthes introduces the argument from design in part III, basing it on an analogy between the universe and a large machine. Just as the machine had a designer, so the universe must have been designed.

Formally, Cleanthes' argument goes something like this:

- Premise 1: The world resembles a machine in some respects.
- Premise 2: We know from experience that machines are created by intelligent designers.
- Conclusion: The world was created by an intelligent designer.

Notice that, if we set aside the more sophisticated mathematical terminology of William Dembski or the contemporary examples from molecular biology of Michael Behe, this is exactly—neither more nor less—the line of reasoning used by modern intelligent design proponents.

Philo, especially in part IV, attacks the argument from several directions. First, he thinks that the analogy between the world and a machine is very weak, and that the world resembles more a naturally generated animal or vegetable than an intelligently designed machine. Because the analogy to a machine is weak, the argument based on it is also considered weak. Second, Philo questions the inductive strength of the argument on two grounds: On the one hand, we have observed only a very limited part of the universe, which provides only a correspondingly limited basis for extrapolation. On the other hand, we know that complexity can be generated without intelligence, as when animals and vegetables generate their own offspring. Third, in part V Philo goes on the counterattack, so to speak, by pointing out to Cleanthes some of the other logically possible conclusions one might draw from Cleanthes' own premises. This is a powerful argument that has been taken into consideration by many modern, mainstream theists, who have increasingly separated their understanding of God from a literal reading of either the Bible or "the book of nature." Given what we know of the world—says Philo—one would have to conclude that God is finite, because the world is also finite, and causes are commensurate to their effects. Moreover, there could actually be a team of gods (because large projects require team effort among humans), and they could be of two sexes, like humans. Furthermore, the imperfections in the world point to a God that is not infinitely powerful. Finally, given the existence of various kinds of evil, the creator cannot simultaneously be omnipotent, omniscient, and benevolent.

In part VI Philo extends his attack on Cleanthes' position by comparing the world to an animal (of which maybe God could be considered the soul) or maybe a vegetable. In part VII Philo suggests that order may result from generation (as when animals and plants produce offspring, which are orderly organic systems) or from instinct, as when a bird builds a nest—that is, an instinctive ordering of natural materials to form a shelter. He also notes that we know of no case in which reason gives rise to generation, but we have plentiful examples the other way around. Philo means that we know from experience that rational beings—that is, humans—have been generated by the natural process of reproduction, yet we never see the converse; that is, we never see reason exercising a similarly (re)productive capacity by producing natural creatures (as Cleanthes maintains that God, a rational, supernatural entity, did when He produced an entire natural cosmos).

The remaining five parts of the dialogue, not reproduced here, contain several other components of the discussion. In part VIII Philo discusses yet another alternative, the Epicurean system of random movements of atoms (which modern creationists still mistakenly insist is analogous to the idea of natural selection—the latter being anything but random, however). In part IX Demea presents a mix of ontological and cosmological arguments for the existence of God, to which Cleanthes gives a typically Humean response that no particular existence is logically necessary (contra the ontological argument), and that even if one grants that something has to exist, why not simply the world instead of a divinity (contra the cosmological argument)? In part X the theme of good and evil is explored in more depth. Here Cleanthes advances the bizarre view (common among fundamentalists today) that there must be evil so that we can appreciate God's benevolence, while Philo and Demea agree that people are brought to God by fear of evil. In part XI Cleanthes suggests that God's benevolence may be finite, and Philo lists four types of evil (imperfections in nature—such as earthquakes—the existence of pain, the fact that animals and people are forced to live frugal and mostly unhappy lives, and the actions of the laws of nature). Finally, in part XII Philo concludes that although it is natural for the mind to see evidence of design in nature, this perception of evidence is more likely a result of the limited power of our reasoning because little about God's character can be discerned from the analysis of nature.

Here, then, are parts IV through VII of Hume's Dialogues Concerning Natural Religion as they appear in the standard Kemp Smith edition.

Part IV

It seems strange to me, said CLEANTHES, that you, DEMEA, who are so sincere in the cause of religion, should still maintain the mysterious, incomprehensible nature of the Deity, and should insist so strenuously, that he has no manner of likeness or semblance to human creatures. The Deity, I can readily allow, possesses many powers and attributes, of which we can have no comprehension: But if our ideas, so far as they go, be not just and adequate, and correspondent to his real nature, I know not what there is in this subject worth insisting on. Is the name, without any meaning, of such mighty importance? Or how do you mystics, who maintain the absolute incomprehensibility of the Deity, differ from sceptics or atheists, who assert, that the first cause of All is unknown and unintelligible? Their temerity must be very great, if, after rejecting the production by a mind; I mean, a mind resembling the human (for I know of no other), they pretend to assign, with certainty, any other specific, intelligible cause: And their conscience must be very scrupulous indeed, if they refuse to call the universal, unknown cause God or Deity; and to bestow on him as many sublime eulogies and unmeaning epithets, as you shall please to require of them.

Who could imagine, replied DEMEA, that CLEANTHES, the calm, philosophical CLEANTHES, would attempt to refute his antagonists, by affixing a nick-name to them; and like the common bigots and inquisitors of the age, have recourse to invective and declamation, instead of reasoning? Or does he not perceive, that these topics are easily retorted, and that anthropomorphite is an appellation as invidious, and implies as dangerous consequences, as the epithet of mystic, with which he has honoured us? In reality, CLEANTHES, consider what it is you assert, when you represent the Deity as similar to a human mind and understanding. What is the soul of man? A composition of various faculties, passions, sentiments, ideas; united, indeed, into one self or person, but still distinct from each other. When it reasons, the ideas, which are the parts of its discourse, arrange themselves in a certain form or order; which is not preserved entire for a moment, but immediately gives place to another arrangement. New opinions, new passions, new affections, new feelings arise, which continually diversify the mental scene, and produce in it the greatest variety, and most rapid succession imaginable. How is this compatible with that perfect immutability and simplicity, which all true theists ascribe to the Deity? By the same act, say they, he sees past, present, and future: His love and his hatred, his mercy and his justice are one individual operation: He is entire in every point of space; and complete in every instant of duration. No succession, no change, no acquisition, no diminution. What he is implies not in it any shadow of distinction or diversity. And what he is, this moment, he ever has been, and ever will be, without any new judgment, sentiment, or operation. He stands fixed in one simple, perfect state; nor can you ever say, with any propriety, that this act of his is different from that other, or that this judgment or idea has been lately formed, and will give place, by succession, to any different judgment or idea.

I can readily allow, said CLEANTHES, that those who maintain the perfect simplicity of the supreme Being, to the extent in which you have explained it, are complete mystics, and chargeable with all the consequences which I have drawn from their opinion. They are, in a word, atheists, without knowing it. For though it be allowed, that the Deity possesses attributes, of which we have no comprehension; yet ought we never to ascribe to him any attributes, which are absolutely incompatible with that intelligent nature, essential to him. A mind, whose acts and sentiments and ideas are not distinct and successive; one, that is wholly simple, and totally immutable; is a mind which has no thought, no reason, no will, no sentiment, no love, no hatred; or in a word, is no mind at all. It is an abuse of terms to give it that appellation; and we may as well speak of limited extension without figure, or of number without composition.

Pray consider, said PHILO, whom you are at present in veighing against. You are honouring with the appellation of atheist all the sound, orthodox divines almost, who have treated of this subject; and you will, at last, be, yourself, found, according to your reckoning, the only sound theist in the world. But if idolaters be atheists, as I think may justly be asserted, and Christian theologians the same; what becomes of the argument, so much celebrated, derived from the universal consent of mankind?

But because I know you are not much swayed by names and authorities, I shall endeavour to show you, a little more distinctly, the inconveniences of that anthropomorphism, which you have embraced; and shall prove, that there is no ground to suppose a plan of the world to be formed in the divine mind, consisting of distinct ideas, differently arranged; in the same manner as an architect forms in his head the plan of a house which he intends to execute.

It is not easy, I own, to see, what is gained by this supposition, whether we judge of the matter by reason or by experience. We are still obliged to mount higher, in order to find the cause of this cause, which you had assigned as satisfactory and conclusive.

If reason (I mean abstract reason, derived from enquiries a priori) be not alike mute with regard to all questions concerning cause and effect; this sentence at least it will venture to pronounce, that a mental world or universe of ideas requires a cause as much as does a material world or universe of objects; and if similar in its arrangement must require a similar cause. For what is there in this subject, which should occasion a different conclusion or inference? In an abstract view, they are entirely alike; and no difficulty attends the one supposition, which is not common to both of them. Again, when we will need force experience to pronounce some sentence, even on these subjects, which lie beyond her sphere; neither can she perceive any material difference in this particular, between these two kinds of worlds, but finds them to be governed by similar principles, and to depend upon an equal variety of causes in their operations. We have specimens in miniature of both of them. Our own mind resembles the one: A vegetable or animal body the other. Let experience, therefore, judge from these samples. Nothing seems more delicate with regard to its causes than thought; and as these causes never operate in two persons after the same manner, so we never find two persons, who think exactly alike. Nor indeed does the same person think exactly alike at any two different periods of time. A difference of age, of the disposition of his body, of weather, of food, of company, of books, of passions; any of these particulars or other more minute, are sufficient to alter the curious machinery of thought, and communicate to it very different movements and operations. As far as we can judge, vegetables and animal bodies are not more delicate in their motions, nor depend upon a greater variety or more curious adjustment of springs and principles.

How therefore shall we satisfy ourselves concerning the cause of that Being, whom you suppose the Author of nature, or, according to your system of anthropomorphism, the ideal world, into which you trace the material? Have we not the same reason to trace that ideal world into another ideal world, or new intelligent principle? But if we stop, and go no farther; why go so far? Why not stop at the material world? How can we satisfy ourselves without going on ad infinitum? And after all, what satisfaction is there in that infinite progression? Let us remember the story of the Indian philosopher and his elephant. It was never more applicable than to the present subject. If the material world rests upon a similar ideal world, as far as abstract reason can judge, it is perfectly indifferent, whether we rest on the universe of matter or on that of thought; nor do we gain any thing by tracing the one into the other. This ideal world must rest upon some other; and so on, without end. It were better, therefore, never to look beyond the present material world. By supposing it to contain the principle of its order within itself, we really assert it to be God; and the sooner we arrive at that divine Being so much the better. When you go one step beyond the mundane system you only excite an inquisitive humour, which it is impossible ever to satisfy.

To say, that the different ideas, which compose the reason of the supreme Being, fall into order, of themselves, and by their own nature, is really to talk without any precise meaning. If it has a meaning, I would fain know, why it is not as good sense to say, that the parts of the material world fall into order, or themselves, and by their own nature? Can the one opinion be intelligible, while the other is not so?

We have, indeed, experience of ideas, which fall into order, of themselves, and without any known cause: But, I am sure, we have a much larger experience of matter, which does the same; as in all instances of generation and vegetation, where the accurate analysis of the cause exceeds all human comprehension. We have also experience of particular systems of thought and of matter, which have no order; of the first, in madness, of the second, in corruption. Why then should we think, that order is more essential to one than the other? And if it requires a cause in both, what do we gain by your system in tracing the universe of objects into a similar universe of ideas? The first step, which we make, leads us on for ever. It were, therefore, wise in us, to limit all our enquiries to the present world, without looking farther. No satisfaction can ever be attained by these speculations, which so far exceed the narrow bounds of human understanding.

It was usual with the peripatetics, you know, CLEANTHES, when the cause of any phenomenon was demanded, to have recourse to their faculties or occult qualities, and to say, for instance, that bread nourished by its nutritive faculty, and senna purged by its purgative: But it has been discovered, that this subterfuge was nothing but the disguise of ignorance; and that these philosophers, though less ingenuous, really said the same thing with the sceptics or the vulgar, who fairly confessed, that they knew not the cause of these phenomena. In like manner, when it is asked, what cause produces order in the ideas of the supreme Being, can any other reason be assigned by you, anthropomorphites, than that it is a rational faculty, and that such is the nature of the Deity? But why a similar answer will not be equally satisfactory in accounting for the order of the world, without having recourse to any such intelligent Creator as you insist on, may be difficult to determine. It is only to say, that such is the nature of material objects, and that they are all originally possessed of a faculty of order and proportion. These are only more learned and elaborate ways of confessing our ignorance; nor has the one hypothesis any real advantage above the other, except in its greater conformity to vulgar prejudices.

You have displayed this argument with great emphasis, replied CLEANTHES: You seem not sensible, how easy it is to answer it. Even in common life, if I assign a cause for any event; is it any objection, PHILO, that I cannot assign the cause of that cause, and answer every new question, which may incessantly be started? And what philosophers could possibly submit to so rigid a rule? Philosophers, who confess ultimate causes to be totally unknown, and are sensible, that the most refined principles, into which they trace the phenomena, are still to them as inexplicable as these phenomena themselves are to the vulgar. The order and arrangement of nature, the curious adjustment of final causes, the plain use and intention of every part and organ; all these bespeak in the clearest language an intelligent cause or Author. The heavens and the earth join in the same testimony: The whole chorus of nature raises one hymn to the praises of its Creator. You alone, or almost alone, disturb this general harmony. You start abstruse doubts, cavils, and objections: You ask me, what is the cause of this cause? I know not; I care not; that concerns not me. I have found a Deity; and here I stop my enquiry. Let those go farther, who are wiser or more enterprising.

I pretend to be neither, replied PHILO: And for that very reason, I should never perhaps have attempted to go so far; especially when I am sensible, that I must at last be contented to sit down with the same answer, which, without farther trouble, might have satisfied me from the beginning. If I am still to remain in utter ignorance of causes, and can absolutely give an explication of nothing, I shall never esteem it any advantage to shove off for a moment a difficulty, which you acknowledge, in its full force, recur upon me. Naturalists indeed very justly explain particular effects by more general causes; though these general causes themselves should remain in the end totally inexplicable; but they never surely thought it satisfactory to explain a particular effect by a particular cause, which was no more to be accounted for than the effect itself.

An ideal system, arranged of itself, without a precedent design, is not a whit more explicable than a material one, which attains its order in a like manner; nor is there anymore difficulty in the latter supposition than in the former.

Part V

But to show you still more inconveniences, continued PHILO, in your anthropomorphism; please do take a new survey of your principles. Like effects prove like causes. This is the experimental argument;[1] and this, you say too, is the sole theological argument. Now it is certain, that the liker the effects are, which are seen, and the liker the causes, which are inferred, the stronger is the argument. Every departure on either side diminishes the probability, and renders the experiment less conclusive. You cannot doubt of this principle; Neither ought you to reject its consequences.

All the new discoveries in astronomy, which prove the immense grandeur and magnificence of the works of nature, are so many additional arguments for a Deity, according to the true system of theism: But according to your hypothesis of experimental theism, they become so many objections, by removing the effect still farther from all resemblance to the effects of human art and contrivance. For if Lucretius, even following the old system of the world, could exclaim,

> *Quis regere immensi summam, quis habere profundi*
> *Indu manu validas potis est moderanter habenas?*
> *Quis pariter coelos omnes convertere? et omnes*
> *Ignibus atheriis terras suffire feraces?*
> *Omnibus inve locis esse omni tempore prasto?*

If Tully esteemed this reasoning so natural, as to put it into the mouth of his Epicurean: "Quibus enim oculis animi intueri potuit vester plato facricam illam tanti operis, qua construi a Deo."

If this argument, I say, had any force in former ages; how much greater must it have at present; when the bounds of nature are so infinitely enlarged, and such a magnificent scene is opened to us? It is still more unreasonable to form our idea of so unlimited a cause from our experience of the narrow productions of human design and invention.

The discoveries by microscopes, as they open a new universe in miniature, are still objections, according to you; arguments, according to me. The farther we push our researches of this kind, we are still led to infer the universal cause of All to be vastly different from mankind, or from any object of human experience and observation.

And what say you to the discoveries in anatomy, chemistry, botany? . . . These surely are no objections, replied CLEANTHES: They only discover new instances of art and contrivance. It is still the image of mind reflected on us from unnumerable objects. Add, a mind like the human, said PHILO. I know of no

---

[1] Hume's use of the word *experimental* here is analogous to what we today call *experiential*—that is, based on sense experience. The "experimental argument," therefore, should be understood as an *experiential* one—that is, an inductive argument in which the premises are based on sensory observations.

other, replied CLEANTHES. And the liker the better, insisted PHILO. To be sure, said CLEANTHES.

Now, CLEANTHES, said PHILO, with an air of alacrity and triumph, mark the consequences. First, by this method of reasoning, you renounce all claim to infinity in any of the attributes of the Deity. For as the cause ought only to be proportioned to the effect, and the effect, so far as it falls under our cognisance, is not infinite; what pretensions have we, upon your suppositions, to ascribe that attribute to the divine Being? You will still insist, that, by removing him so much from all similarity to human creatures, we give into the most arbitrary hypothesis, and at the same time weaken all proofs of his existence.

Secondly, you have no reason, on your theory, for ascribing perfection to the Deity, even in his finite capacity; or for supposing him free from every error, mistake, or incoherence in his undertakings. There are many inexplicable difficulties in the works of nature, which, if we allow a perfect Author to be proved a priori, are easily solved, and become only seeming difficulties, from the narrow capacity of man, who cannot trace infinite relations. But according to your method of reasoning, these difficulties become all real; and perhaps will be insisted on, as new instances of likeness to human art and contrivance. At least, you must acknowledge, that it is impossible for us to tell, from our limited views, whether this system contains any great faults, or deserves any considerable praise, if compared to other possible, and even real systems. Could a peasant, if the *Aeneid* were read to him, pronounce that poem to be absolutely faultless, or even assign to it its proper rank among the productions of human wit; he, who had never seen any other production?

But were this world ever so perfect a production, it must still remain uncertain, whether all the excellencies of the work can justly be ascribed to the workman. If we survey a ship, what an exalted idea must we form of the ingenuity of the carpenter, who framed so complicated, useful, and beautiful a machine? And what surprise must we entertain, when we find him a stupid mechanic, who imitated others, and copied an art, which, through a long succession of ages, after multiplied trials, mistakes, corrections, deliberations, and controversies, had been gradually improving? Many worlds might have been botched and bungled, throughout an eternity, ere this system was struck out: Much labour lost: Many fruitless trials made: And a slow, but continued improvement carried on during infinite ages in the art of world-making. In such subjects, who can determine, where the truth; nay, who can conjecture where the probability, lies; amidst a great number which may be imagined?

And what shadow of an argument, continued PHILO, can you produce, from your hypothesis, to prove the unity of the Deity? A great number of men join in building a house or ship, in rearing a city, in framing a commonwealth: Why may not several Deities combine in contriving and framing a world? This is only so much greater similarity to human affairs. By sharing the work among several, we may so much farther limit the attributes of each, and get rid of that extensive power and knowledge, which must be supposed in one Deity, and which, according to you, can only serve to weaken the proof of his existence. And if such foolish, such vicious creatures as man can yet often unite in framing and executing one plan, how much more those deities or demons, whom we may suppose several degrees more perfect!

To multiply causes, without necessity, is indeed contrary to true philosophy: But this principle applies not to the present case. Were one Deity antecedently proved by your theory, who were possessed of every attribute requisite to the production of the universe; it would be needless, I own (though not absurd) to suppose any other Deity existent. But while it is still a question, whether all these attributes are united in one subject, or dispersed among several independent Beings: By what phenomena in nature can we pretend to decide the controversy? Where we see a body raised in a scale, we are sure that there is in the opposite scale, however concealed from sight, some counterpoising weight equal to it: But it is still allowed to doubt, whether that weight be an aggregate of several distinct bodies, or one uniform united mass. And if the weight requisite very much exceeds any thing which we have ever seen conjoined in any single body, the former supposition becomes still more probable and natural. An intelligent Being of such vast power and capacity, as is necessary to produce the universe, or, to speak in the language of ancient philosophy, so prodigious an animal, exceeds all analogy, and even comprehension.

But further, CLEANTHES, men are mortal, and renew their species by generation; and this is common to all living creatures. The two great sexes of male and female, says MILTON, animate the world. Why must this circumstance, so universal, so essential, be excluded from those numerous and limited Deities? Behold then the theogony of ancient times brought back upon us.

And why not become a perfect anthropomorphite? Why not assert the Deity or Deities to be corporeal, and to have eyes, a nose, mouth, ears, etc.? EPICURUS maintained, that no man had ever seen reason but in a human figure; therefore the gods must have a human figure. And this argument, which is deservedly so much ridiculed by Cicero, becomes, according to you, solid and philosophical.

In a word, CLEANTHES, a man, who follows your hypothesis, is able, perhaps, to assert, or conjecture, that the universe, sometime, arose from something like design: But beyond that position he cannot ascertain one single circumstance, and is left afterwards to fix every point of his theology, by the utmost licence of fancy and hypothesis. This world for aught he knows, is very faulty and imperfect, compared to a superior standard; and was only the first rude essay of some infant Deity, who afterwards abandoned it, ashamed of his lame performance; it is the work only of some dependent, inferior Deity; and is the object of derision to his superiors: it is the production of old age and dotage in some superannuated Deity; and ever since his death, has run on at adventures, from the first impulse and active force, which it received from him . . . You justly give signs of horror, DEMEA, at these strange suppositions: But these, and a thousand more of the same kind, are CLEANTHES'S suppositions, not mine. From the moment the attributes of the Deity are supposed finite, all these have place. And I cannot, for my part, think, that so wild and unsettled a system of theology is, in any respect, preferable to none at all.

These suppositions I absolutely disown, cried CLEANTHES: They strike me, however, with no horror; especially, when proposed in that rambling way in which they drop from you. On the contrary, they give me pleasure, when I see, that, by the utmost indulgence of your imagination, you never get rid of the hypothesis of design in the universe; but are obliged, at every turn, to have recourse to it. To this concession I adhere steadily; and this I regard as a sufficient foundation for religion.

Part VI

It must be a slight fabric, indeed, said DEMEA, which can be erected on so tottering a foundation. While we are uncertain, whether there is one Deity or many; whether the Deity or Deities, to whom we owe our existence, be perfect or imperfect, subordinate or supreme, dead or alive; what trust or confidence can we repose in them? What devotion or worship address to them? What veneration or obedience pay them? To all the purposes of life, the theory of religion becomes altogether useless: And even with regard to speculative consequences, its uncertainty, according to you, must render it totally precarious and unsatisfactory.

To render it still more unsatisfactory, said PHILO, there occurs to me another hypothesis, which must acquire an air of probability from the method of reasoning so much insisted on by CLEANTHES. That like effects arise from like causes: This principle he supposes the foundation of all religion. But there is another principle of the same kind, no less certain, and derived from the same source of experience; that where several known circumstances are observed to be similar, the unknown will also be found similar. Thus, if we see the limbs of a human body, we conclude, that it is also attended with a human head, though hid from us. Thus, if we see, through a chink in a wall, a small part of the sun, we conclude, that, were the wall removed, we should see the whole body. In short, this method of reasoning is so obvious and familiar, that no scruple can ever be made with regard to its solidity.

Now if we survey the universe, so far as it falls under our knowledge, it bears a great resemblance to an animal or organized body, and seems actuated with a like principle of life and motion. A continual circulation of matter in it produces no disorder: A continual waste in every part is incessantly repaired: The closest sympathy is perceived throughout the entire system: And each part or member, in performing its proper offices, operates both to its own preservation and to that of the whole. The world, therefore, I infer, is an animal, and the Deity is the SOUL of the world, actuating it, and actuated by it.

You have too much learning, CLEANTHES, to be at all surprised at this opinion, which, you know, was maintained by almost all the theists of antiquity, and chiefly prevails in their discourses and reasonings. For though sometimes the ancient philosophers reason from final causes, as if they thought the world the workmanship of God; yet it appears rather their favourite notion to consider it as his body, whose organization renders it subservient to him. And it must be confessed, that as the universe resembles more a human body than it does the works of human art and contrivance; if our limited analogy could ever, with any propriety, be extended to the whole of nature, the inference seems juster in favour of the ancient than the modern theory.

There are many other advantages too, in the former theory, which recommended it to the ancient theologians. Nothing more repugnant to all their notions, because nothing more repugnant to common experience, than mind without body; a mere spiritual substance, which fell not under their senses nor comprehension, and of which they had not observed one single instance throughout all nature. Mind and body they knew, because they felt both: An order, arrangement, organization, or internal machinery in both they likewise knew, after the same manner: And it could not but seem reasonable to transfer this experience to the universe, and to have, both of them, order and arrangement naturally inherent in them, and inseparable from them.

Here therefore is a new species of anthropomorphism, CLEANTHES, on which you may deliberate; and a theory which seems not liable to any considerable difficulties. You are too much superior surely to systematical prejudices, to find any more difficulty in supposing an animal body to be, originally, of itself, or from unknown causes, possessed of order and organization, than in supposing a similar order to belong to mind. But the vulgar prejudice, that body and mind ought always to accompany each other, ought not, one should think, to be entirely neglected; since it is founded on vulgar experience, the only guide which you profess to follow in all these theological inquiries. And if you assert, that our limited experience is an unequal standard, by which to judge of the unlimited extent of nature; you entirely abandon your own hypothesis, and must thenceforward adopt our mysticism, as you call it, and admit of the absolute incomprehensibility of the divine nature.

This theory, I own, replied CLEANTHES, has never before occurred to me, though a pretty natural one; and I cannot readily, upon so short an examination and reflection, deliver any opinion with regard to it. You are very scrupulous, indeed, said PHILO; were I to examine any system of yours, I should not have acted with half that caution and reserve, in starting objections and difficulties to it. However, if any thing occur to you, you will oblige us by proposing it.

Why then replied CLEANTHES, it seems to me that, though the world does, in many circumstances, resemble an animal body; yet is the analogy also defective in many circumstances, the most material: No organs of sense; no seat of thought or reason; no one precise origin of motion and action. In short, it seems to bear a stronger resemblance to a vegetable than to an animal; and your inference would be so far inconclusive in favour of the soul of the world.

But in the next place, your theory seems to imply the eternity of the world; and that is a principle which, I think, can be refuted by the strongest reasons and probabilities. I shall suggest an argument to this purpose, which, I believe, has not been insisted on by any writer. Those, who reason from the late origin of arts and sciences, though their inference wants not force, may perhaps be refuted by considerations derived from the nature of human society, which is in continual revolution between ignorance and knowledge, liberty and slavery, riches and poverty; so that it is impossible for us, from our limited experience, to foretell with assurance what events may or may not be expected. Ancient learning and history seem to have been in great danger of entirely perishing after the inundation of the barbarous nations; and had these convulsions continued a little longer, or been a little more violent, we should not probably have now known what passed in the world a few centuries before us. Nay, were it not for the superstition of the Popes, who preserved a little jargon of LATIN, in order to support the appearance of an ancient and universal church, that tongue must have been utterly lost; in which case, the Western world, being totally barbarous, should not have been in a fit disposition for receiving the GREEK language and learning, which was conveyed to them after the sacking of CONSTANTINOPLE. When learning and books had been extinguished, even the mechanical arts would have fallen considerably to decay; and it is easily imagined, that fable or tradition might ascribe to them a much later origin than the true one. This vulgar argument, therefore, against the eternity of the world, seems a little precarious.

But here appears to be the foundation of a better argument. LUCULLUS was the first that brought cherry-trees from ASIA to EUROPE; though that tree thrives so well in many EUROPEAN climates, that it grows in the woods without any culture. Is it possible, that, throughout a whole eternity, no EUROPEAN had ever passed into ASIA, and thought of transplanting so delicious a fruit into his own country? Or if the tree was once transplanted and propagated, how could it ever afterwards perish? Empires may rise and fall; liberty and slavery succeed alternately; ignorance and knowledge give place to each other; but the cherry-tree will still remain in the woods of GREECE, SPAIN and ITALY, and will never be affected by the revolutions of human society.

It is not two thousand years since vines were transplanted into FRANCE; though there is no climate in the world more favourable to them. It is not three centuries since horses, cows, sheep, swine, dogs, corn, were known in AMERICA. Is it possible, that, during the revolutions of a whole eternity, there never arose a COLUMBUS, who might open the communication between EUROPE and that continent? We may as well imagine, that all men would wear stockings for ten thousand years, and never have the sense to think of garters to tie them. All these seem convincing proofs of the youth, or rather infancy, of the world; as being founded on the operation of principles more constant and steady than those by which human society is governed and directed. Nothing less than a total convulsion of the elements will ever destroy all the EUROPEAN animals and vegetables, which are now to be found in the Western world.

And what argument have you against such convulsions? replied PHILO. Strong and almost incontestable proofs may be traced over the whole earth, that every part of this globe has continued for many ages entirely covered with water. And though order were supposed inseparable from matter, and inherent in it; yet may matter be susceptible of many and great revolutions, through the endless periods of eternal duration. The incessant changes, to which every part of it is subject, seem to intimate some such general transformations; though at the same time, it is observable, that all the changes and corruptions, of which we have ever had experience, are but passages from one state of order to another; nor can matter ever rest in total deformity and confusion. What we see in the parts, we may infer in the whole; at least, that is the method of reasoning on which you rest your whole theory. And were I obliged to defend any particular system of this nature (which I never willingly should do), I esteem none more plausible than that which ascribes an eternal, inherent principle of order to the world; though attended with great and continual revolutions and alterations. This at once solves all difficulties; and if the solution, by being so general, is not entirely complete and satisfactory, it is, at least, a theory, that we must, sooner or later, have recourse to, whatever system we embrace. How could things have been as they are, were there not an original, inherent principle of order somewhere, in thought or in matter? And it is very indifferent to which of these we give the preference. Chance has no place, on any hypothesis, sceptical or religious. Everything is surely governed by steady, inviolable laws. And were the inmost essence of things laid open to us, we should then discover a scene, of which, at present, we can have no idea. Instead of admiring the order of natural beings, we should clearly see, that it was absolutely impossible for them, in the smallest article, ever to admit of any other disposition.

Were any one inclined to revive the ancient Pagan Theology, which maintained, as we learned from Hesiod, that this globe was governed by 30,000 Deities, who arose from the unknown powers of nature: You would naturally object, CLEANTHES, that nothing is gained by this hypothesis, and that it is as easy to suppose all men and animals, beings more numerous, but less perfect, to have sprung immediately from a like origin. Push the same inference a step farther; and you will find a numerous society of Deities as explicable as one universal Deity, who possesses, within himself, the powers and perfections of the whole society. All these systems, then, of scepticism, polytheism, and theism, you must allow, on your principles, to be on a like footing, and that no one of them has any advantages over the others. You may thence learn the fallacy of your principles.

Part VII

But here, continued PHILO, in examining the ancient system of the soul of the world, there strikes me, all on a sudden, a new idea, which, if just, must go near to subvert all your reasoning, and destroy even your first inferences, on which you repose such confidence. If the universe bears a greater likeness to animal bodies and to vegetables, than to the works of human art, it is more probable that its cause resembles the cause of the former than that of the latter, and its origin ought rather to be ascribed to generation or vegetation than to reason or design. Your conclusion, even according to your own principles, is therefore lame and defective.

Pray open up this argument a little farther, said DEMEA. For I do not rightly apprehend it, in that concise manner in which you have expressed it.

Our friend, CLEANTHES, replied PHILO, as you have heard, asserts, that since no question of fact can be proved otherwise than by experience, the existence of a Deity admits not of proof from any other medium. The world, says he, resembles the works of human contrivance: Therefore its cause must also resemble that of the other. Here we may remark, that the operation of one very small part of nature, to wit man, upon another very small part, to wit that inanimate matter lying within his reach, is the rule by which CLEANTHES judges of the origin of the whole; and he measures objects, so widely disproportioned, by the same individual standard. But to waive all objections drawn from this topic; I affirm, that there are other parts of the universe (besides the machines of human invention) which bear still a greater resemblance to the fabric of the world, and which therefore afford a better conjecture concerning the universal origin of this system. These parts are animals and vegetables. The world plainly resembles more an animal or a vegetable, than it does a watch or a knitting-loom. Its cause, therefore, it is more probable, resembles the cause of the former. The cause of the former is generation or vegetation. The cause, therefore, of the world, we may infer to be some thing similar or analogous to generation or vegetation.

But how is it conceivable, said DEMEA, that the world can arise from any thing similar to vegetation or generation?

Very easily, replied PHILO. In like manner as a tree sheds its seed into the neighbouring fields, and produces other trees; so the great vegetable, the world, or this planetary system, produces within itself certain seeds, which, being scattered into the surrounding chaos, vegetate into new worlds. A comet, for instance is the seed of a world; and after it has been fully ripened, by passing from sun to sun, and star to star, it is at last tossed into the unformed elements, which everywhere surround this universe, and immediately sprouts up into a new system.

Or if, for the sake of variety (for I see no other advantage), we should suppose this world to be an animal; a comet is the egg of this animal; and in like manner as an ostrich lays its egg in the sand, which, without any farther care, hatches the egg, and produces a new animal; so . . .

I understand you, says DEMEA: But what wild, arbitrary suppositions are these? What data have you for such extraordinary conclusions? And is the slight, imaginary resemblance of the world to a vegetable or an animal sufficient to establish the same inference with regard to both? Objects, which are in general so widely different; ought they to be a standard for each other?

Right, cries PHILO: This is the topic on which I have all along insisted. I have still asserted, that we have no data to establish any system of cosmogony. Our experience, so imperfect in itself, and so limited both in extent and duration, can afford us no probable conjecture concerning the whole of things. But if we must need fix on some hypothesis; by what rule, pray, ought we to determine our choice? Is there any other rule than the greater similarity of the objects compared? And does not a plant or an animal, which springs from vegetation or generation, bear a stronger resemblance to the world, than does any artificial machine, which arises from reason and design?

But what is this vegetation and generation of which you talk? said DEMEA. Can you explain their operations, and anatomize that fine internal structure, on which they depend? As much, at least, replied PHILO, as CLEANTHES can explain the operations of reason, or anatomize that internal structure, on which it depends. But without any such elaborate disquisitions, when I see an animal, I infer, that it sprang from generation; and that with as great certainty as you conclude a house to have been reared by design. These words, generation, reason, mark only certain powers and energies in nature, whose effects are known, but whose essence is incomprehensible; and one of these principles, more than the other, has no privilege for made being a standard to the whole of nature.

In reality, DEMEA, it may reasonably be expected, that the larger the views are which we take of things, the better will they conduct us in our conclusions concerning such extraordinary and such magnificent subjects. In this little corner of the world alone, there are four principles, reason, instinct, generation, vegetation, which are similar to each other, and are the causes of similar effects. What a number of other principles may we naturally suppose in the immense extent and variety of the universe, could we travel from planet to planet and from system to system, in order to examine each part of this mighty fabric? Any one of these four principles above mentioned (and a hundred others which lie open to our conjecture) may afford us a theory, by which to judge of the origin of the world; and it is a palpable and egregious partiality, to confine our view entirely to that principle, by which our own minds operate. Were this principle more intelligible on that account, such a partiality might be somewhat excusable: But reason, in its internal fabric and structure, is really as little known to us as instinct or vegetation; and perhaps even that vague, indeterminate word, nature, to which the vulgar refer everything, is not at the bottom more inexplicable. The effects of these principles are all known to us from experience: But the principles themselves, and their manner of operation, are totally unknown: Nor is it less intelligible, or less conformable to experience to say, that the world arose by vegetation from a seed shed by another world, than to say that it arose from a divine reason or contrivance, according to the sense in which CLEANTHES understands it.

But methinks, said DEMEA, if the world had a vegetative quality, and could sow the seeds of new worlds into the infinite chaos, this power would be still an additional argument for design in its Author. For whence could arise so wonderful a faculty but from design? Or how can order spring from any thing, which perceives not that order which it bestows?

You need only look around you, replied PHILO, to satisfy yourself with regard to this question. A tree bestows order and organization on that tree which springs from it, without knowing the order: an animal in the same manner on its offspring; a bird on its nest; and instances of this kind are even more frequent in the world than those of order, which arise from reason and contrivance. To say that all this order in animals and vegetables proceeds ultimately from design is begging the question; nor can that great point be ascertained otherwise than by proving a priori both that order is, from its nature, inseparably attached to thought, and that it can never, of itself, or from original unknown principles, belong to matter.

But farther, DEMEA; this objection, which you urge, can never be made use of by CLEANTHES, without renouncing a defence which he has already made against one of my objections. When I enquired concerning the cause of that supreme reason and intelligence, into which he resolves every thing; he told me, that the impossibility of satisfying such enquiries could never be admitted as an objection in any species of philosophy. We must stop somewhere, says he; nor is it ever within the reach of human capacity to explain ultimate causes, or show the last connections of any objects. It is sufficient, if the steps, so far as we go, are supported by experience and observation. Now that vegetation and generation, as well as reason, are experienced to be principles of order in nature, is undeniable. If I rest my system of cosmogony on the former, preferably to the latter, it is at my choice. The matter seems entirely arbitrary. And when CLEANTHES asks me what is the cause of my great vegetative or generative faculty, I am equally entitled to ask him the cause of his great reasoning principle. These questions we have agreed to forbear on both sides; and it is chiefly his interest on the present occasion to stick to this agreement. Judging by our limited and imperfect experience, generation has some privileges above reason: For we see every day the latter arise from the former, never the former from the latter.

Compare, I beseech you, the consequences on both sides. The world, say I, resembles an animal, therefore it is an animal, therefore it arose from generation. The steps, I confess, are wide; yet there is some small appearance of analogy in each step. The world, says CLEANTHES, resembles a machine, therefore it is a machine, therefore arose from design. The steps are here equally wide, and the analogy less striking. And if he pretends to carry on my hypothesis, a step farther, and to infer design or reason from the great principle of generation, on which I insist; I may, with better authority, use the same freedom to push farther his hypothesis, and infer a divine generation or theogony from his principle of reason. I have at least some faint shadow of experience, which is the utmost that can ever be attained in the present subject. Reason, in innumerable instances, is observed to arise from the principle of generation, and never to arise from any other principle.

Hesiod, and all the ancient mythologists, were so struck with this analogy, that they universally explained the origin of nature from an animal birth, and copulation. PLATO too, as far as he is intelligible, seems to have adopted some such notion in his TIMAEUS.

The BRAHMINS assert, that the world arose from an infinite spider, who spun this whole complicated mass from his bowels, and annihilates afterwards the whole or any part of it, by absorbing it again, and resolving it into his own essence. Here is a species of cosmogony, which appears to us ridiculous; because a spider is a little contemptible animal, whose operations we are never likely to take for a model of the whole universe. But still here is a new species of analogy, even in our globe. And were there a planet wholly inhabited by spiders (which is very possible), this inference would there appear as natural and irrefragable as that which in our planet ascribes the origin of all things to design and intelligence, as explained by CLEANTHES. Why an orderly system may not be spun from the belly as well as from the brain, it will be difficult for him to give a satisfactory reason.

I must confess, PHILO, replied CLEANTHES, that of all men living, the task which you have undertaken, of raising doubts and objections, suits you best, and seems, in a manner, natural and unavoidable to you. So great is your fertility of invention, that I am not ashamed to acknowledge myself unable, in a sudden, to solve regularly such out-of-the-way difficulties as you incessantly start upon me: Though I clearly see, in general, their fallacy and error. And I question not, but you are yourself, at present, in the same case, and have not the solution so ready as the objection; while you must be sensible, that common sense and reason are entirely against you; and that such whimsies as you have delivered, may puzzle, but never can convince us.

– APPENDIX B –

Bryan's Last Speech

The text of this speech was obtained from the library at Bryan College, in Dayton, Tennessee. William Jennings Bryan—one of the prosecutors of John Scopes at the 1925 trial (see Chapter 1)—had planned to make this speech his closing argument in the trial but did not actually get to deliver it. According to Ed Larson's book, *Summer for the Gods*, Bryan had subsequently planned to use this as a stump speech while traveling across the country, but he delivered it only a couple of times, dying before he could begin his cross-country tour.

May it please the court, and gentlemen of the jury:

Demosthenes, the greatest of ancient orators, in his "oration on the crown," the most famous of his speeches, began by supplicating the favor of all the gods and goddesses of Greece. If, in a case which involved only his own fame and fate, he felt justified in petitioning the heathen gods of his country, surely we, who deal with the momentous issues involved in this case, may well pray to the ruler of the universe for wisdom to guide us in the performance of our several parts in this historic trial.

Let me in the first place, congratulate our cause that circumstances have committed the trial to a community like this and entrusted the decision to a jury made up largely of the yeomanry of the state. The book in issue in this trial contains on its first page two pictures contrasting the disturbing noises of a great city with the calm serenity of the country. It is a tribute that rural life has fully earned.

I appreciate the sturdy honesty and independence of those who come into daily contact with the earth, who, living near to nature, worship nature's God, and who, dealing with the myriad mysteries of earth and air, seek to learn from revelation about the Bible's wonder-working God. I admire the stern virtues, the vigilance and the patriotism of the class from which the jury is drawn, and am reminded of the lines of Scotland's immortal bard, which, when changed but slightly, describe your country's confidence in you:

O, Scotia, my dear, my native soil!
For whom my warmest wish to heaven is sent,
Long may thy hardy sons of rustic toil
Be blest with health, and peace, and sweet content.

And, Oh, may heaven their simple lives prevent
From luxury's contagion, weak and vile
Then, howe'er crowns and coronets be rent
A virtuous populace may rise the while,
And stand, a wall of fire, around their much loved isle.

Let us now separate the issues from the misrepresentations, intentional or unintentional, that have obscured both the letter and the purpose of the law.

This is not an interference with freedom of conscience. A teacher can think as he pleases and worship God as he likes, or refuse to worship God at all. He can believe in the Bible or discard it; he can accept Christ or reject Him. This law places no obligations or restraints upon him. And so with freedom of speech; he can, so long as he acts as an individual, say anything he likes on any subject.

This law does not violate any right guaranteed by any constitution to any individual. It deals with the defendant, not as an individual, but as an employee, an official or public servant, paid by the state, and therefore under instructions from the state.

The right of the state to control the public schools is affirmed in the recent decision in the Oregon case, which declares that the state can direct what shall be taught and also forbid the teaching of anything "manifestly inimical to the public welfare." The above decision goes even farther and declares that the parent not only has the right to guard the religious welfare of the child, but is in duty bound to guard it. That decision fits this case exactly. The state had a right to pass this law, and the law represents the determination of the parents to guard the religious welfare of their children.

It need hardly be added that this law did not have its origin in bigotry. It is not trying to force any form of religion on anybody. The majority is not trying to establish a religion or to teach it—it is trying to protect itself from the effort of an insolent minority to force irreligion upon the children under the guise of teaching science. What right has a little irresponsible oligarchy of self-styled "intellectuals" to demand control of the schools of the United States, in which 25,000,000 children are being educated at an annual expense of nearly $2,000,000,000?

Christians must, in every state of the Union, build their own colleges in which to teach Christianity; it is only simple justice that atheists, agnostics and unbelievers should build their own colleges if they want to teach their own religious views or attack the religious views of others.

The statute is brief and free from ambiguity. It prohibits the teaching, in the public schools, of "any theology that denies the story of divine creation as taught in the Bible," and teaches, "instead, that man descended from a lower order of animals." The first sentence sets forth the purpose of those who passed the law. They forbid the teaching of any evolutionary theory that disputes the Bible record of man's creation and, to make sure that there shall be no misunderstanding, they place their own interpretation on their language and specifically forbid the teaching of any theory that makes man a descendant of any lower form of life.

The evidence shows that the defendant taught, in his own language as well as from a book outlining the theory, that man descended from lower forms of life. Howard Morgan's testimony gives us a definition of evolution that will become known throughout the world as this case is discussed.

Howard, a 14-year-old boy, has translated the words of the teacher and the textbook into language that even a child can understand. As he recollects it, the defendant said "a little germ of one-cell organism has formed in the sea; this kept evolving until it got to be a pretty good sized animal, then came on to be a land animal, and it kept evolving, and from this was man."

There is no room for difference of opinion here, and there is no need for expert testimony. Here are the facts, corroborated by another student, Harry Helton, and admitted to be true by counsel for defense. White, superintendent of schools, testified to the use of Hunter's Civic Biology, and to the fact that the defendant not only admitted teaching evolution, but declared that he could not teach it without violating the law. Robinson, the chairman of the school board, corroborated the testimony of Superintendent White in regard to the defendant's admissions and declaration. These are the facts; they are sufficient and undisputed; a verdict of guilty must follow.

But the importance of this case requires more. The facts and arguments presented to you must not only convince you of the justice of conviction in this case, but, while not necessary to a verdict of guilty, they should convince you of the righteousness of the purpose of the people of the state in the enactment of this law.

The state must speak through you to the outside world and repel the aspersions cast by counsel for the defense upon the intelligence and the enlightenment of the citizens of Tennessee. The people of this state have a high appreciation of the value of education. The state constitution testifies to that in its demand that education shall be fostered and that science and literature shall be cherished. The continuing and increasing appropriations for public instruction furnish abundant proof that Tennessee places a just estimate upon the learning that is secured in its schools.

Religion is not hostile to learning. Christianity has been the greatest patron learning has ever had. But Christians know that "the fear of the Lord is the beginning of wisdom." Now, just as it has been in the past, and they therefore oppose the teaching of guesses that encourage godlessness among the students.

Neither does Tennessee undervalue the service rendered by science. The Christian men and women of Tennessee know how deeply mankind is indebted to science for benefits conferred by the discovery of the laws of nature and by the designing of machinery for the utilization of these laws. Give science a fact and it is not only invincible, but it is of incalculable service to man.

If one is entitled to draw from society in proportion to the service that he renders to society, who is able to estimate the reward earned by those who have given to us the use of steam, the use of electricity, and enable us to utilize the weight of water that flows down the mountainside? Who will estimate the value of the service rendered by those who invented the radio? Or, to come more closely to our home life, how shall we recompense those who gave us the sewing machine, the tractor, the threshing machine, the automobile and the method now employed in making artificial ice? The department of medicine also opens an unlimited field for invaluable service.

Typhoid and yellow fever are not feared as they once were. Diphtheria and pneumonia have been robbed of some of their terrors, and a high place on the scroll of fame still awaits the discoverer of remedies for arthritis, tuberculosis and other dread diseases to which mankind is heir.

Christianity welcomes truth from whatever source it comes, and is not afraid that any real truth from any source can interfere with the divine truth that comes by Inspiration from God Himself. It is not scientific truth to which Christians object, for true science is classified knowledge, and nothing therefore can be scientific unless it is true.

Evolution is not truth; it is merely an hypothesis—it is millions of guesses strung together. It had not been proven in the days of Darwin; he expressed astonishment that with two or three million species it had been impossible to trace any species to any other species. It had not been proven in the days of Huxley, and it has not been proven up to today. It is less than four years ago that Prof. Bateson came all the way from London to Canada to tell the American scientists that every effort to trace one species to another had failed—every one.

He said he still had faith in evolution, but had doubts about the origin of species. But of what value is evolution if it cannot explain the origin of species? While many scientists accept evolution as if it were a fact, they all admit, when questioned, that no explanation has been found as to how one species developed into another.

Darwin suggested two laws, sexual selection and natural selection. Sexual selection has been laughed out of the classroom and natural selection is being abandoned, and no new explanation is satisfactory even to scientists. Some of the more rash advocates of evolution are wont to say that evolution is as firmly established as the law of gravitation or the Copernican theory. The absurdity of such a claim is apparent when we remember that anyone can prove the law of gravitation by throwing a weight into the air, and that anyone can prove the roundness of the earth by going around it, while no one can prove evolution to be true in any way whatever.

Chemistry is an insurmountable obstacle in the path of evolution. It is one of the greatest of the sciences; it separates the atoms—isolates them and walks about them so to speak. If there were in nature a progressive force, an eternal urge, chemistry would find it. But it is not there.

All of the 92 original elements are separate and distinct; they combine in fixed and permanent proportions. Water is H_2O, as it has been from the beginning. It was here before life appeared and has never changed; neither can it be shown that anything else has materially changed.

There is no more reason to believe that men descended from some inferior animal than there is to believe that a stately mansion had descended from a small cottage. Resemblances are not proof, they simply put us on inquiry.

As one fact, such as the absence of the accused from the scene of the murder, outweighs all resemblances that a thousand witnesses could swear to, so the inability of science to trace any of the millions of species to another species, outweighs all the resemblances upon which evolutionists rely to establish man's blood relationship with the brutes.

But while the wisest scientists can not prove a pushing power, such as evolution is supposed to be, there is a lifting power that any child can understand. The plant lifts the mineral up into a higher world, and the animal lifts the plants up into a world still higher. So, it has been reasoned by analogy, man rises, not by a power within him, but only when drawn upward by a higher power.

There is spiritual gravitation that draws all souls toward heaven, just as surely as there is physical force that draws all matters on the surface of the earth towards the earth's center. Christ is our drawing power; he said, "I, if I be lifted from the earth, will draw all men unto Me," and his promise is being fulfilled daily all over the world.

It must be remembered that the law under consideration in this case does not prohibit the teaching of evolution up to the line that separates man from the lower form of animal. The law might well have gone further than it does and prohibit the teaching of evolution in lower forms of life; the law is a very conservative statement of the people's opposition to an anti-Biblical hypothesis. The defendant was not content to teach what the law permitted; he, for reasons of his own, persisted in teaching that which was forbidden for reasons entirely satisfactory to the law makers.

Many of the people who believe in evolution do not know what evolution means. One of the science books taught in the Dayton high schools has a chapter on "The evolution of machinery." This is a very common misuse of the term. People speak of the evolution of the telephone, the automobile, and the musical instrument. But these are merely illustrations of man's power to deal intelligently with inanimate matter; there is no growth from within in the development of a machinery.

Equally improper is the use of the word "evolution" to describe the growth of a plant from a seed, the growth of a chicken from an egg, or the development of any form of animal life from a single cell.

All these give us a circle, not a change from one species to another.

Evolution—the evolution involved in this case, and the only evolution that is a matter of controversy anywhere—is the evolution taught by the defendant, set forth in the books now prohibited by the new state law, and illustrated in the diagram printed on page 194 of Hunter's Civic Biology.

The author estimates the number of species in the animal kingdom at 518,900. These are then divided into 18 classes, and each class indicated on the diagram by a circle, proportioned in size to the number of species in each class and attached by a stem to the trunk of the tree. It begins at protozoa and ends with mammals.

Passing over the classes with which the average man is unfamiliar, let me call your attention to a few of the larger and better known groups. The insects are numbered at 360,000, over two-thirds of the total number of species in the animal world. The fishes are numbered at 13,000, the amphibians at 1,400, the reptiles at 3,500, and the birds at 13,000, while 3,500 mammals are crowded together in a little circle that is barely higher than the bird circle. No circle is reserved for man alone.

He is, according to the diagram, shut up in the little circle entitled "mammals," with 3,499 other species of mammals. Does it not seem a little unfair not to distinguish between man and lower forms of life? What shall we say of the intelligence, not to say religion of those who are so particular to distinguish between fishes and reptiles and birds, but put a man with an immortal soul in the same circle with the wolf, the hyena, and the skunk? What must be the impressions made upon children by such a degradation of man?

In the preface of this book, the author explains that it is for children, and adds that "the boy or girl of average ability upon admission to the secondary school is not a thinking individual." Whatever may be said in favor of teaching evolution to adults, it surely is not proper to teach it to children who are not yet able to think.

The evolutionist does not undertake to tell us how protozoa, moved by interior and resident forces, sent life up through all the various species, and can not prove that there was actually any such compelling power at all. And yet, the school children are asked to accept their guesses and build a philosophy of life upon them. If it were not so serious a matter, one might be tempted to speculate upon the various degrees of relationship that, according to evolutionists, exists between man and other forms of life.

It might require some very nice calculation to determine at what degree of relationship the killing of a relative ceases to be murder and the eating of one's kin ceases to be cannibalism. But it is not a laughing matter when one considers that evolution not only offers no suggestion as to a creator but tends to put the creative act so far away [as] to cast doubt upon creation itself. And, while it is shaking faith in God as a beginning, it is also creating doubt as to heaven at the end of life.

Evolutionists do not feel it is incumbent upon them to show how life began or at what point, in their long drawn out scheme of changing species, man became endowed with hope and promise of immortal life.

God may be a matter of indifference to the evolutionists, and a life beyond may have no charm for them, but the mass of mankind will continue to worship their Creator and continue to find comfort in the promise of the Saviour that he has gone to prepare a place for them. Christ has made of death a narrow, star-lit strip between the companionship of yesterday and the reunion of tomorrow, and evolution strikes out the stars and deepens the gloom that enshrouds the tomb.

If the results of evolution were unimportant, one might require less proof in support of the hypothesis, but before accepting a new philosophy of life, built upon a materialistic foundation, we have reason to demand something more than a guess; "we may well suppose" is not a sufficient substitute for "thus saith the Lord."

If you, your honor, and you, gentlemen of the jury, would have an understanding of the sentiment that lies back of the statute against the teaching of evolution, please consider these facts: First, as to the animals to which evolutionists would have us trace our ancestry. The following is Darwin's family tree, as you will find it set forth on pages 180–181 of his "Descent of Man."

> *"The most ancient progenitors in the kingdom of vertebrata, at which we were able to obtain an obscure glance, apparently consisted of a group of marine animals, resembling the larvae of existing asidians. These animals probably gave rise to a group of fishes, as lowly organized as the lancelot; and from these the canoids, and other fishes like the lepidosiren, must have developed. From such fish a very small advance would carry us on to the amphibians. We have seen that birds and reptiles were once intimately connected together; and the monotremata now connect mammals with reptiles in a slight degree. But no one can at present say what line of descent the three higher and related classes, namely, mammals, birds and reptiles, were derived from the two lower classes, namely amphibians and fishes.*
>
> *In the class of mammals the steps are not difficult to conceive which led from the ancient monotremata to the ancient marsupials; and from these to the early progenitors of the placental mammals. We may thus ascend to the lemuridae; and the interval is not very wide from these to the simiadae. The simiadae then branched off into two great stems, the new world and the old world monkeys; and from the latter, at a remote period, man, the wonder and glory of the universe, proceeded. Thus we have given to man a pedigree of prodigious length, but not, it may be said, of noble quality."*

Darwin, on page 171 of the same book, tries to locate his first man, that is, the first man to come down out of the trees, in Africa. After leaving man in company with gorillas and chimpanzees, he says: "But it is useless to speculate on the subject." If he had only thought of this earlier, the world might have been spared much of the speculation that his brute hypothesis has excited.

On page 79 Darwin gives some fanciful reasons for believing that man is more likely to have descended from the chimpanzee than from the gorilla. His speculations are an excellent illustration of the effect that the evolutionary hypothesis has in cultivating the imagination. Professor J. Arthur Thomson says that the "idea of evolution is the most potent thought- economizing formula the world has yet known." It is more than that; it dispenses with thinking entirely and relies on the imagination.

On page 141 Darwin attempts to trace the mind of man back to the mind of lower animals. On page 118 and 114 he endeavors to trace man's moral nature back to the animals. It is all animal, animal, animal, with never a thought of God or religion.

Our first indictment against evolution is that it disputes the truth of the Bible account of man's creation and shakes faith in the Bible as the Word of God. This indictment we prove by comparing the processes described as evolutionary with the text of Genesis. It not only contradicts the Mosaic record as to the beginning of human life, but it disputes the Bible doctrine of reproduction according to kind—the greatest scientific principle known.

Our second indictment is that the evolutionary hypothesis, carried to its logical conclusion, disputes every vital truth of the Bible. Its tendency, natural, if not inevitable, is to lead those who really accept it, first to agnosticism and then to atheism. Evolutionists attack the truth of the Bible, not openly at first, but by using weasel-words like "poetical," "symbolical," and "allegorical" to suck the meaning out the inspired record of man's creation.

We call as our first witness Charles Darwin. He began life as a Christian. On page 39, volume 1, of the life and letters of Charles Darwin, by his son, Francis Darwin, he says, speaking of the period of 1828 to 1831, "I did not then in the least doubt the strict and literal truth of every word in the Bible." On page 412 of volume 2, of the same publication, he says, "when I was collecting facts for 'The Origin' my belief in what is called a personal God was firm as that of Doctor Puzey himself."

It may be a surprise to your honor, and to you, gentlemen of the jury, as it was to me, to learn that Darwin spent three years at Cambridge studying for the ministry.

This was Darwin as a young man, before he came under the influence of doctrine that man was from a lower order of animals. The change wrought in his religious views will be found in a letter written to a German youth in 1879, and printed on page 277 of volume I of the life and letters above referred to. The letter begins:

> *"I am much engaged, an old man, and out of health, and I can not spare time to answer your questions fully, nor indeed can they be answered. Science has nothing to do with Christ, except insofar as the habit of scientific research makes a man cautious in admitting evidence. For myself, I do not believe that there ever has been any revelation. As for a future life, every man must judge for himself between conflicting vague probabilities."*

Note that "science has nothing to do with Christ, except insofar as the habit of scientific research makes a man cautious in admitting evidence," stated plainly, that simply means that "the habit of scientific research" makes one cautious in accepting the only evidence that we have of Christ's existence, mission, teaching, crucifixion and resurrection, namely the evidence found in the Bible.

To make this interpretation of his words the only possible one, he adds "for myself, I do not believe that there ever has been any revelation." In rejecting the Bible as a revelation from God he rejects the Bible's conception of God, and he rejects also the supernatural Christ of whom the Bible, and the Bible alone, tells. And, it will be observed, he refuses to express any opinion as to a future life.

Now let us follow with his son's exposition of his father's views as they are given in extracts from a biography written in 1876. Here is Darwin's language as quoted by his son:

> *"During these two years (October, 1838, to January, 1839) I was led to think much about religion. Whilst on board the Beagle I was quite orthodox, and I remember being heartily laughed at by several of the officers (though themselves orthodox) for quoting the Bible as an unanswerable authority on some point of morality. When thus reflecting I felt compelled to look for a first cause, having an intelligent mind, in some degree analogous to man; and I deserved not to be called an atheist. This conclusion was strong in my mind about the time, as far as I can remember, when I wrote the 'Origin of Species.' It is since that time that it has very gradually, with many fluctuations, become weaker. Then arises the doubt, can the mind of man, which has, as I fully believe, been developed from a mind as low as that possessed by the lowest animals, be trusted when it draws such grand conclusions?*
>
> *I can not pretend to throw the least light on such abstruse problems. The mystery of the beginning of all things is insolvable by us; and I, for one, must be content to remain an agnostic."*

When Darwin entered upon his scientific career he was "quite orthodox and quoted the Bible as an unanswerable authority on some point of morality."

Even when he wrote "Origin of Species," the thought of "a first cause, having an intelligent mind, in some degree analogous to man," was strong in his mind. It was after that time that very gradually, with many fluctuations, his belief in God became weaker. He traces this decline for us and concludes by telling us that he can not pretend to throw the least light on such abstruse problems—the religious problems above referred to. Then comes the flat statement that he "must be content to remain an agnostic," and, to make clear what he means by the word agnostic, he says that "the mystery of the beginning of all things is insolvable by us"—not by him alone but by everybody. Here we have the effect of evolution upon its most distinguished exponent; it led him from an orthodox Christian, believing every word of the Bible and in a personal God, down and down to helpless and hopeless agnosticism.

But there is one sentence upon which I reserve comment—it throws light upon its downward pathway: "Then arises the doubt, can the mind of man, which has, as I fully believe, been developed from a mind as low as that possessed by the lowest animals, be trusted when it draws such grand conclusions?"

Here is the explanation; he drags man down to the brute levels, and then, judging man by brute standards he questions "whether man's mind can be trusted to deal with God and immortality."

How can any teacher tell his students that evolution does not tend to destroy his religious faith? How can an honest teacher conceal from his students the effect of evolution upon Darwin himself? And is it not stranger still that preachers who advocate evolution never speak of Darwin's loss of faith, due to his belief in evolution? The parents of Tennessee have reason enough to fear the effect of evolution upon the mind of their children. Belief in evolution can not bring those who hold such belief any compensation for the loss of faith in God, trust in the Bible and belief in the supernatural character of Christ. It is belief in evolution that has caused so many scientists and so many Christians to reject the miracles of the Bible, and then give up, one after another, every vital trust in Christianity. They finally cease to pray and sunder the tie that binds them to their Heavenly Father.

The miracle should not be a stumbling block to anyone. It raises but three questions: First, could God perform a miracle? Yes, the God who created the universe can do anything he wants to do with it. He can temporarily suspend any law that he has made or he may employ higher laws that we do not understand.

Second: Would God perform a miracle? To answer that question in the negative one would have to know more about God's plans and purposes than a finite mind can know and yet some are so wedded to evolution that they deny that God would perform a miracle merely because a miracle is inconsistent with evolution.

If we believe that God can perform a miracle and might desire to do so, we are prepared to consider with open mind the third question, namely: Did God perform the miracles recorded in the Bible? The same evidence that establishes the authority of the Bible establishes the truth of miracles performed.

Now let me read of one of the most pathetic confessions that has come to my notice. George John Romanes, a distinguished biologist, sometimes called the successor of Darwin, like Darwin, was reared in the orthodox faith, and like Darwin, was led away from it by evolution.

For 25 years he could not pray. Soon after he became an agnostic, he wrote a book entitled, "A Candid Examination of Theism," publishing it under the assumed name "Physicus." In his book he says:

> *"And for so much as I am far from being able to agree with those who affirm that the twilight doctrine in the 'new faith' is a desirable substitute for*

> *the waning splendor of 'the old' I am not ashamed to confess that with this virtual negation of God the universe to me has lost its soul of loveliness; and although from hence the precept 'work while it is day' will doubtless but gain an intensified force from the terribly intensified meaning of the words that 'the night cometh when no man can work,' yet when at times I think, as think at times I must, of the appalling contrast between the hallowed glory of that creed which once was mine, and the lonely mystery of existence as now I find it—at such times I shall ever feel it impossible to avoid the sharpest pang of which my nature is susceptible."*

Do these evolutionists stop to think of the crime they commit when they take faith out of the hearts of men and women and lead them out into a starless night? What pleasure can they find in robbing a human being of "the hallowed glory of that creed" that Romanes once cherished, and in substituting the "lonely mystery of existence" as he found it? Can the fathers and mothers of Tennessee be blamed for trying to protect their children from such a tragedy?

If anyone has been led to complain of the severity of the punishment that hangs over the defendant, let him compare this crime and its mild punishment with the crimes for which a greater punishment is ascribed. What is the taking of a few dollars from one in day or night in comparison with the crime of leading one away from God and away from Christ?

He who spake as never man spake, thus describes the crimes that are committed against the young: "It is impossible but that offenses will come: but woe unto him through whom they come. It were better for him that a millstone were hanged about his neck and he be cast into the sea than he should offend one of these little ones."

Christ did not overdraw the picture. Who is able to set a price upon the life of a child—a child into whom a mother poured her life and for whom a father has labored? What may a noble life mean to the child itself, to the parents and to the world?

And, it must be remembered that we can measure the effect on only that part of life which is spent on earth; we have no way of calculating the effect on that infinite circle of life [of] which existence here is but a small arc. The soul is immortal and religion deals with the soul; the logical effect of the evolutionary hypothesis is to undermine religion and thus affect the soul. I recently received a list of questions that were to be discussed in a prominent eastern school for women. The second question in the list read: "Is religion an obsolescent function that should be allowed to atrophy quietly, without arousing the passionate prejudice of outworn superstitions?" The real attack of evolution, it will be seen, is not upon orthodox Christianity or even upon Christianity, but upon religion—the most basic fact in man's existence and the most practical thing in life.

James H. Leuba, a professor of psychology at Bryn Mawr college, Pennsylvania, published a few years ago a book entitled, "Belief in God and Immortality." In this book he relates how he secured the opinions of scientists as to the existence of a personal God and a personal immortality. He issued a volume entitled, "American Men of Science," which he says, included the names of "practically every American who may properly be called a scientist."

There are 5,500 names in the book. He selected 1,000 names as representative of the 5,500, and addressed them personally. Most of them, he said, were teachers in schools of higher learning. The names were kept confidential. Upon the answer received, he asserts that over half of them doubt or deny the existence of a personal God and a personal immortality, and he asserts that unbelief being greatest among the most prominent. Among biologists, believers in a personal God numbered less than 31 per cent while believers in a personal immortality numbered only 37 per cent.

He also questioned the students in nine colleges of high rank and from 1,000 answers received, 97 per cent of which were from students between 18 and 20, he found that unbelief increased from 15 percent in the Freshman class up to 40 to 45 per cent among the men who graduated. On page 280 of this book we read "the students' statistics show that young people enter college, possessed of the beliefs still accepted, more or less perfunctorily, in the average home of the land, and gradually abandon the cardinal Christian beliefs." This change from belief to unbelief he attributed to the influence of the persons "of high culture under whom they studied."

The people of Tennessee have been patient enough; they acted none too soon. How can they expect to protect society, and even the church, from the deadening influence of agnosticism and atheism if they permit the teachers employed by taxation to poison the mind of the youth with this destructive doctrine? And remember, that the law has not heretofore required the writing of the word "poison" on poisonous doctrines. The bodies of our people are so valuable that the druggists and physicians must be careful to properly label all poisons; why not be as careful to protect the spiritual life of our people from the poisons that kill the soul?

There is a test that is sometimes used to ascertain whether one suspected of mental infirmity is really insane. He is put into a tank of water and told to dip the tank dry while a stream of water flows into the tank. If he has not sense enough to turn off the stream he is adjudged insane. Can parents justify themselves if, knowing the effect of belief in evolution, they permit irreligious teachers to inject skepticism and infidelity in the minds of their children?

Do bad doctrines corrupt the morals of students? We have a case in point. Mr. Darrow, one of the most distinguished criminal lawyers in our land, was engaged about a year ago in defending two rich men's sons who were on trial for as dastardly a murder as was ever committed. The older one, "Babe" Leopold, was a brilliant student, 19 years old. He was an evolutionist and an atheist. He was also a follower of Nietzsche, whose books he had devoured and whose philosophy he had adopted. Mr. Darrow made a plea for him, based upon the influence that Nietzsche's philosophy had exerted on the boy's mind.

Here are extracts from his speech:

> *"Babe took to philosophy . . . He grew up in this way; he became enamored of the philosophy of Nietzsche. Your honor, I have read almost everything that Nietzsche ever wrote. A man of wonderful intellect; the most original philosopher of the last century. A man who made a deeper imprint on philosophy than any other man within a hundred years. In a way he has reached more people, and still he has been a philosopher of what we might call the intellectual cult.*
>
> *He wrote one book called 'Beyond Good and Evil,' which was a criticism of all moral precepts, as we understood them, and a treatise that the intelligent was beyond good and evil; that the laws for good and the laws for evil did not apply to anybody who approached the superman. He wrote on the will to power. I have just made a few short extracts from Nietzsche that show the things that he (Leopold) has read, and these are short and almost taken at random. It is not how this would affect you. It is not how it would affect me. The question is, how it would affect the impressionable, visionary, dreamy mind of a boy—a boy who should never have seen it—too early for him."*
>
> *Quotations from Nietzsche: 'Why so soft, oh my brethren? Oh why so soft, so unresisting and yielding? Why is there so much disavowal and abnegation, in your heart? Why is there so little faith in your looks? For all creators are hard and it must seem blessedness unto you to press your hand upon millenniums and upon wax. This new table, ah, my brethren, I put over you; become hard. To be obsessed by moral consideration presupposes a very low grade of intellect. We should substitute for morality the will to our own end and consequently to the means to accomplish that. A great man, a man whom nature has built up and invented in a grand style, is colder, harder, less cautious and more free from the fear of public opinion. He does not possess the virtues which are compatible with respectability, with being respected, nor any of these things which are counted among the virtues of the herd.'"*

Mr. Darrow says: That the superman, a creation of Nietzsche, has permeated every college and university in the civilized world.

> *"There is not any university in the world where the professor is not familiar with Nietzsche, not one . . . Some believe it and some do not believe it. Some read it as I do and take it as a theory, a dream, a vision, mixed with good and bad, but not in any way related to human life. Some take it seriously . . . There is not a university in the world of any high standing where the professors do not tell you about Nietzsche and discuss him or where the books are not there.*
>
> *If this boy is to blame for this, where did he get it? Is there any blame at-*

> *tached because somebody took Nietzsche's philosophy seriously and fashioned his life upon it? And there is no question in this case but what that is true. Then who is to blame? The university would be more to blame than he is; the scholars of the world would be more to blame than he is. The purposes of the world . . . are more to blame than he is. Your honor, it is hardly fair to hang a 19-year-old boy for the philosophy that was taught him at the university. It does not meet my ideas of justice and fairness to visit upon his head the philosophy that has been taught by university men for 25 years."*

In fairness to Mr. Darrow, I think I ought to quote two more paragraphs. After this bold attempt to excuse the student on the ground that he was transformed from a well-meaning youth into murderer by the philosophy of an atheist, and on the further ground that his philosophy was in the libraries of all the colleges and discussed by the professors—some adopting the philosophy and some rejecting it—on these two grounds, he denied that the boy should be held responsible for the taking of human life. He charges that the scholars in the universities were more responsible than the boy, and that the universities were more responsible than the boy, because they furnished such books to the students, and then he proceeds to exonerate the universities and scholars, leaving nobody responsible. Here is Mr. Darrow's language:

> *"Now I do not want to be misunderstood about this. Even for the sake of saving the lives of my clients, I do not want to be dishonest and tell the court something that I do not honestly think in this case. I do not think that the universities are to blame. I do not think they should be held responsible. I do think however, that they are too large and that they should keep a closer watch, if possible, upon the individual.*
>
> *But you can not destroy thought, because forsooth, some brain may be deranged by thought. It is the duty of the university as I conceive it, to be the greatest storehouse of the wisdom of the ages, and to have its students come there and learn, choose. I have no doubt but that it has meant the death of many, but that we can not help."*

This is a damnable philosophy, and yet it is the flower that blossoms on the stalk of evolution. Mr. Darrow thinks the universities are in duty bound to feed out this poisonous stuff to their students, and when the students become stupefied by it and commit murder, neither they nor the universities are to blame. I protest against the adoption of any such a philosophy in the state of Tennessee. A criminal is not relieved from responsibility merely because he found Nietzsche's philosophy in a library which ought not to contain it. Neither is the university guiltless if it permits such corrupting nourishment to be fed to the souls that are entrusted to its care. But, going a step farther, would the state be blameless if it permitted the universities under its control to be turned into training schools for murder? When we get back to the root of this question, we will find that the legislature not only had a right to protect the students from the evolutionary hypothesis, but was in duty bound to do so.

While on this subject, let me call your attention to another proposition embodied in Mr. Darrow's speech. He said that Dicky Loeb, the younger boy, had read trashy novels, of the blood and thunder sort. He even went so far as to commend an Illinois statute which forbids minors reading stories of crime. Here is what Mr. Darrow said: "We have a statute in this state, passed only last year, if I recall it, which forbids minors reading stories of crime. Why? There is only one reason; because the legislature in its wisdom, thought it would have a tendency to produce these thoughts and this life in the boys who read them."

If Illinois can protect her boys, why can not this state protect the boys of Tennessee? Are the boys of Illinois any more precious than yours?

But to return to philosophy of an evolutionist. Mr. Darrow said: "I say to you seriously that the parents of Dickey Loeb are more responsible than he, and yet few boys had better parents." Again, he says, "I know that one of two things happened to this boy; that this terrible crime was inherent in his organism, and came from some ancestor, or that it came through his education and his training after he was born." He thinks the boy was not responsible for anything; his guilt was due, according to his philosophy, either to heredity or to environment. But let me complete Mr. Darrow's philosophy based on evolution. He says: "I do not know what remote ancestor may have sent down the seed that corrupted him, and I do not know through how many ancestors it may have passed until it reached Dickey Loeb. All I know is, it is true, and there is not a biologist in the world who will not say I am right."

Psychologists who build upon the evolutionary hypothesis teach that man is nothing but a bundle of characteristics inherited from brute ancestors. That is the philosophy which Mr. Darrow applied in this celebrated criminal case. "Some remote ancestor"—he does not know how remote—"sent down the seed that corrupted him." You cannot punish the ancestor—he is not only dead but, according to the evolutionists, he was a brute and may have lived a million years ago. And he says that all the biologists agree with him. No wonder so small a per cent of the biologists, according to Leuba, believe in a personal God.

This is the quintessence of evolution, distilled for us by one who follows that doctrine to its logical conclusion. Analyze this dogma of darkness and death. Evolutionists say that back in the twilight of life a beast, name and nature unknown, planted a murderous seed and that the impulse that originated in that seed throbs forever in the blood of the brute's descendants, inspiring killings innumerable, for which murderers are not responsible because coerced by a fate fixed by the laws of heredity. It is an insult to reason and shocks the heart. That doctrine is as deadly as leprosy; it may aid a lawyer in a criminal case, but it would, if generally adopted, destroy all sense of responsibility and menace the morals of the world. A brute, they say, can predestine a man to crime, and yet they deny that God-incarnated flesh can release a human being from his bondage or save him from ancestral sins. No more repulsive doctrine was ever proclaimed by man; if all the biologists of the world teach this doctrine—as Darrow says they do—then may Heaven defend the youth of our land from their impious babblings.

Our third indictment against evolution is that it diverts attention from pressing problems of great importance to trifling speculation. While one evolutionist is trying to imagine what happened in the dim past, another is trying to pry open the door of the distant future. One recently grew eloquent over ancient worms, and another predicted that 75,000 years hence everyone will be bald and toothless. But those who endeavor to clothe our remote ancestors with hair and those who endeavor to remove the hair from the heads of our remote descendants ignore the present with its imperative demands. The science of "how to live" is the most important of all the sciences. It is desirable to know the physical sciences, but it is necessary to know how to live. Christians desire that their children shall be taught all the sciences, but they do not want them to lose sight of the Rock of Ages while they study the age of rocks; neither do they desire them to become so absorbed in measuring the distance between the stars that they will forget Him who holds the stars in His Hand.

While not more than two percent of our population are college graduates, these, because of enlarged powers, need a "heavenly vision," even more than those less learned, both for their own restraint and to assure society that their enlarged powers will be used for the benefit of society and not against the public welfare.

Evolution is deadening to spiritual life of a multitude of students. Christians do not desire less education, but they desire that religion shall be entwined with learning so that our boys and girls will return from college with their hearts aflame with love of God and love of fellow men, and prepared to lead in the altruistic work that the world so sorely needs. The cry in the business world, in the industrial, even in the religious world—is for consecrated talents—for ability plus a passion for service.

Our fourth indictment against the evolutionary hypothesis is that, by paralyzing the hope of reform, it discourages those who labor for the improvement of man's condition. Every upward-looking man or woman seeks to lift the level upon which mankind stands, and they trust that they will see beneficent changes during the brief span of their own lives. Evolution chills their enthusiasm by substituting aeons for years. It obscures all beginnings in the mists of endless ages. It is represented as a cold and heartless process, beginning with time and ending in eternity, and acting so slowly that even the rocks can not preserve a record of the imaginary changes through which it is credited with having carried an original germ of life that appeared sometime from somewhere. Its only program for man is scientific breeding, a system under which a few supposedly superior intellects, self-appointed, would direct the mating and the movements of the mass of mankind—an impossible system! Evolution, disputing the miracle, and ignoring the spiritual in life, has no place for the regeneration of the individual. It recognizes no cry of repentance and scoffs at the doctrine that one can be born again.

It is thus the intolerant and unrelenting enemy of the only process that can redeem society through the redemption of the individual. An evolutionist would never write such a story as the Prodigal Son; it contradicts the whole theory of evolution. The two sons inherited from the same parents and, through their parents, from the same ancestors, proximate and remote. And these sons were reared at the same fireside and were surrounded by the same environment during all the days of their youth; and yet they were different.

If Mr. Darrow is correct in the theory applied to Loeb, namely, that his crime was due either to inheritance or to environment, how will he explain the difference between the elder brother and wayward son? The evolutionist may understand from observation, if not by experience, even though he cannot explain, why one of these boys was guilty of every immorality, squandered the money that the father had laboriously earned, and brought disgrace upon the family name; but his theory does not explain why a wicked young man underwent a change of heart, confessed his sin, and begged for forgiveness. And because the evolutionist cannot understand this fact, one of the most important in the human life, he cannot understand the infinite love of the heavenly Father who stands ready to welcome home any repentant sinner, no matter how far he has wandered, how often he has fallen, or how deep he has sunk in sin.

Your honor has quoted from a wonderful poem written by a great Tennessee poet, Walter Malone. I venture to quote another stanza which puts into exquisite language the new opportunity which a merciful God gives everyone who will turn from sin to righteousness:

"Tho' deep in mire wring not your hands and weep,
I lend my arm to all who say 'I can.'
No shamefaced outcast ever sank so deep,
But he might rise and be a man."

There are no lines like these in all that evolutionists have ever written. Darwin says that science has nothing to do with the Christ who taught the spirit embodied in the words of Walter Malone, and yet this spirit is the only hope of human progress. A heart can be changed in the twinkling of an eye, and, a change in the life follows a change in the heart. If one heart can be changed, then a world can be born in a day. It is the fact that inspires all who labor for man's betterment. It is because Christians believe in individual regeneration and in the regeneration of society through the regeneration of individuals that they pray: "Thy Kingdom come, Thy will be done in earth as it is in heaven." Evolution makes a mockery of the Lord's prayer!

To interpret the words to mean that the improvement desired must come slowly through unfolding ages—a process with which each generation could have little to do—is to defer hope, and hope deferred makes the heart sick.

Our fifth indictment of the evolutionary hypothesis is that, if taken seriously and made the basis of a philosophy of life, it would eliminate love and carry man back to a struggle of tooth and claw. The Christians who have allowed themselves to be deceived into believing that evolution is a beneficent, or even a rational process, have been associating with those who either do not understand its implications or dare not avow their knowledge of these implications. Let me give you some authority on this subject. I will begin with Darwin, the high priest of evolution, to whom all evolutionists bow.

On pages 149 and 150, in "The Descent of Man," already referred to, he says:

> *"With savages, the weak in body or mind are soon eliminated; and those that survive commonly exhibit a vigorous state of health. We civilized men, on the other hand, do our utmost to check the process of elimination; we build asylums for the imbecile, the maimed and the sick; we institute poor laws; and our medical men exert their utmost skill to save the life of everyone to the last remnant—There is reason to believe that vaccination has preserved thousands who, from a weak constitution, would formerly have succumbed to smallpox. Thus the weak members of civil society propagate their kind. No one who has attended to the breeding of domestic animals will doubt that this must be highly injurious to the race of man. It is surprising how soon a want of care, or care wrongly directed, leads to the degeneration of a domestic race, but, excepting in the case of man himself, hardly anyone is so ignorant as to allow his worst animals to breed.*
>
> *The aid which we feel impelled to give to the helpless is mainly an incidental result of the instinct of sympathy, which was originally acquired as part of the social instincts, but subsequently rendered in the manner previously indicated more tender and more widely diffused. Nor could we check our sympathy, even at the urging of hard reason, without deterioration to the noblest part of our nature . . . We must, therefore, bear the undoubtedly bad effects of the weak surviving and propagating their kind."*

Darwin reveals the barbarous sentiment that runs through evolution and dwarfs the moral nature of those who become obsessed with it. Let us analyze the quotation just given. Darwin speaks with approval of the savage custom of eliminating the weak so that only the strong will survive and complains that "we civilized men do our utmost to check the process of elimination."

How inhuman such a doctrine as this! He thinks it injurious to "build asylums for the imbecile, the maimed, and the sick," or to care for the poor. Even the medical men come in for criticism because they "exert their utmost skill to save the life of everyone to the last moment." And then note his hostility to vaccination, because it has "preserved thousands who, from a weak constitution would, but for vaccination, have succumbed to smallpox!" All of the sympathetic activities of civilized society are condemned because they enable "the weak members to propagate their kind." Then he drags mankind down to the level of the brute and compares the freedom given to man unfavorably with the restraint that we put on barnyard beasts.

The second paragraph of the above quotation shows that his kindly heart rebelled against the cruelty of his own doctrine. He says that we "feel impelled to give to the helpless," although he traces it to a sympathy which he thinks is developed by evolution; he even asserts that we could not check this sympathy "even at the urging of hard reason," without deterioration of the noblest part of our nature. "We must therefore bear" what he regards as "the undoubtedly bad effects of the weak surviving and propagating their kind." Could any doctrine be more destructive of civilization? And what a commentary on evolution! He wants us to believe that evolution develops a human sympathy that finally becomes so tender that it repudiates the law that created it and thus invites a return to a level where the extinguishing of pity and sympathy will permit the brutal instincts to again do their progressive work.

Let no one think that this acceptance of barbarism, as the basic principle of evolution, died with Darwin. Within three years a book has appeared whose author is even more frankly brutal than Darwin. The book is entitled "The New Decalogue of Science," and has attracted wide attention.

One of our most reputable magazines has recently printed an article by him defining the religion of a scientist. In his preface he acknowledges indebtedness to 21 prominent scientists and educators, "nearly all of them doctors" and "professors."

One of them who has recently been elevated to the head of a great state university read the manuscript over twice and made many valuable suggestions. The author describes Nietzsche, who, according to Mr. Darrow, made a murderer out of Babe Leopold, as the bravest soul since Jesus.

He admits Nietzsche was "gloriously wrong," but he affirms that Nietzsche was "gloriously right in his fearless questioning of the universe and of his own soul."

In another place the author says: "Most of our morals today are jungle products." And then he affirms that: "It would be safer, biologically, if they were more so." Now, after these two samples of his views, you will not be surprised when I read you the following:

> *"Evolution is a bloody business, but civilization tries to make it a pink tea. Barbarism is the only process by which man has ever organically progressed and civilization is the only process by which he has ever organically declined.*
>
> *Civilization is the most dangerous enterprise on which man ever set out. For when you take man out of the bloody, brutal, but beneficent hand of natural selection you place him at once in the soft, daintily gloved, but far more dangerous hand of artificial selection.*
>
> *And unless you call science to your assistance and make this artificial selection as efficient as the rude methods of nature, you bungle the whole task."*

This aspect of evolution may amaze some of the ministers who have not been permitted to enter the inner circle of the iconoclasts whose theories menace all the ideals of civilized society. Do these ministers know that evolution is a "bloody business"? Do they know that barbarism is the only process by which man has ever organically progressed, and "that civilization is the only process by which he has ever organically declined"?

Do they know that the bloody, brutal hand of natural selection is beneficent and the artificial selection "found in civilization is dangerous"? What shall we think of the distinguished educators and scientists who read the manuscript before publication and did not protest against this pagan doctrine?

To show that this is a world-wide matter, I now quote from a book issued from the press in 1918, seven years ago. The title of the book is "The Science of the Power ," and its author, Benjamin Kidd, being an Englishman, could not have any national prejudice against Darwin. On pages 46 and 47 we find Kidd's interpretation of evolution:

> *"Darwin's presentation of the evolution of the world as the product of natural selection in never-ceasing war, as a product that is to say, of a struggle in which the individual efficient in the fight for his own interests was always the winning type touched the profoundest depths of the psychology of the west.*
>
> *The idea seemed to present the whole order of progress in the world as the result of a purely mechanical and materialistic process resting on force. In so doing it was a conception which reached the springs of that heredity born of the unmeasured ages of conquest out of which the western mind has come. Within half a century the 'Origin of Species' had become the Bible of the doctrine of the omnipotence of force."*

Kidd goes so far as to charge that "Nietzsche recited the interpretation of the popular Darwinism, delivered with the fury and intensity of genius." And yet Nietzsche denounced Christianity as the "doctrine of the degenerate," and mercy as "the refuge of weaklings."

Kidd says that Nietzsche gave Germany the doctrine of Darwin's efficient animal in the voice of his sermon, and that Bernhardi and the military textbooks in due time gave Germany the doctrine of the superman translated into the national policy of the superstate aiming at world power.

And what else but the spirit of evolution can account for the popularity of the selfish doctrine, "each one for himself, and the devil take the hindmost," that threatens the very existence of the doctrine of brotherhood?

In 1900—25 years ago, while an international peace congress was in session at Paris, the following editorial appeared in *L'Univers:*

> *"The spirit of peace has fled the earth because evolution has taken possession of it. The plea for peace in past years has been inspired by faith in the divine nature and the divine origin of man; men were then looked upon*

> *as children of one father and war therefore was fratricide. But now that men are looked upon as children of apes, what matters it whether they were slaughtered or not?"*

When there is poison in the blood, no one knows on what part of the body it will break out, but we can be sure that it will break out unless the blood is purified.

One of the leading universities of the south, (I love the state too well to mention its name), publishes a monthly magazine entitled, "Journal of Social Forces." In the January issue of this year a contributor has a lengthy article on "Zoology and Ethics," in the course of which he says:

> *"No attempt will be made to take up the matter of the good or evil of sexual intercourse among humans aside from the matter of conscious procreation, but as an historian it might be worth while to ask the exponents of the impurity complex to explain the fact that without exception the great herds of cultural affluescence have been those characterized by a large amount of freedom in sex relations and that those of the greatest cultural degradation and decline have been accompanied with greater sex repression and purity."*

No one charges or suspects that all or any large percentage of the advocates of evolution sympathize with this loathsome application of evolution to social life, but it is worth while to inquire why those in charge of a great institution of learning allow such filth to be poured out for the stirring of the passions of its students.

Just one more quotation: "The Southeastern Christian Advocate" of June 25, 1925, quotes five eminent college men of Great Britain as joining in answer to the question: "Will civilization survive?"

Their reply is that "Greatest danger to our civilization is the abuse of the achievements of science. Mastery over the forces of nature has endowed the twentieth century man with a power which he is not fit to exercise, unless the development of morality catches up with the development of technique, humanity is bound to destroy itself."

Can any Christian remain indifferent? Science needs religion to direct its energies and to inspire with lofty purpose those who employ the forces that are unloosed by science. Evolution is at war with religion because religion is supernatural; it is therefore the relentless foe of Christianity, which is a revealed religion.

Let us, then, hear the conclusion of the whole matter—Science is a magnificent material force, but it is not a teacher of morals. It is perfect machinery, but it adds no moral restraints to protect society from the misuse of the machine. It can also build gigantic intellectual ships, but it constructs no moral rudders for the control of storm-tossed human vessels.

It not only fails to supply the spiritual element needed, but some of its unproven hypotheses rob the ship of its compass and thus endanger its cargo.

In war, science has proven itself an evil genius; it has made war more terrible than it ever was before. Man used to be content to slaughter his fellowmen on a single plane—the earth's surface. Science has taught him to go down into the water and shoot up from below and to go up into the clouds and shoot down from above, thus making the battlefield three times as bloody as it was before, but science does not teach brotherly love. Science has made war so hellish that civilization was about to commit suicide; and now we are told that newly discovered instruments of destruction will make the cruelties of the late war seem trivial in comparison with the cruelties of wars that may come in the future. If civilization is to be saved from the wreckage threatened by intelligence not consecrated by love, it must be saved by the moral code of the meek and lowly Nazarene. His teachings, and His teachings, alone, can solve the problems that vex the heart and perplex the world.

The world needs a Savior more than it ever did before, and there is only one Name under heaven given among men whereby we must be saved. It is this Name that evolution degrades, for, carried to its logical conclusion, it robs Christ of the glory of a virgin birth, of the majesty of His deity and mission and of the triumph of His resurrection. It also disputes the doctrine of the atonement.

It is for the jury to determine whether this attack upon the Christian religion shall be permitted in the public schools of Tennessee by teachers employed by the state and paid out of the public treasury. This case is no longer local, the defendant ceases to play an important part. The case has assumed the proportions of a battle-royal between unbelief that attempts to speak through so-called science and the defenders of the Christian faith, speaking through the legislators of Tennessee. It is again a choice between God and Baal; it is also a renewal of the issue in Pilate's court. In that historic trial—the greatest in history—force, impersonated by Pilate, occupied the throne. Behind it was the Roman government, mistress of the world, and behind Roman government were the legions of Rome. Before Pilate, stood Christ, the Apostle of Love. Force triumphed; they nailed Him to the tree and those who stood around mocked and jeered and said, "Christ is dead." But from that day the power of Caesar waned and the power of Christ increased. In a few centuries the Roman government was gone and its legions forgotten; while the crucified Lord has become the greatest fact in history and the growing figure of all time.

Again force and love meet face to face, and the question, "What shall I do with Jesus?" must be answered. A bloody, brutal doctrine—evolution—demands, as the rabble did nineteen hundred years ago, that He be crucified. That cannot be the answer of this jury representing a Christian state and sworn to uphold the laws of Tennessee. Your answer will be heard throughout the world; it is eagerly awaited by a praying multitude. If the law is nullified, there will be rejoicing wherever God is repudiated, the Savior scoffed at and the Bible ridiculed. Every unbeliever of every kind and degree will be happy. If, on the other hand, the law is upheld and the religion of the school children protected, millions of Christians will call you blessed and, with hearts full of gratitude to God, will sing again that grand old song of triumph:

"Faith of our fathers, living still,
In spite of dungeon, fire and sword;
Oh, how our hearts beat high with joy,
Whene'er we hear the glorious word:
Faith of our fathers—holy faith,
We will be true to thee till death!"

– NAME INDEX –

– SUBJECT INDEX –